COLLOQUIES ON THE SIMPLES & DRUGS OF INDIA

COLLOQUIES
ON THE
SIMPLES & DRUGS OF INDIA

BY

GARCIA DA ORTA

NEW EDITION (LISBON, 1895) EDITED AND ANNOTATED BY THE CONDE DE FICALHO

TRANSLATED WITH AN INTRODUCTION AND INDEX

BY

SIR CLEMENTS MARKHAM, K.C.B., F.R.S.
HON. FELLOW OF THE ASIATIC SOCIETY OF BENGAL;
HON. MEMBER OF THE LISBON GEOGRAPHICAL SOCIETY MEMBER OF THE PORTUGUESE SOCIETY OF HISTORICAL STUDIES

GYAN

Published by

GYAN

Gyan Publishing House
5, Ansari Road
Daryaganj, New Delhi-110002
Phone: 011-47034999, 9811692060
E-mail: books@gyanbooks.com

Distribution Network
gyanbooks.com
India, USA, Canada, UK, Australia, France

ISBN : 978-81-212-2398-0 (PB)
First Published, 1913

2nd Impression 2021

Printed at: Gyan Press, Delhi.

Colloquies on the Simples & Drugs of India
Author: Sir Clements Markham

COLLOQUIES

ON THE

SIMPLES & DRUGS OF INDIA

BY

GARCIA DA ORTA

NEW EDITION (LISBON, 1895) EDITED AND ANNOTATED
BY THE CONDE DE FICALHO

TRANSLATED WITH AN INTRODUCTION AND INDEX
BY

SIR CLEMENTS MARKHAM, K.C.B., F.R.S.
HON. FELLOW OF THE ASIATIC SOCIETY OF BENGAL; HON. MEMBER OF THE LISBON GEOGRAPHICAL SOCIETY
MEMBER OF THE PORTUGUESE SOCIETY OF HISTORICAL STUDIES

LONDON
HENRY SOTHERAN AND CO.
43 PICCADILLY, W., AND 140 STRAND, W.C.
MCMXIII

DEDICATION

MY DEAR BIRDWOOD,

I have never forgotten that 28th of September 1860 when you welcomed me in the Town Hall at Bombay, and gave the Chinchona plants a temporary home under its portico. I found you full of interest and enthusiasm in all matters relating to the good of the people of India. That interest and that enthusiasm have never flagged throughout a long and useful life, and during a life-long devotion to a great cause. For these reasons I ask you to accept this Dedication, for you are the Garcia da Orta of British India, and much more.—Ever your old Friend and Admirer,

CLEMENTS R. MARKHAM.

JULY 20, 1913.

INTRODUCTION

THE great work of Garcia da Orta on the drugs and simples of India is well known, at least by name, to all who are interested in the botany and pharmacy of our Indian Empire. The work is interesting because it contains the first mention of many important plants, of their uses, and the first descriptions of the treatment of diseases—cholera, for instance—and of several customs and habits of the natives.

Garcia da Orta was born in about 1490 or perhaps a few years later, at Elvas,[1] so famous for its plums, near the Spanish frontier and on the way to Badajos. His father appears to have been an owner of houses and other property at Elvas, named Jorge da Orta, who, besides Garcia, had a son, Francisco of Portalegre, who married Caterina Lopes. Francisco had a son, Jorge, nephew of Garcia da Orta, who was a surgeon, and the heir to his uncle. The elder Jorge also had a daughter, Garcia's sister, married to the Bachiller Gabriel Luis.[2]

In the boyhood of Garcia da Orta there was residing on

[1] Barbosa Machado, quoted by Ficalho. Machado wrote *Bibliotheca Lusitana.*

[2] *Estudos e Notas Elvenses por a Thomaz Pires* (Elvas). (Editor, Antonio Jose Torres Carvalho, 1905.)

his estate of Labruja, near Elvas, Dom Fernao de Sousa, of a noble family descended from a natural son of King Alfonso III. by a very beautiful Moor, daughter of the Cadi of Faro. The two families were on friendly terms, and under the auspices of the Sousa family young Garcia was sent to the Spanish Universities, studying both at Salamanca and at Alcala de Henares, from 1515 to 1525.[1] After his return to Portugal in 1526 he was for some years a village doctor at Castello de Vide, near his native town of Elvas. His patrons, in 1532, got him appointed lecturer in the Lisbon University, and he held that appointment from 1532 to 1534. In the latter year he undertook to go out to India as a physician with Martin Affonso de Sousa, second son of the Lord of Labruja—Orta calls him his "amo." The fleet of five ships left the Tagus on March 12, 1534, reaching Goa in September.

At that time Nuño da Cunha was Governor of Portuguese India, 1529–1538. He was succeeded by Garcia da Noronha as Viceroy, 1538–1540. Then followed Estevan, nephew of Vasco da Gama, 1540–1542; and Martin Affonso de Sousa, the friend and patron of Garcia da Orta, was Governor from 1542 to 1545.

Garcia da Orta saw a great deal of active service as physician to his friend, and became personally well acquainted with the countries bordering on the west coast of India. He was at Diu when it was ceded to the Portuguese by Bahadur Shah, the King of Cambay, and he accompanied Sousa in a march across Kattiawar nearly to Ahmedabad, with the army of Bahadur Shah. He afterwards formed a great friendship with Bahram Nizam Shah, whose capital was at Ahmednagar, visiting him, and acting as his physician. Garcia da Orta also accompanied Sousa in a campaign from Cochin against the Zamorin, and in Ceylon. This seems to be the extent of his

[1] Information from Orta himself (see pp. 1-5). The oldest entries of matriculations at Salamanca only go back to 1546, of degrees to 1525. The Alcala books, now at the University of Madrid, only go back to 1548.

personal knowledge. He does not appear to have been at Bijapur or Bijayanagar in the Deccan, though he often mentions those places, and he knew nothing of Bengal, Berar, or the kingdom of Delhi. The great physician had a house and garden with many medicinal herbs at Goa, and in about 1554 he was granted a long lease of the island of Bombay,[1] which he sublet. His tenant was Simao Toscano, who, with the rent, brought him presents of mangos and other fruits to Goa. He was in practice for many years, and after his friend Sousa left India he was physician to the Viceroy, Pedro Mascarenhas, 1554–1555. Garcia da Orta was always adding to the great amount of erudition he brought with him to India. He had a most extensive and accurate knowledge of the writings of all who had gone before him, and his sound common sense enabled him, with confidence, to separate facts from fables. He knew personally all the plants within his reach from which drugs were derived. For the rest he was indefatigable in his enquiries from native physicians, and in his examinations of Yogis from the kingdom of Delhi, and of traders and others from all parts—Deccanis, Guzeratis, Bengalis, Cingalese, Moors, Persians, Arabs, and Malays.

In 1558 Dom Constantino de Braganza, the brother of the Duke, came out as Viceroy, bringing with him the licentiate Dimas Bosque as his physician. This learned person[2] became a friend of Dr. Garcia da Orta, and it is more than probable that the old physician, who had then been practising in India for upwards of a quarter of a century, was induced by Dimas Bosque to undertake the production of a great work on the

[1] See page 193. What is now the island of Bombay included several islets separated by channels in Orta's time. Bombay Island was only one of them. Gerson da Cunha, who had studied the early documents, came to the conclusion that the Bombaim of Orta was scarcely a tenth part of the present island. See his "Origin of Bombay" in an extra number (1900) of the *Journal of the Bombay Branch of the Royal Asiatic Society*.

[2] See account of him in the Annotated Index. Dimas Bosque wrote the preface to Orta's work.

drugs and simples of his adopted country. All his friends must have felt that his vast knowledge ought not to die with him.

The work was finished in the time of Dom Francisco Coutinho, Count of Redondo, who was Viceroy from 1561 to 1564,[1] to whom it was dedicated. Luis Camoens, the immortal poet, was an intimate friend of Garcia da Orta at this time. Camoens had been banished to Macao, but he returned to Goa in 1561, and was engaged in writing *Os Lusiados.* Wilhelm Stork, in his *Life* of Camoens, says that the poet passed many agreeable and instructive hours in the house of the learned old man, admiring his collections and examining his extensive library, finding in both materials for the last two cantos of *Os Lusiados.* This is quite likely, for the poet composed an ode to the Viceroy in which, after an eloquent address to the Count of Redondo, he exclaims :—

> Favorecei a antigua
> Sciencia que já Achilles estimou
> Olhai que vos obrigua
> Verdes que em vosso tempo se mostrou
> O fruto d'aquella ORTA onde florecem
> Prantas novas que os doutos nao conhecem
>
> Olhai que em vossos annos
> Produze huma ORTA insigne varias ervas
> Nos campos Lusitanos
> As quaes aquellas doutas e protervas
> Medea e Circe nunca conheceram
> Posta que as leis da Magica excederam
>
> E vede carreguado
> Deannos, letras, e longua experiencia
> Hum velho que insinado
> Das gangeticas Musas na sciencia
> Podaliria subtil e arte syluestre
> Vence o velho Chiron de Achilles mestre.[2]

[1] He died at Goa on February 19, 1564.

[2] This sonnet by Camoens only appears in the work of Garcia da Orta. It has not been included in any of the collections of the sonnets of Camoens, and was probably

Orta means a *garden* in Portuguese, and the poet plays upon the old physician's name.

The work was completed and published at Goa by "Johannes de Endem" on April 10, 1563. The title was *Coloquios dos simples e drogas he cousas medicinais da India compostos pello Doutor Garcia da Orta.* This first edition is excessively rare. It is full of typographical errors, being the third book ever printed in India.[1] There was a second edition published at Goa.

Clusius (Charles de l'Escluze) made a sort of *résumé* or epitome in Latin in 1567 (Plantin, Antwerp). It is very different from the original.[2] The same may be said of the Italian translation by Annibal Briganti (Venice, 1582), and of the French translation by Antoine Colin (Lyons, 1619). Orta is here called "Du Jardin." Their translations are from the Latin of Clusius, not from the *Coloquios* of Orta, which they never saw.

F. A. de Varnhagen printed an edition of the *Coloquios* at Lisbon in 1872. It contains many *lacunae* and imperfections, is without notes, and only attempted the identification of one plant, in which it is quite wrong.

The standard edition of Garcia da Orta is by Count Ficalho, in two volumes, the first published in 1891, the

written at Goa on the spur of the moment. Throughout there is a play on the words *Orta* the physician's name, and *Orta* a garden, so that a literal translation cannot convey the sense. The general meaning may be rendered as follows :—

"The lore which Achilles once valued I studied with thee, you opened my eyes to its charm. In your garden of herbs each flower, each tree, were seen in your time by your friend. The fruits of that garden collected from far, were unknown to the learned of old. See how, in thine age thy wisdom and care brought many new simples to light. Unknown to the ancients, but revealed to our sage are the plants in thy garden of herbs. You have opened to us an inspiring page. To thy neighbours like magic it seems. Taught of yore by the muses of Ganges and Ind, full of learning, as of years, in all that is known of the true healing art, old Chiron must bow before thee."

[1] The first was a Catechism by St. Francis Xavier (1557). The second was a *Compendio espiritual* by Dr. Pereira, the first Archbishop of Goa (1561).

[2] 8vo, pp. 250. There were editions in 1574, 1579, 1593, 1605.

second in 1895. The text has been preserved with great care. Count Ficalho was a very accomplished botanist, and in his full and admirable notes to each Colloquy he displays a surprising amount of erudition, having evidently consulted every work that has been written since on the subject, whether in the form of separate books, or of papers in the *Transactions* of Societies. It is quite a model for faithful and thorough editing. Count Ficalho also wrote the *Life and Times of Garcia da Orta*, now very scarce.[1]

Following after the frontispiece, the work of Garcia da Orta contains the approval of the Count Viceroy dated November 5, 1562. Then there is a dedication to the author's old friend and master, Martin Affonso da Sousa, and a sonnet to the same. The prologue or preface is by Dimas Bosque, and finally comes the celebrated ode by Camoens.

The work itself contains chapters on fifty-seven drugs and simples. These chapters are written in colloquies between Garcia da Orta and a Spanish doctor named Dr. Ruano, who is supposed to be an old college friend of Orta, but is clearly an imaginary person.[2] In the Fifty-sixth Colloquy Dimas Bosque, certainly a real person, joins the party. It was not uncommon at that time for authors to put their works into dialogue form.

Count Ficalho, in his *Life* of Orta, admirably explains the spirit and intention of the dialogues, in the following passage:

"The two interlocutors are the two characters united in Garcia da Orta, the two sides of his spirit placed in front one of the other. Dr. Ruano, the man of the schools, the former student of Salamanca, erudite, ready with quotations, with Dioscorides and Pliny at his finger ends. Dr. Orta, the traveller and observer, who, in the face of all the quotations,

[1] *Garcia da Orta e o seu tempo* (Lisbou, 1886), *por o Conde de Ficalho.* This is another very excellent piece of literary work (8vo, pp. 392).

[2] In Spanish and Portuguese the word FULANO is used for any one whose name is unknown, *Fulano tal*, same as *So-and-so*, or *Thingamy*. In Portuguese SICRANO and BELTRANO are also used, and apparently, in former times, RUANO—"the man in the street."

says tranquilly, 'I have seen it.' It is enough for us to note to which of these two entities Orta attaches his own name for evidence as to which of the two he prefers. From this situation, admirably conceived and maintained with much talent, the most interesting controversies result, which bring out, in the clearest light, the spirit of the work."

The work is of great value, owing to its giving the first descriptions of plants and drugs. In addition it contains a great deal of interesting matter. There is some account of the politics of Guzerat and the Deccan at that time, of the greatness of China and of its sea-borne trade, of the controversy about the Spice Islands between the Spaniards and Portuguese, of an expedition to Ilha de Vacas and Jafnapatam, and of other events of the time. There is also an account of Indian names of chessmen. There is the fable of Parizataco, and of the camphor tree. The effects of bhang are described, the method of using betel and the etiquette connected with it, the various uses of the cocoa-nut, an interesting account of Diu and Bassein, some stories about elephants, and about fights between cobras and mungoos. There is an amusing story of a theft from a lady when under the influence of stramonium, and the first account of the treatment of a case of cholera; with other episodes of the same kind. There is also the second description of the cave of Elephanta. I believe the first was written by Dom João de Castro, Viceroy 1545–48. We also get a glimpse of the old bachelor's establishment and habits, and of his intelligent servant girl Antonia. So that the work is by no means confined to a description of drugs and their uses.

Garcia da Orta is believed to have died in about 1570 at Goa, at a good old age, having been thirty-six years practising as a physician in India.

His work still lives. In the *Pharmographia* of Hanbury and Fluckiger, one of the latest and best works on *materia medica*, Garcia da Orta is referred to seven times as an authority.[1] Dr. Ullerspergen of Munich has published an appreciative work on the Colloquys. Gerson da Cunha, in his *Origin of Bombay*, writes of Orta with unbounded praise, and gives some translated passages from his work. In Sir Henry Yule's Glossary, Garcia da Orta is frequently quoted. Sir Henry says of his work: "A most valuable book, full of curious matter and good sense."

The Spanish work of Christoval Acosta followed almost immediately on that of Orta, and is copied wholesale from it. In a preface by the Licentiate Juan Costa, Professor of Rhetoric at Salamanca, this is denied in the following words: "Acosta did not compose the work at his ease, but in durance of sad captivity which he suffered in Africa, Asia, and China. Although his work owes much to the diligence of Garcia da Orta, yet I have compared the two works, and I can say that Orta only sketches the first lines, and that Acosta gave the living colours, and made perfect what Orta had commenced." This may be rhetoric. It is not truth.

Acosta was a native of Burgos and brother of the better-known Jesuit, Josef Acosta, author of the *History of the East and West Indies*. The title of the botanist's work is *Trata de las drogas y medicinas de las Indias Orientales con sus plantas por Christoval Acosta* (Burgos, 1578), dedicated to the *Illustre Senado de Burgos*, his native town. It is a fat little quarto, describing sixty-nine plants and other sources of drugs and medicines. The great merit of the work is that there are full-page illustrations to forty-six plants with

[1] 1. Name and description of the *brindones* of Orta, *Garcinia indica*.
2. Orta's *mamelos de Benguala* or Bengal quince, or *beli Fructus Beloe*.
3. Orta's account of *Cortex Margosae*.
4. The story about the lady who was robbed when under the influence of *Datura alba*, told by Orta.
5. Orta's account of the use of bhang.
6. *Fructus Cardamomi*. Orta on the trade.
7. Camphor, with reference to Orta's account of Chinese trade.

the roots, very well drawn. There are 448 pages of text. It is not all copied from Orta, but the greater part certainly is. Acosta occasionally makes independent remarks, and there are a few plants not in Orta. Mr. Hanbury noticed that the *Semen Tiglii* was first described by Acosta. He was certainly a great traveller, and he gave himself the surname of "Africanus," his more famous brother, the learned Jesuit and writer on Peru and Mexico, being "Americanus." Acosta's portrait faces some odes in his honour, as a frontispiece.

At the end of this Introduction there is a list of the plants in the Colloquies of Garcia da Orta, with the modern scientific names and the names in Acosta, denoting those that are illustrated in Acosta's work. These lists are followed by one of the money, weights, and measures mentioned by Garcia da Orta.

At the end of the book there is an Index of the authorities with biographical notices of the more important mentioned, of Indian Princes and others; an Index of the names of plants and minerals mentioned in the work; also an Index of place-names.

The drawings of plants by Acosta have been used to illustrate the Colloquies of Garcia da Orta.

Garcia da Orta frequently refers to plants and drugs being hot and dry, cold and moist, in the first, second, or third degrees.

Sir George Birdwood has furnished me with the following very interesting explanatory note on this point:—

Da Orta's repeated qualification of drugs refers to the theory of the constitution of the human body and of its diseases and their treatment held in Europe from the earliest Greek and Roman period down to the 17th and 18th centuries, and to the present day throughout native Asia. The constituents of all things were fire, air, water, and earth; fire was hot and dry, air hot and moist, water cold and moist, and earth cold and dry; and of the bodily humours "composed" thereof, blood was regarded as hot and moist, phlegm as cold and moist, red bile as hot and dry, and black bile as cold and dry. The excess of these humours, as the cause of diseases, had to be treated by drugs of their opposite qualities. In India, at least, this in practice means that a drug is either hot or cold, and that it is given either to promote or repress aphrodisia. Chaucer, in *The Canterbury Tales*, writes of the Doctor of Physic:—

> He knew the cause of everich maladye,
> Were it of hoot, or cold, or moiste, or drye,
> And where engendred, and of what humours.
> He was a verrey parfait practisour!

I have to thank Sir George Birdwood for seeing to the scientific names of plants in the footnotes, and Dr. Dalgado of Estoril for much valuable advice and assistance; and for obtaining for me the very scarce work on the *Life and Times of Garcia da Orta*, by Count Ficalho, and the *Origin of Bombay*, by Gerson da Cunha.

CLEMENTS R. MARKHAM.

JULY 20, 1913.

LIST OF PLANTS

No. of Colloquy.	Orta's Names.	Scientific Name.	Chapter in Acosta.	Acosta's Names.	Acosta's Illustration—+. None—0.
2	Aloes . . .	Alöe socotrina	25	Azicear . .	+ whole plant.
3	Amber . . .		26	Ambar . . .	0
4	Amomo . . .	Amomum Cardamomum	28	Amomo . .	0
5	Arbore triste . .	Nyctanthes arbor tristis	27	Arbol triste . .	+ branches.
6	Anacardo . .	Semecarpus Anacardium (Linn.)	30	Anacardo . .	0
7	Altiht . . .	Ferula fœtida	62	Assa fetida . .	0
8	Bangue . . .	Cannabis indica	61	Bangue . .	+ whole plant.
9	Benjuy . . .	Styrax Benzoin			
10	Ber . . .	Garcinia indica	15	Mançanas de la India . . .	+ leaves and fruit.
11	Calamo . . .	Acorus Calamus	63	Calamo aromatico .	+ plant.
12	Camphor . .	Dryobalanops aromatica	33	Canfora . .	0
13	Cardamom . .	Elettaria Cardamomum	64	Cardamom . .	0
14	Cassia fistula . .	Cathartocarpus fistula	17	Caña fistola . .	0
15	Canela . . .	Ravensara aromatica	1	Canela . . .	+ leaves and berries.
16	Coqus (cocos) . .	Cocos nucifera	14	Cocos Palma . .	0
17	Costo . . .	Saussurea Lappa of Clarke	65	Costo . . .	0
18	Crisocolo. . .	Biborate of soda			
19	Cubebras . .	Cubeba officinalis	18	Cubebas . .	0

LIST OF PLANTS (*continued*)

No. of Colloquy.	Orta's Names.	Scientific Name.	Chapter in Acosta.	Acosta's Names.	Acosta's Illustration—+. None—0.
20	Datura . .	Datura alba	11	Datura . . .	+ tree-flowers.
21	Marfim . .	Ivory		Elephant . .	+
22	Faufel . .	Areca Catechu	12	Avellana Indica .	+ tree.
23	Folio Indo .	Cinnamomum Tamala	19	Folio Indo . .	0
24	Galanga . .	Alpinia Galanga	7	Galanga . .	+ whole plant and roots.
25	Cravo . .	Caryophyllus aromaticus	3	Clavos . . .	+ the tree.
26	Gengivre . 5	Zingiber officinale	36	Gengibre . .	+ whole plant.
27	Herbas contra camaras . 5	Hollarhena anti-dysenterica			
28	Jaca . . 3½	Artocarpus integrifolia	37	Jaca . . .	+ fruit and leaves.
29	Lacre . . 11	Coccus lacca	16	Lacre . . .	0
30	Linaloes . . 13	Aquilaria Agallocha			
31	Cate . . 7½	Acacia Catechu	20	Cate . . .	0
32	Maça e Noz . 5	Myristica fragrans	4-5	Nuez Moscada Macer . .	+ the tree. + branch.
33	Manna Purgativa 3	Cotoneaster nummularia	66	Manna . . .	0
34	Mangas . . 10	Mangifera indica	49	Mangas . .	+ leaves and berries.
35	Margarita or Aljofar . 5½	Meleagrina margaretifera			
36	Mungo e Melem de India . . 11½	Cucumis Melo Citrullus vulgaris			
37	Mirabolanos . 6	Terminalia Chebula	41	Mirabolanos . .	0
38	Mangostaes . 1	Nephellium Litchi Garcinia Mangostana			
39	Negundo or Sambali . . 2	Vitex Negundo	42	Negundo . .	+ branches, male and female.

LIST OF PLANTS (*continued*)

No. of Colloquy.	Orta's Names.	Scientific Name.	Chapter in Acosta.	Acosta's Names.	Acosta's Illustration—+. None—0.
40	Nimbo . . 1	Melia Azadirachta	43	Nimbo . . .	0
41	Amfiam (Opio). 4	Papaver somniferum	68	Opio . . .	0
42	Pao da Cobra . 5	Pauwolfia serpentina	55	Palo de culebra .	+ root and leaves.
43	Diamão . . 11				
44	Pedras Preciosas 8				
45	Pedra Bezar . 4		21	Piedra Bezahar .	
46	Pimenta . . 9½	Piper nigrum Piper longum	2	Pimienta . .	+ leaves and berries.
47	Raiz da China . 12	Guaiacum officinale	10	Palo de la China .	+ leaves and roots.
48	Ruibarbo . . 2	Rheum officinale	44	Reobarbaro . .	0
49	Sandalo . . 7	Pterocarpus santalinus	22	Sandalo . .	0
50	Espiquenardo . 7	Nardostachys Jatamansi	23	Spicanardi . .	0
51	Espodio . . 6½	Bambusa arundinacea	46	Spodio . . .	+ stalk and leaves.
52	Esquinanto . 5½	Andropogon lauiger	24	Schinantho . .	0
53	Tamarindos . 6	Tamarindus indica	8	Tamarindo . .	+ leaves and beans.
54	Turbit . . 16½	Ipomoea Turpethum	47	Turbit . . .	0
55	Thure e Mirra 3	Boswellia carterii, and Balsamodendro (Commiphora) Myrrha			
56	Tutia . . 1½	a mineral medicine			
57	Zedoaria . . 5½	Curcuma aromatica			
58	Algunas cousas novas . . 14				
59	Betre . 13/231	Piper Betle			

LIST OF PLANTS (*continued*)

In Orta, not in Acosta.	In Acosta, not in Orta.
Benjuy	Higuara de las Indias
Crisocola	Durion + mentioned by Orta but not seen (under Datura). Seen and drawn by Acosta
Linaloes	Yerbaviva +
Mungo	Yerba mimosa +
Mangosteen	Piñares de Maluco
Tutia	Yerba de Maluco +
Betle	Palo de Maluco
Pearl	Caius +
Precious stones	Moringa +
	Ananas + (pine apple)
	Charameis +
	Sargasso + (gulf weed)
	Carcapuli +
	Añil

MONEY, WEIGHTS, AND MEASURES OCCURRING IN THE WORK OF GARCIA DA ORTA

A REPORT on the money, weights, and measures of India and adjacent countries was made by the King's accountant in 1554.

Money.

CEITAL—⅞ of a ceital = 1 real.

CRUZADO or VENEZIANO DE OURO—about 4 shillings. In Goa its value was 7 TANGUAS or 420 reis; at Cochin, in the time of Martin Affouso de Sousa, 426 reis. It varied with the price of pepper.

DINHEIRO—same as a CEITAL.

LEAES—or LEAL at Goa 1⅛ rei.

LIVRA—36 reis.

PARDAO—about a shilling or 5 TANGUAS (300 reis) at Goa. PARDAO DE OURO = 6 TANGUAS at Goa and Diu; at Cochin, 360 reis.

REAL.

REI—a milresi or 1000 reis, equivalent to 3s. 4d.

TANGUA—half a PARDAO. In Goa a TANGUA was 72 reis, a TANGUA DE PRATA 72 reis, a TANGUA BRANCA 115⅛.

VINTEM—about a penny.

Measures.

PALMO—span, 0·22 m.

COVADO—1½ ft., 0·66 m.

Weights.

ARRATA, ARRATEL—1 lb.

ARROBA—14·688 kilos.

BAR—4 quintals.

CANADA—2½ lbs.

CANDIL—(for rice and wheat) at Goa, 20 mao; (for rice and wheat) at Diu, 8 paras; at Cochin, 14 paras.

CELANUA—25 carats.

GANTA—148 in a candy.

GRAO DE TRIGO—160 = 40 carats.

MAO-MAN (MAUND)—27 lbs. In Diu, 24¾ lbs.; in Goa, 8 to 12 canadas.

MANGELI—a carat in Tamil.

OITAVA—0·00359 kilo.

ONÇA—oz.

QUILATE—a carat.

QUINTAL—58·752 kilos.

RATTI—96 rattis = 100 grains, 1·75 grains Troy.

TROLLA—3 trollas = 540 grains.

FIRST COLLOQUY

In the which is introduced to the reader, Dr. Ruano, one well known to the author in Salamanca and Alcalá, and who came to India in a ship whereof his brother-in-law was purser; and came only to learn about the drugs of India and about all other simples in that country: and how he arrived at Goa and heard of the author there, they being formerly known to each other; and how he went to the author to declare the object of his voyage, and how the author answered him.

ORTA

WELL! we have now adopted the same life that you proposed to enter upon after we had completed our studies; but why have you come to India? It will be well if you can tell me any way in which I could serve you here, for, from this time, I will try to do so.

RUANO

You should know that I came because I had a share in the ship, of which my brother-in-law was the purser; but my presence, as well as your own, in this land, may well be excused, for I come with a great desire to know about the medicinal drugs (such as are called the drugs of pharmacy in Portugal) and other medicines of this country, as well as the fruits and spices. I further wish to learn their names in different languages, and the trees or herbs from which they are taken. I also desire to know how the native physicians use them; and to learn what other plants and fruits there are belonging to this land, which are not medicinal; and what customs will be met with; for all such things may be described as having been seen by you or by other persons worthy of credit.

ORTA

In all these things I will serve you, and will tell you what is true, but I fear that what I may say will be found to be unworthy of record, for so eminent a scholar, who has risen so high in speculation, will only be satisfied by very rare things.

RUANO

If those things content you they will also content me. For it may well be that a thing, which you know well, will for that reason not have much value in your eyes; while I, who did not before know about it, will value it highly, as is reasonable. For some physicians, who came to Spain, neither gave me the information I want, nor satisfied my understanding. But I know that when you talk to me, all you say is worthy of being written down. For this, I have a book in which I write down my enquiries in *a b c* order.

ORTA

I say, my dear Sir, that as your curiosity incites you to desire a knowledge of all that arises here on the subject you have referred to, I equally desire to tell you all I know. As our friendship is so great and so old, that which I communicate to you is to be understood to be to the best of my knowledge, and without any adulation or flattery. On these conditions I promise to serve you and to tell you what I know, and presently you can make such observations as your knowledge suggests, and refer to what seems doubtful; and thus we can discover the truth.

RUANO

In what you say I find what is very agreeable to me, and we shall sleep, making ready to carry out our intention to institute enquiries to-morrow.

SECOND COLLOQUY

OF THE ALOES

RUANO

IT seems to me that the time has now come for you to reply to my enquiries, and as a certain order greatly assists the memory it will be well to commence alphabetically with some names which need elucidation.

ORTA

What you propose respecting alphabetical order appears to me not to be good, for the reason that it may happen that the things discussed at first may be of little value or not much noted, or amusing to read. Just as I have often heard it said that the greatest sins should be the first to be confessed to the priest; or that the best reasons should be given first when we read lessons; or when things have to be sought, the most important must be asked for first.

RUANO

Rather, my dear Sir (saving your better judgment), I disagree with you on many points, because the beginnings of prayers do not move the feelings and affections as much as the rest, and because the end is fixed in the memory more than what came first, neither is a doctrine dwelt upon very subtilely at the beginning, without the promise to make the argument more convincing afterwards.

ORTA

Nevertheless, what you have said has not convinced me. Some day a little book may be printed, making a joke of me, or showing up my errors and badly arranged reasons. Some people in reading, not finding in the beginning anything that they like, without considering further, will give this book to the four winds, covering me with a thousand curses and vituperations, and, what is worse, will direct against me invectives and other kinds of abuse when I do not deserve it.

RUANO

Your writings will meet with no other evil than the malignancy of readers, for they are true, and many are unknown to physicians in Spain, much less to those in the rest of Europe. I have already enquired of physicians in Spain, and they gave me no more reason than that they knew all. Of these men some were learned, but time going on, their thoughts were turned more to enriching themselves than to philosophizing; though as the philosopher says,[1] "to philosophize is better than to enrich oneself." But they prefer first to enrich themselves and then to philosophize. I say that as you have drawn out this treasure, I want your work for myself alone, and for a very few other people that I know in Spain, and some of our disciples, if you will not mind their knowing it; and some of your disciples are as learned as you and I. We could learn from them because they think little of practice and much of the schools. You and I will do the opposite. For now you have neither masters nor preceptors in Salamanca or Alcalá, for all are either dead or banished from Spain. Turning

[1] Aristotle, *Topics* (libro 3). Note by da Orta.

to our investigations, tell me the names of the aloe in all the languages known to you, how it is used, and which is the best, for Pliny and Dioscorides highly praise that of this country.

ORTA

There is little that is noteworthy to be said about the ALOE,[1] nevertheless I will comply with your wish. I say that ALOES or ALOA is Latin and Greek. The Arabs call it CEBAR, and the Guzeratis and Deccanis AREÁ; the Canarese, who are the inhabitants of this sea-coast, call it CATECOMER, the Spaniards ACIBAR, and the Portuguese AZEVRE. After the herb has been dried with smoke, it is called *Herva-babosa* by the Portuguese. There is a great quantity of this herb in Cambay, Bengal, and many other parts; but that of Socotra has much the highest value, and there is a trade with it to Turkey, Persia, Arabia, and throughout Europe. They call it *Socotrine aloes.* That island is 128 leagues from the mouth of the straits whence as much may be seen of Arabia as of Ethiopia, for the entrance of the strait has Arabia on one side and Ethiopia on the other. It is not here that a city exists, as is stated by Laguna, for there is not a city in the island, but only people with large flocks. Nor is there pavement or ground to receive a tear that may fall. For they neither have a city, nor in the island is there any civilization, nor is there any mistake about the great abundance there of this herb. Owing to the little care of the negroes of that land, in not separating the herbs from this *Herva-babosa*, they get mixed, and for this reason one does not appear to be so good as the other. Also you must not believe that it is better on the top than in the middle, and worst at the bottom, nor is it full of sand if it is prepared with care, for all is good. Nor can it be mistaken for Gum Arabic or

[1] *Aloë socotrina*, and *A. Perryi*, Baker, Liliaceae. Aloes was known to the Greeks as a product of Socotra from about the first century B.C. See Pliny, xxvii. 4 (5); Juvenal, vi. 181 ("Plus aloës quam mellis habet"); and is the *sibr* of the Arabs, the *alia* of the Hindus, the *kumarika* of the Cingalese, and the *alivah* of the Malays.

Acacia (as Pliny and Dioscorides affirm), for in our country there is little Gum Arabic or Acacia, and to speak the truth, no one has sent to enquire, from persons of good faith, respecting what I have related. It may well be that this AZEVRE is falsified in other countries.

RUANO

How is it known that the aloes of Socotra, called by some writers *Suco-cetrino,* is the best? How do you know that they can discern the Persian, Arabian, and Turkish one from the other at Ormuz, whither they take it to sell, as they say.

ORTA

A rich merchant well known to common fame, and well read in their literature, named Khoja Perculim,[1] served as secretary to the governors. One day I asked him how it was called in Turkish, Persian, and Arabic. He told me that the word was CEBAR in all those languages. On making further enquiries, he said that the best of all came from Socotra, but that they were brought from many other parts of India to Ormuz, Aden, and Jiddah, whence they are taken to Cairo and on to Alexandria, the port of the Nile. The merchants know very well what is from Socotra, and what from Cambay and other parts; and they value that from Socotra four times as high as any other kind. After this I went to see NIZAMOXA,[2] who is one of the principal kings in the Deccan. He is called the NIZAMALUCO.[2] He is also well read in his own literature. He always has physicians of Turkey and Persia to whom he gives large rents. These men knew the difference perfectly. They told me that they could tell the kind from Socotra because its parts joined well one with another, while the other kinds of AZEVRE did not connect well because the parts came

[1] A Parsee. When Bahadur Shah ceded Baçaim to Nuna da Cunha in 1534, Khoja Perculim served as interpreter, and he then became known to Garcia da Orta.

[2] Nizam-Shah. Nizam-ul-Mulk.

from different plants. This is a thing perfectly well known, they added, and the king, their master, always has the Socotra kind brought to him. So that there are not two species, or three species in use, but one only. This I understand, though you may hold that locality causes variety in species. Certainly there are bad and good, and the plants differ in degrees of goodness. But this diversity in quality does not prevent the parts from mixing well, for they are all one species. Some doctors call it SUCO-CETRINO, but not many. I do not myself see the reason, but in truth it is so called.

RUANO

But how do we explain the statements of Pliny and Dioscorides, who say that the best of all is that from India, and that other kinds are from Alexandria or Arabia?

ORTA

To this I answer you that you are not to understand merely that the kind coming from India is the best. It was first brought to India from Socotra. But, as I have already told you, they also export AZEVRE from Cambay and Bengal to Ormuz, Aden, and to Juda (as we have corrupted the name, for the natives called it Gida[1]). All these are always brought from other parts, but, as I say, that of Socotra is the best, above all the rest. For who buys devils, sells devils.

RUANO

However, Mesué says better, that one is brought from Socotra, another from Persia, another from Armenia, and another from Arabia.

ORTA

Mesué does not do better, but only less erroneously than the others. For in truth what goes from here to Portugal,

[1] Jidda.

and I have seen all, has been brought from Socotra. When your doctors may have said that it came from Alexandria, it is to be understood that in times past quantities of drugs were brought to Ormuz, thence to Bussora, and thence to Aden and Gida, whence, by land in camel caravans, they reached Suez and Alexandria, the port of the Nile. Thence the galleys of Venice carried them to that port, and they were distributed and sold all over Europe. It was not because there was AZEVRE in Alexandria.

RUANO

If there is not AZEVRE in Alexandria, you will also say that there is not RUIBARBO. Therefore that author is wrong who says that one cannot make a dollar for purging or deoppilating, there being such a quantity of RUIBARBO in Alexandria.

ORTA

This doctor means the quantity of RUIBARBO that comes to Alexandria from other parts.

RUANO

Touching the names, there is one a little doubtful and not in Mateo Silvatico. He says it is called SABER[1] or CANTHAR or REAMAL; but he may be mistaken, not being an Arab, and what shall we say of Serapio who called it SABER?

ORTA

It was not called CEBAR only, and in course of time it became SABER. For this the translator is not alone in fault, but the lapse of time also, which alters everything. In Arabic it is certainly CEBAR.

[1] *Sibr* is the Arabic for the drug aloes, obtained from *Aloë socotrina* and other species of Aloë. See footnote, page 6. Here rhubarb [*rhawand Chini*] and gamboge [*osara rhawand*] seem to be confused with aloes.

SECOND COLLOQUY

Ruano

Is it used among the Indians?

Orta

The physicians of Persia, Arabia, and Turkey use this medicine because they know Avicenna thoroughly. They call him *Abolahi*, and his five books "*Canum*." They know Rasis whom they call *Benzacaria*, and Halirodoam and Mesué, although it is not the work that we used. They also have all the works of Hypocras, Galen, Aristotle and Plato, though not so complete as from the Greek fountain-head. The Gentio[1] physicians also use it for purgatives, in kidney diseases, colics, and also for healing wounds. They have, in their pharmacy, a medicine called Mocebar made of aloe and myrrh, which they also call Bola. It is much used by them to cure horses, and to kill maggots in wounds. But among us bad aloes are called Cabalino, as a modern doctor writes, saying that the worst is used at the farrieries; and that the kind Cabalino is neither used to cure beast nor men, only Socotrino. From what Serapiam says on the authority of Alcamzi, we should understand that Cabalino can be used with less harm for farriers' purposes and for wounds. A Gentio physician of Sultan Bahadur, King of Cambay, uses, as an ordinary and efficacious medicine, pieces of leaves of Herva-babosa boiled, with salt between them. He takes eight ounces of this infusion with which he causes four or five motions, without injury or annoyance to him who takes it. Here, in this city of Goa, I took this herb, pounded and mixed with milk, to be given to those who suffer from disease in the kidneys or bladder, or to be administered in any other way. It proved very efficacious as a cure, and now we use this medicine and derive benefit from it. We also used

[1] A gentile or heathen, the Portuguese name for Hindus in contradistinction to Moors or Muhammadans. The English corrupted the word into "Gentoo."

AZEVRE for the injuries to the legs of birds, a remedy well known to falconers; and to reduce inflammation of the gums. Mateolo Senes does not appear to be well informed, where he says that the herb is better to look at than to use as medicine.

RUANO

All you have said is not wanting in interest, but I am told that when you taste HERVA-BABOSA it is bitter, and has a very strong smell.

ORTA

Reading in Antonio Musa and other modern writers I find it stated that the bitterness does not exist in the HERVA-BABOSA of our land, and I have often tried this myself. I found it very bitter, and nearer the root it became more bitter, while at the upper point there was no bitterness; but a disagreeable smell from all parts, so that the statement of Antonio Musa that the kind from Socotra is more bitter is false. For I have myself tried the Indian kind, and have ordered that of Socotra to be tried, finding that both have a very bitter taste. It has not been tried in Spain. And further, I may tell you that I find, in Silvatico and the Plateario, that all things that are bitter are useful, the more bitter they are the better, except the aloes. It appears that Antonio Musa is of a contrary opinion, and it appears to me that Musa is right, knowing that bitterness preserves from putrefaction, and is very good in other ways.

RUANO

Resolve this doubt for me. If the medicines containing aloes are taken while fasting, and food is taken afterwards, should there be a long interval before the food is given?

ORTA

Do not ask me this, for you know it better than I do.

RUANO

Nevertheless I want to hear your opinion, and to know the course you adopt.

ORTA

Galen orders five pills of the size of a grain of corn to be given for pains in the head, and Pliny says that it is a very good medicine, taken a short time after drinking, and then a little good food. It is a very good prescription, and also used by the Moorish physicians of this country. But as the aloes are weakening, the medicine will not work until the system is strengthened by good nourishing food, but little in quantity, as I said, so as to digest well. The medicine ought to be taken fasting and not after a meal lest it should corrupt the food. Each one of these points is supported by reasons and texts, and all can be shown to agree. As to the question how the food suits, so as to mix with the medicine, you know better than I, and will excuse me for not saying more on the subject.

RUANO

It grows more in places near the sea, as Dioscorides tells us?

ORTA

I have travelled all over India, more than 200 leagues of road, and I saw this HERVA-BABOSA in every place.

RUANO

They tell me that it yields a gum.

ORTA

It has no gum except sometimes on the leaves like some kinds of viscous water, of which no use is made.

RUANO

Ruelio says that the pills of Rasis, which are prescribed

for the plague, compounded by Rufo, contain aloes, myrrh, amoniaco, temiama, and wine. Why do these Maumetistas[1] leave out amoniaco, temiama, and wine; and increase the açafram?

ORTA

You seem very much attached to these modern authors who, in order to praise the Greeks, speak evil of Arabs and of some Moors born in Spain, and others of Persia, calling them "Maumetistas" and barbarians (which they hold to be the worst epithet there is in the world), especially the Italians, as well as the Greeks, who are not those who are now called Rumis, and the Turks, a cruel and ill-conditioned race who, at present, persecute Christians more than any other people. I must tell you that I do not deny that the medicine of Rufo is what they say and is very good. I also grant that the pills of Rasis (which we use) are well tried and excellent. The açafram is put into them because it is a cordial and purgative, and for many other virtues which it possesses.

RUANO

It seems that you make a difference between Rumis and Greeks, and I have always understood that these two words mean the same thing?

ORTA

Although your question does not refer to medicine, I answer that they are very different, the Turks being of the province of Anatolia (formerly called Asia Minor) and the Rumis are of Constantinople and its empire.

RUANO

How do you know this: by books or by what has been told you by certain persons?

[1] Muhammadans.

SECOND COLLOQUY

Orta

I have often asked a white soldier whether he was a Turk, when I have been campaigning in the wars of these Indian kings, and he has answered no that he was a Rumi. Others, when I have asked them whether they were Rumis, have said no that they were Turks. When I asked them what was the difference between one and the other, they told me that I could not understand because I did not know the names of the countries, nor could I understand the language. Finding myself in the house of that excellent worthy, Martin Affonso de Sousa (whom I was serving), he showed me a *Platina*[1] containing a legend in the life of San Silvestre, where we find it written that when Constantine, leaving Rome to the Pope, went to Constantinople, he granted to it the privilege of being called Rome, and that the country should be called Romeos. The *Platina* says that they have those names to this day.

Ruano

I have much pleasure in hearing these things although they do not refer to the subject of our colloquies. But to return to the aloes, you told me that we could answer Menardo and other moderns who find fault with Mesué, Serapiam, and Avicenna, because they say that the aloe opens the veins and is bad for the hemorrhoids, and because these Arabs say that it is less purgative when mixed with honey. For they affirm that it is less harmful to the stomach than other soluble medicines. Menardo and these others say that it does not only cause hemorrhage, but closes the hemorrhoids, and that it is not pernicious to the stomach, but on the contrary does good, not doing any harm whatever. Mixed with honey it is more soluble than other soluble medicines. The first opinion

[1] Bartolomeo de Sacchi, Platina, 1442–1481; author of *In vitas summorum pontificum ad Sixtum IV.*, 1479.

is proved by Galen and many other authorities, and the second by the fact that honey is a cause of solubility, for two solutions are more purgative than one.

Orta

I have told you that I do not feel obliged to reply to questions referring to matters which are better known in Spain, where one can read many things that are written every day while practising, and conferring with many learned physicians. But I, out here, am not consulted by any one, and the want of counsel is not made up for by books. However, in answer to your first point, I tell you that Antonio Musa speaks as a man without feeling, for he has no reverence for any master, and concedes the accuracy of the first point affirmed by Mesué, that it opens a hemorrhage, and he says that he has seen this many times in his practice. I also say that I have had experience, and seen the effects many times, and that it causes great pains with the hemorrhage. All this you can do with the aloes, through its bitter quality, opening the veins and stimulating the expulsive virtue. In this way it purges the bile remaining in the stomach and in the navel, as Dioscorides and Serapiam say. As to closing the veins, it is proved by the authority of Jacob de Partibus that it retains outside and opens from within. This authority refers to many medicines that, taken within, give occasion for an operation, and applied on the outside have different effects, such as an onion which sustains when used inwardly, and applied outwardly causes wounds and ulcers. As regards the second point, objected to by Mesué, that it purges less when taken with honey, I tell you that both are soluble, both aloes and honey, and more soluble together than aloes by itself, the less soluble is weakened, which is the honey. The third point is where he reprehends Mesué because he says that it is less harmful to the stomach, being comforting to the stomach. To this I say that it is to be understood that it gives comfort to the stomach by an accident, which the physicians call *per accidens*, namely, drawing out the bad humours of the stomach without doing any harm, or at least very little. In this way we have to understand the authorities referred to by Menardo and other modern writers.

RUANO

On all the points you have referred to, you have thoroughly satisfied me, and much more in what you have said touching the effects of the aloes. The first qualities, warmth, cold, humidity, dryness, are caused to a certain degree, and less warmth remits something which, in degree, causes more warmth. The second and third qualities are aperient and dioretic, more strong and intense, being more dioretic, united with other things becoming less so. Thus the aloes being more purgative, when united with honey become less purgative, and all become less soluble. Hence it comes that it purges more with ten grains of *escamonea*[1] than with five drachms of solution, and an ounce of *cassia fistula*. In one drachm of rhubarb there enters more *escamonea* than in twelve grains. I have tried this many times, and can give no other reason. And now tell me whether they have aloes *metallico* around Jerusalem.

ORTA

I enquired respecting this of some Jews who came here and said they were inhabitants of Jerusalem. Some of them were sons of physicians, others of apothecaries. They all told me that it was untrue, and that there was none in Palestine; and here we may make an end of the aloes, if you are satisfied.

RUANO

Before you conclude I must offer you many thanks. But

[1] Scammony, the gum resin from the root of *Convolvulus Scammonia* of the Levant, and now obtained also from the root of *Ipomora orizabensis* of Mexico.

I would also ask you first to solve a doubt as to the way in which they take the pills and purgatives in this country, and how long a patient is kept from food after taking them. I ask this to see whether the followers of Avicenna who cure kings in this country, have the same custom as we have in Spain.

ORTA

I say that the pills are taken in the way we take them, as also are the liquid purgatives. The time is at break of day, without eating or drinking anything for five hours. These are taken to quiet the stomach according to the rule of Avicenna, consisting of two drachms of ALMECEGA dissolved in rose water, the stomach being rubbed with the bile of a cow, then wet cloths are put over the abdomen to promote motion and stimulate the expulsive virtue. The motion is very satisfactory. At the end of the five hours the patient drinks three ounces of chicken broth, but nothing else, sleeps a little and drinks some rose water. After sleeping he purges well. They say that the broth and rose water fortify nature, but that it is not good to eat much as it impedes digestion and the purging is less effective. Enquiring if this treatment is given to all that require purging, they replied that this was the usual practice among learned physicians, but they allege no authority whatever for it.

RUANO

They have a good deal of reason in what they do and practise, for the bile as a solutive applied exteriorly has a biting effect in compelling expulsion; and not to eat fowl is the express text of Avicenna, where he says that he who has to take medicine should take much in the morning, and eat in the afternoon, after three hours, four ounces of bread and a little wine and water, six hours after that he should go into a bath, remaining there a convenient time, and then have another meal. This is the text translated into the Portuguese language, according to the version of Belunense. However these Moors have not adopted this practice without authority, nor is their custom without reason. But Mateus de Gadi gives a different version of this text, and applies it chiefly to the treatment of sciatica, for (saving better judgment) it may be applied to many diseases. And do they use the bath recommended in the text?

ORTA

Yes they use it, but not on the same day. The bath is taken on another day afterwards. A bath is prescribed to Brahmins, Baneanes,[1] and all Gentios who never eat without first washing the body, and the Moors bathe every third day at the least.

RUANO

Why do you take the text amended by Belunense? I ask whether you find that translation correct.

ORTA

I have tried it many times, and read the ordinary translated text, having the Arabic of Avicenna before me. It agreed with what I said, and the translation containing the corrections of Belunense. As it is now the hour for dinner, we will not converse more, but after dinner we will talk of the amber.[2]

[1] A Hindu trader, especially from Guzerat, spelt *Banyan* by the English.

[2] Acosta copies the whole of this Colloquy on the Aloe. He gives a plate. It is his chapter xxv., headed "Azuiar."

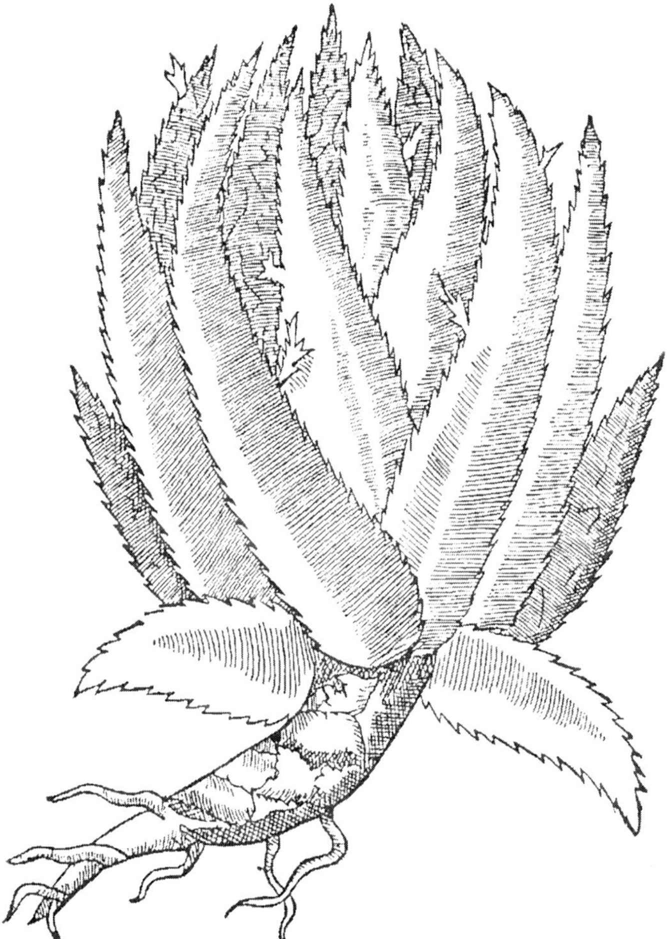

Plate I.—Azuiar.

THIRD COLLOQUY

AMBER

RUANO

I SHOULD first like to know about ALJOFAR.

ORTA

And I formerly wished to have much of it, both large and perfect, that I might study it. In the chapter on MARGARITA we shall find for what it is necessary and useful. Now we will speak of AMBER, for it also is a medicine, and it is worth more to have plenty of it than to know how it is generated.

RUANO

Tell me the whole truth about it, and set forth the facts.

ORTA

The Arabs call it AMBAR[1] and the Latins AMBARUM according to the Latin variation of words. Other nations and languages, as I know, give it the same name or with very little variation.

[1] *Amber* in Arabic means "perfume," and the word is applied by the Arabs to our Ambergris; while Amber, they call *inkitrium*, a corruption of the Greek ἤλεκτρον.

RUANO

What reason do you give me that it has the same name among all people?

ORTA

There are certain names that do not vary, or vary very little in all the languages that I know, such as AMBAR, LIMÃO, LARANJA, SABAM, and some others. But by many LIMÃO is called LINBON, and LARANJA turns into NARANJA[1] and so with many others.

RUANO

How is it generated and what is it?

ORTA

Some say that it is the foam of the whale, and affirm that it is the excrement or the foam of a sea animal, others that it comes from a fountain issuing from the bottom of the sea, and this appears the best explanation, and most conformable to the truth. Avicenna and Serapiam believed that it generated in the sea, in the same way that fungi generate on rocks and trees, and that when the sea is tempestuous it gives out stones which turn into amber. This opinion is also more conformable to the truth than other theories of Avicenna. For when the wind is high much amber is cast on the Comoro Islands, Emgoxa, Mozambique, and all that coast. And because the Maldive Islands are to the east, more reaches those islands when the wind comes from the west.

RUANO

But I am not quite clear why that long chain of islands is called the Maldives.

ORTA

In the matter of the names of lands and seas and regions

[1] The Lemon and Orange are both natives of the Indo-Persian Caucasus region, and the names of both are Indo-Persian, almost unchanged, that is *limu* and *narang* respectively. We say "*an* orange" instead of "*a* norange," just as we say "*a* newt" for "*an* ewt."

are we not a good deal misled from not knowing that the foreign languages supply the etymology of the names? I must tell you that I have heard it said that the natives do not call it Maldiva but Nalediva. In the Malabar language *nale* means four and *diva* island. So that in that language the word signifies "four islands," while we, corrupting the name, call it Maldiva. In the same way we call an island, which is 12 leagues from Goa, Angediva because it is a group of five islands, and this is the meaning in the Malabar language in which *ange* means five. These derivations are well known, and I, therefore, do not wish to make a proof by demonstration.[1]

RUANO

I rejoice much to know this for it satisfies one's understanding, so that when they talk of this I can give a more worthy account. And now we can proceed with the amber.

ORTA

Avicenna and Serapiam say that some amber is swallowed by a fish called AZEL which dies as soon as it has eaten it, and its body goes floating on the sea. The men of that region take grains and draw them out so that they penetrate to the amber which is not good, except some which have reached the backbone. This, they say, is good and pure, and this according to the length of time that it has been in the stomach or against the backbone.

RUANO

And what do you think of that? Is it likely?

ORTA

No, for I have made enquiries and no one has said that they have seen it.

[1] Yule refers to Orta's derivation, but considers that it is certainly wrong. But he adds, "hard as it may be to say what is the right one."—*Glossary*, p. 546.

RUANO

That reason does not seem conclusive on all points, and, as you are learned and no longer young, you should give another.

ORTA

I say that the unreasoning animals, by a natural instinct, seek the food that is suitable for them and not that which is poisonous, except when what is injurious gets mixed with that which is wholesome—in the same way as we deceive the rats when the food they know to be good is mixed with red arsenic. So that we need not suppose that the fish goes to seek for the amber which will kill it. I say further that amber is one of the principal cordials, and though it may be poisonous to the fish, it also has virtues which are the reverse of fatal. These reasons, though not actually demonstrable, appear to me to be plausible.

RUANO

And to me conclusive as you have not seen any persons worthy of belief, with different experience. I would ask now what you have heard respecting amber which is worth so much, where the best is found, and for what it is used in these parts.

ORTA

First I must tell you of a great error of Avenrrois,[1] who says that amber is a kind of camphor which originates in the fountains of the sea, and not on the surface water, and that the best is called by the Arabs ASCAP. I asked the physicians of NIZAMOXA (vulgarly called NIZAMALUCO) what amber this could be, but they were unable to satisfy me as they do not possess the works of Avenrrois and Abenzoar. But that the statement of Avenrrois is false and unworthy of the great philosopher is clear. First in saying that camphor originates in the sea, for camphor is cold and dry in the third degree, while amber is warm and dry in the second degree, whence it

[1] Averroes [*i.e.* Abul Walid ibn-Rushd] here obviously refers to Ambergris.

is clear that they cannot be included in the same genus. In conclusion I tell you that as there are lands where there is red earth such as ochre or Armenian *bola*,[1] and others where there is loam or *cardea* (?), so may there not be islands or lands where there is amber, where the ground may be spongy or in some other condition. That this may be true is proved by the great quantity of it, for I have seen pieces as big as a man, and one piece 90 *palmos* in circumference and 18 long. Some people declare that they found an island of amber, noted its position, and returned to the land whence they came. Wishing to go back for the amber, they collected provisions sufficient for the voyage, but were never able to find the island. Perhaps it was God's will that they should not discover it by reason of the castles of vanity they would raise on finding it, and because of the small thanks they had offered up to Him for the discovery. Also because these men might be saved having little of this world's goods, and not if they were wealthy. For God, who is merciful, knew best what was good for them. In the year 1555 I found beyond Cape Comorin a piece of near 30 quintals, and thinking that the find was tar I held it very cheap. However on being divided among many persons it returned to its accustomed price. The locality where I found it was in front of the Maldive Islands. It is manifest that this is true because the pieces come sometimes in the beaks of birds, and at others in the shells of various molluscs, for they stick to the amber, and the birds have it lodged on them, the cleanest being the best. This that I tell you is the most accurate information that can be had.

RUANO

Is it found in other parts such as the coasts of Ethiopia ?

[1] Bolarmenico ; a red silicate of alumina ; the μίλτος of Homer, *Iliad*, ii. 637 ; the "rubrica" of Pliny, xxxiii. 7 (38) ; the "Bolum Armenum" of the old "Dispensatories" ; the *Bolearmonike* of "The Nomenclator" of 1585 ; our "red-ochre."

AMBER

ORTA

Some is found in Timor, but seldom and in small quantity. They tell me that it is also found in Brazil, and in the year 1530 a piece was found at Setubal. But no account need be taken of these finds, as they seldom occur and in small quantities.

RUANO

Now tell me why it may not be the foam of the whale or its evacuation.

ORTA

This notion cannot be accepted because the whale oil, that I have seen, has an abominable smell, and not like the amber. Besides there are whales in many parts, and none yield amber, either in Spain or Galiza. For the same reason it cannot be from the foam of the sea, for where the waves are heard in contest with the wind there is foam, and those who say that a fish eats it are confuted and proved to have a false notion. This is what the Arabs say, for the Greeks did not find this simple, or only Aecio.

RUANO

Which is the best kind to choose?

ORTA

The kind that is nearest to white is the best, for instance the grey amber, or that with alternate grey and white streaks, and especially when it is light. The proof is that putting a pin into it, more oil is cast into the hole, and this is the best. The black amber is very bad, and I had a piece of it for a small price, which had very little smell. It was mixed with ALMISCRE[1] to make beads. It mixed very badly making many cracks. That which is as white as an ostrich's egg is very bad according to Serapio. I have not seen or heard of any one

[1] The *salip misri* of Egypt, Persia, and India, or tubers of various orchids of the genera *Orchis* and *Eulophia*, which in accordance with the doctrine of "signatures" are all reputed powerful aphrodisiacs; and strong glues.

who has seen it. If any one has seen it, it must have been adulterated with plaster.

RUANO

Menardo says, in his *Letuario de gemis*, that amber is a new thing, and it is esteemed according to its price. Further in his *Letuario di ambra*, he says that an electuary composed of it is very precious, much used by women and old men. It seems to be a contradiction of this doctor to say that it does not cost so much as the *Letuario de gemis*. With the amber it makes a very beautiful mixture which is much used and prized by the people of this India, and not only by us. And first, with reference to what I have said, explain to me some names which are in Serapiam and Avicenna. For instance, Serapiam says that there is much amber in the lands of Zing.

ORTA

It must be the parts of Sofala for ZINGUE or ZANQUE, among the Persians and Arabs, means *cafre*[1] or black. Because all that coast of Ethiopia is inhabited by negroes. Serapiam calls it ZINGUE. Avicenna also makes mention of Melinde, calling it ALMENDELI. What they call SELACHITICUM is Ceilão, one of the most famous islands in the world, possessed by the King our Lord, and not very far from the Maldives. It is not a city, as Laguna says, but an island with many cities. The greatest quantity of amber comes from Sofala as far as Brava. There is also some on the coast of Arabia, and a greater quantity, it is said, on the coast of Ethiopia.

RUANO

Is it much esteemed among the Indians and Moors of this land?

ORTA

Among the rich and powerful, yes! not used for food, but

[1] Arabic, *kafir*, in the plural *kafra*, "black," as in Caffraria; and Persian, *zenj*, "blacks," as in Zanguebar. It was the Orientals themselves who originated the "phrase," in its opprobrious sense, which we translate, "D—d Nigger"; as also that which we translate, "I don't care a dam(n)."

for medicine in conformity with the rule of Avicenna, and according to the quantity, for an ounce of the best is worth more than precious stones.

RUANO

What is the size of the largest piece you have seen in this land?

ORTA

I have seen a piece which weighed more than fifteen pounds, but I am told that there are much larger pieces in Ethiopia. I do not know for what price they are sold, but I am certain that pieces reaching the Nizamoxa will be bought at a good price, judging from the value that is put upon large pieces. This amber is not only worth much among the Moors, it also has a high value with the Gentios, and what is marvellous, it has a still higher value among the Chinese. For our Portuguese take it to them, and sell it at the rate of 1500 *crusados* for a *cate*, which is twenty ounces. However, our people are taking such a quantity that it will soon be cheaper, each time being worth less according to the cupidity of those who bring it.

RUANO

How do the Chinese know that it is a good medicine, having bought it at such a high price?

ORTA

Diogo Pereira, who is a nobleman very well known in those parts, told me that the Chinese believe all that we hold concerning the engendering of amber, and say the same thing word for word. They declare that it is very efficacious for the ailments of women, and that it is good for the heart, the brain, and the stomach. And now, leaving the smell of the amber,[1] let us pass on to the amomo.

[1] Acosta devotes his chapter xxvi. to Amber, copying what Garcia da Orta says, slightly condensed.

FOURTH COLLOQUY

AMOMO

Ruano

THERE arises so much doubt as to what the *amomum*[1] is, that some writers would take the *acoro* for it, seeing that Galen gives it the same virtues; but there is also much doubt about *acoro*. They say that the *amomum* enters into *tiriaca*. For this reason Mateolo Senense weeps for the perdition of the human race in having lost the *amomum*, for, without it, help cannot be given in curing the infirmities of men. This writer says that he does not hold for certain that this drug enters into the *tiriaca*[2] of Andronicus. For this he is blamed and reprehended by some writers because they had affirmed that under some headings he stated that it did enter into *tiriaca* and under others, according to them, he had said the contrary. Nor does Mateolo offer us a remedy, but laments

[1] "Cardomums," derived from *Amomum Cardamomum, Elettaria Cardamomum*, and other species of the Zingiberaceae.

[2] *Theriaca, i.e.* "treacle" [from French *triacle*, and *thériaque*], an antidote to bites of snakes, and to poisons generally, and prescribed by Galen for the Emperor Marcus Aurelius Antoninus who took a small dose every day. Galen's prescription is lost.

the loss, adding that neither can that which is called the rose of Jericho[1] be also the *amomum*. For this he gives several good reasons and amends many texts. What you have heard related will never be finished. You can see that yourself, or you may see it from Laguna and others. Certainly, according to many writers this *amomum* enters into *tiriaca*, and it is not good to experiment with unknown medicine. I have a strong desire to learn whether the *amomum* is in this land, and whether the Moorish physicians have it who cure the kings, believing it to be *pes columbinus*,[2] for this is a great error, as the above authors have proved.

ORTA

If in this land I have seen the simples that are in your land of Europe I will draw out your doubt, for I will tell you all that is known of the matter in India. These modern writers say that *tiriaca* cannot be made for want of *amomum*. I asked an apothecary who was Spanish in language, but Jew by his false religion, who said he was from Jerusalem, what *amomum* was. He told me that it was *hamama* in Arabic, which means "dove's foot," and he had known it very well; but that he had seen it in his own country, and not in this, and of this there could be no doubt. Some years afterwards I went to visit Nizamoxa, when I asked his physicians whether they had *amomum*. They said that it was not found in their land. But among other medicines that were brought to the king from Turkey, Persia, and Arabia, for which high prices were paid by reason of the need there was for them to make up prescriptions, *amomum* also came. Among these prescriptions there was a *mitridato*.[3] They gave me some of the *amomum*, which I took

[1] *Anastatica Hierochuntina*, the *Kaf Maryam* or "[Virgin] Mary's Flower" of Palestine, which expanded during the birth of the infant Jesus, the Lord Christ.

[2] *Geranium dissectum* is the "Dove's Foot" of modern botanists.

[3] The "Theriaca," prescribed for the daily use of Mithridates Eupator, King of Pontus, B.C. 20.

to Goa, showed it to the apothecaries, and compared it with some drawings of the simples of Dioscorides. It all appeared to be conformable to the drawing and, though it was dried, it seemed to be the same as the "*dove's foot.*"

RUANO

I do not find this argument to be convincing, for this they would call the "*cow's tongue*" of Avicenna, which I doubt.

ORTA

All the names that we have in Avicenna have been translated literally from the Arabic. It is called *cow's tongue, bird's tongue, dog's tongue,* and *capillus veneris.*[1] The infirmities are also called with reference to the name. The elephantiasis is *daul alfil,* which means foot of elephant. Hydrophobia is *maraz alquelbe,* which means sickness of a dog. Whence I know that *dove's foot* is intended by Avicenna to indicate *amomum.* It is the same as regards many names derived from Avicenna, and we Spaniards imitate the Arabs in this.

RUANO

What use does this king put the *amomum* to?

ORTA

It enters into *mitridato,* which composition is much used from fear of poison, and they have this medicine sealed under their hand. For these kings, or rather tyrants, of this land are much in the habit of poisoning their brothers. Talking with this king one day I tried to discover how they made *tiriaca.* He said that it came in a barrel with a man to make proof. He would, if genuine, buy all the *tiriaca,* weighing out for it the same weight of gold, if he should make proof, which would amount to 2000 *pardaos,* which price is equal to one Spanish crown. Certainly if the devil does not take it first for the partnership of Mafamede, he would keep his word.

[1] The *Adiantum Capillus Veneris* of modern botanists is a fern.

RUANO

Tiriaca is cheaper in Europe, and it is wonderful what a small value it has owing to the quantity there is. You will see there other medicines respecting which there are doubts among us, that is respecting a knowledge of them.

ORTA

Such as *eupatorio* and *mexquetera mexir.*

RUANO

You certainly know that they have not the medicines you mention in this land.

ORTA

It may well be that that is so, for the apothecaries of India gain more by practice than by sale of drugs, and as the profit is small, I neither go to the continent nor to BALAGUATE[1] to seek *herva cidreira, lingoa de vacca, fumus terrae, tamarisco,* or *esparegos,* of which medicines we are in need, but I saw them there. I also saw violets sown in the king's garden. Here in Goa we use, instead, certain flowers from trees, very different from our violets. I do not consent to the use of these, except as lotions, and I order to be made of violets, in conserves, a lotion called *xarope violado,*[2] brought from Ormuz or from Portugal.

RUANO

More curious are our apothecaries in Spain with their poverty, for the love of money grows as the money itself increases.[3]

[1] The country above the Western *ghats,* or stepped hills of Western India, *i.e.* the tableland of peninsular India.

[2] Violet syrup.

[3] Acosta devotes a very short chapter to Amomo, condensing what Orta says (xxviii.), and gives an illustration.

FIFTH COLLOQUY

ANACARDO

Ruano

I WANT to know about Anacardo.[1] The Greek name is derived from the name of the *heart* which is *cor*. I am surprised that there is no mention of this medicine among the ancient Greeks.

Orta

You need not be surprised, for the modern Greeks give this name for the reason you have mentioned. It was a medicine used by Arab writers, and this was not the reason that they changed its name. For they call it Balador, and you find it written in a doubtful manner in their books, so that the word has become corrupt. The Indians call it Bybo, and we Portuguese Fava de Malaqua, because it appears like our large beans on the tree on which it grows, and almost like the beans we saw first at Malaqua.[2] According to what some say there is abundance

[1] *Semecarpus Anacardium*, Linn., Anacardiaceae; the "Marking Nut" of Anglo-Indians.

[2] The fleshy receptacles on which the seeds rest are roasted and eaten. The green fruit pounded into a pulp makes good bird-lime. The tree is 50 feet high with leaves entire, obovate, rounded at the apex.—Drury.

of this medicine[1] in Cananore and Calicut, and in all the other parts of India that I know, namely Cambaya and the Deccan.

RUANO

Antonio de Lebrixa, in his dictionary, gives *anacardus* as a herb often mentioned by Galen.

ORTA

It is true that Lebrixa says this, and he was very learned and curious, but he was deceived as regards the Greek name. Without hearing more he said that Galen made the statement. He was careless, and you need not be surprised at this, for the good Homer sleeps sometimes. Serapio also quotes Galen, who never saw the ANACARDO, and further says that it sometimes kills, which is contrary to our experience; for in this country they use it in milk for asthma and also for worms. When it is green they make a conserve of it with salt, called ACHAR, and sell it in the market as we do olives. When it is dried they use it, like caustic for scrofula, and all India uses it for marking cloths,[2] mixed with lime. Avicenna says that the ANACARDO has a fruit like the kernel of tamarinds, and that the pith is like that of an almond, which does no harm. But lower down he says that it is counted among the deadly poisons. Yet he is more clear than Serapiam, who finds things that put him in doubt, and the contradiction is more clear, for in one place he says that it does no apparent harm, and in another that it is counted among the deadly poisons.

RUANO

Where he says that it does no harm, he means that it does no apparent harm at first, though finally it is fatal.

[1] The acrid juice of the shell is used for rheumatic pains, aches, and sprains.

[2] It is still generally used for marking cotton cloths. See footnote, page 32.

ORTA

Also this may save the question, that it is not poison for everybody, for many Indians eat it, and it is used as a caustic when it is dried.

RUANO

In what degree do you place it—warm and dry ?

ORTA

They place it in the fourth, warm and dry, others in the second part of the third; but neither of these satisfy me, for when green it is clearly not so warm and dry. It, therefore, does not appear reasonable to make it as warm and dry as other spices, such as pepper, which is placed in the third degree. Nor is it found to be vermilion, but a clear black, and for this there cannot be another excuse; but the Sicilian kind will be warm and drier and has a colour more like vermilion.

RUANO

I am quite in agreement with what you say. It seems to be a very good preparation with sour milk for asthma, understanding by sour milk, the milk that has been drawn from its butter, and this agrees with Avicenna.[1]

[1] Acosta treats of *Anacardo* in his chapter xxx., quoting Orta, but gives no Illustration.

SIXTH COLLOQUY

ARBOR TRISTE[1]

Ruano

I COMMENCE, in the name of God, with the drugs and simples which are neither known nor ever seen by us. What is that tree which smells so sweet when it is placed in the sun until it blooms? Tell me whether it is used as medicine or as food. For me I could not wish to experience a more pleasant smell than when I come suddenly to where this tree is growing.

Orta

I have not seen this plant in any other part of India except Goa, and they say that it came here from Malacca. It would be a very good thing if it could be raised elsewhere, for it was grown near Goa and took very well. But, as I say, I am uncertain whence it came.

Ruano

Tell me the name and use of these flowers which I only know from their scent.

[1] *Nyctanthes Arbor tristis*, Oleaceae.

SIXTH COLLOQUY

ORTA

The scent is not so useful, because the flowers in that flower bed, called MOGORY,[1] smell sweeter than the orange flower, and are more valued in giving an agreeable scent to food, in the way that dishes are made to smell sweet in Spain with orange flowers. We use the water of MOGORY for the same purpose. But I have not seen water from the plants used respecting which you enquire, and it may be that they do not yield good water, for their virtue is very superficial, as is the case with the cloves they have in Portugal. We use these flowers to give a tinge to the dishes, like *açafram*.[2] The corollas are yellow and give a strong tinge. In the Goa language their name is PARIZATACO; in Malay, SINGADI.

RUANO

Is the tinge from these flowers slight, like that given to dishes by *açafram* in Spain?

ORTA

No, for it causes a slightly bitter taste.

RUANO

And whence do they say that the *açafram* of this country comes?

ORTA

It is from some roots which grow here, whose virtue I will state presently.

RUANO

Can you make distilled water from the MOGORY flowers you praise so highly?

[1] *Jasminum Sambac.*

[2] *Açafram* = Saffron.

ORTA

Yes. You see them in this flower bed, and you have seen their water which I have, and which is put into a quill to clean the teeth, and which you also have highly praised.

RUANO

Until now I always thought that it was orange-flower water. The people of this country seem much addicted to scents, and owing to this it is said that they are inclined to Venus.

ORTA

Some plants are left for food while others are used for their scent, such as sandal wood, which is very commonly used to anoint the body, and the LINALOE, AMBER, ALMISQUE, and ALGALIA, much used because the price is not so high. This is because there are many cats in many parts of India. This ALGALIA is used for pains of "cold humour" (*humor frio*), anointing the part that hurts with it. There are other flowers which are much used in this country called CHAMPI,[1] which have a very strong smell. It is like a white lily, but not so soft. Know that the kings I have seen cover the floors of their houses all night, and for the greater part of the day, with these flowers and roses. They also have various flowers painted in colours, which look very well. Here at night they take their pleasure and receive the presents from the poor, which consist of these flowers and roses. The expenditure on these flowers is such that I am assured that, in BISNAGUER, the scents bring a revenue to the king of 5000 *pardaos*. It is still more worthy of note that in ORMUZ the

[1] *Michelia Champaca* (Magnoliaceae). This tree is venerated by the Hindus, and dedicated to Vishnu. It is celebrated for the exquisite perfume of its flowers. The name *Champaca* is said to be derived from *Champa*, an island between Cambodia and Cochin China where the tree grows.—DRURY.

labourers who obtain food by working as carriers, buy scents to anoint their bodies at night, and leave the food. As you like to hear the legends and fables of these gentiles, I may mention that this tree was the daughter of a great lord named PARIZATACO. She became enamoured of the sun, who abandoned her after their intercourse, and she killed herself and was burnt (as is the custom in that country). From her ashes grew this tree, the flowers of which abhor the sun and do not appear in its presence. It would seem as if Ovid had been in these parts, for he composed his fables after this manner.

RUANO

Certainly it gives rise to wonder that these are flowers of night and not of the day. Do not trouble yourself to tell me the size and appearance of the tree, for I saw that it was about the size of an olive with leaves like the plum tree. But as this is not medicinal, let us go on to consider the *asafetida* and *anil*.[1]

[1] Acosta's chapter xxvii., on the "Arbor Triste," is partly from Orta and partly original. He gives a plate of the leaves and flowers.

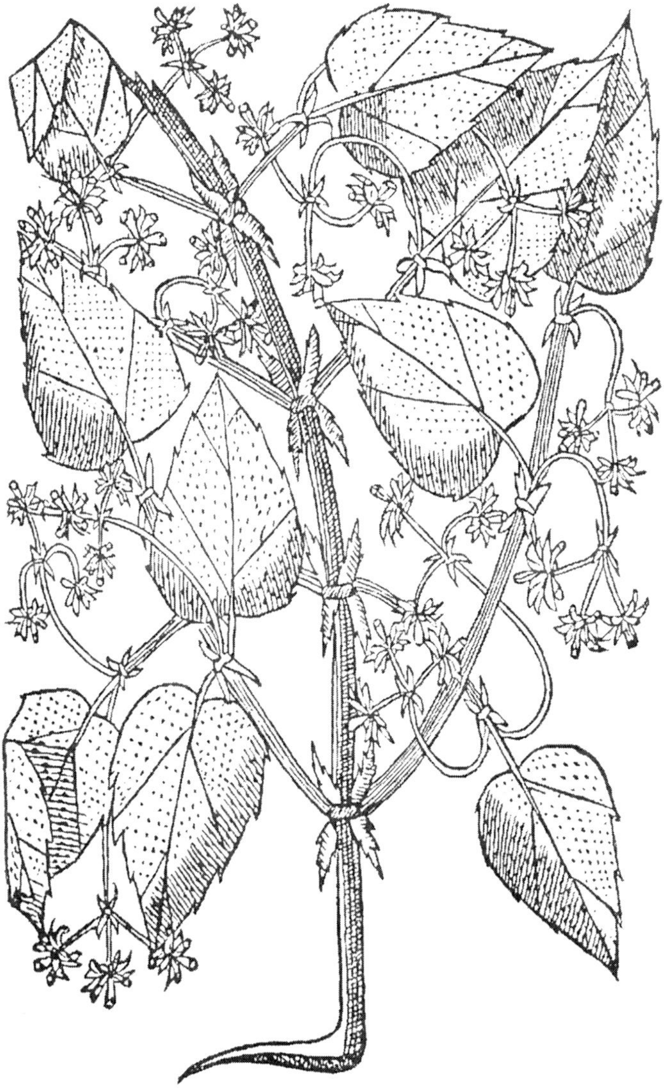

Plate II.—Arbor Triste.

SEVENTH COLLOQUY

ALTIHT, ANJUDEN, ASSAFETIDA, DOLCE, ODORATA, ANIL

RUANO

WE should know about the plant which is called ALTIHT,[1] ANJUDEN,[2] ASSAFETIDA, DOLCE, and ODORATA. Between it and *Laserpicium* the doctors make some difference.

ORTA

I have found more confusion in those names than you have, and this was because they never knew how to tell me the way it was made, nor the tree from which the gum was extracted. They told me that some came from Coraçone[3] to Ormuz, and from Ormuz to India, and also Guzerat. I was also informed that it came from the kingdom of Dely,[4] a very cold country, which is bounded by Coraçone on the other side, and by the region of Chiruan,[5] as Avicenna points out. Without doubt this gum is called ALTIHT in Arabic; some have ANTIT, and any Arab to whom it is shown

[1] Arabic form *hiltit* حلتيت : used by Edrisi.

[2] Or *angadan*, another Arabic name of the plant.

[3] Khorasan. [4] Delhi.

[5] Shirvan, on the west side of the Caspian.

would give one of those names. The Indians call it IMGO,[1] or IMGARA, and the tree from which it is extracted is ANJUDEN or ANGEIDAN.[2] But as this merchandize comes from far inland it is difficult to find out, for certain, what the tree is. Avicenna gives it several names, and this may be because it has one name in one country and a different one in another; for instance, in one ALTIHT, in another ALMHARUT,[2] for it is known that there are diverse languages in the countries whence it comes.

RUANO

What was the reason that the translator turned it into ASSA?

ORTA

I do not believe that the translator wrote ASSA, but LASE, and the word was corrupted into the other form through the lapse of time.

RUANO

First, we must see that ASSA (*fetida*) is the same as LASER or LASERPICIUM. I must tell you that ALTIHT does not appear to me to be the name of the tree, but the juice of the ALCACUZ thickened and hardened; and this was the belief of Gerardo Cremonense in the chapter, "*Da falta do coito en Rasis.*"

ORTA

Gerardo Cremonense was not a good Arabic scholar. He was an Andalusian, while the proper language in which Avicenna wrote was that used in Syria and Mesopotamia, as well as in Persia and Tartary where Avicenna was. They call this language Arabic, and that of our Moors Mogaraby, which

[1] The Sanskrit *hinga* and *hingula*, from *han* = "to kill." Asa-foetida = "stinking mastic," *aza* being the Persian for mastic. The earlier European names of the drug were ὀπός Μηδικός and "Laser Mēdicum"; also "Sterx Diaboli" and "Cibus Deorum," as people disliked it intensely—or liked it. It is yielded by the root of *Ferula foetida*, Regel, and other umbelliferous plants; and possibly by certain composite plants.

[2] *Mahruth* صروث referring specially to the root.

is as much as to say Western Moors. For *garby* in Arabic means "west," and *ma* "of the," so that Gerardo is not very far wrong in this. I say that ALTIHT does not mean anything but the tree of the ASSA FETIDA, and the gum has often been mistaken for the tree.

RUANO

And if ALTIHT is not ASSA DULCIS, what is ASSA DULCIS?

ORTA

ASSA DULCIS is not named by Arabian nor Greek nor Latin doctors of any authority, and if any one uses the name he is in error. For ALCACUZ is called CUZ in Arabic, and its juice boiled is called ROBALCUZ by the Arabs, which the Castilians have corrupted into RABACUZ; so that ROBALCUZ is a name composed of ROB which in Arabic is juice made thick, AL the genitive of the article, as much as to say, "the thick juice of ALCACUZ"; hence we do not call this juice ASSA DULCIS.

RUANO

This derivation seems to me to be right, but before I ask you why LASERPICIUM is ASSA, I want you to tell me how Avicenna belongs to the land of the Tartars, and why the language of our Moors is not so good as that of Syria and Arabia.

ORTA

Avicenna was a native of these parts, and not of Spain. The physicians of Persia and Turkey who cured that king, I have already mentioned to you, told me that Avicenna was of a city called Bochara which is in the Uzbeque province, a part of Tartary, as we call it, or of the Moguoras as it is known to the natives. Andreas Belunensis calls that country Persia, but this is an undue extension of Persia, for Persia is a small kingdom. Afterwards I met some discreet and well-informed merchants who had lived at Ormuz for a long time, and I asked them the position of Bochara. They told me that it was in the country of Uzbeque, and that there was a great deal of *manna* in it. I was told the same by Coge Perculim, a learned man who was in Goa. As the son-in-law of Belunense said that Avicenna was a person of weight and learning, I enquired if he was a king. They answered that he was not, but that he was *guazil*,[1] which with us is as much as to say a magistrate, a noble.

RUANO

This appears to me to be true, for we, by the chronicles of Spain, know who the kings of Cordova and Seville were at that time, and we do not find his name. But I believe that he was an important person whoever he may have been.

ORTA

Answering your other question I may say that it is a difficult thing to prove that one language is better than another. These physicians and the other men of letters, they call Mullas, say that the works of Avicenna, of Galen, and the Greek philosophers, and of the false prophet, were written in the language of Syria. The other language of our Africa they call Barbara, and our Moors are known to them as MOGARABY. For this reason the Moors of Persia and Arabia call those of our lands, known to us as ALGARVES, by the name of ALGARBY, which means Moors of the west, for to us Algarve is to the westward. I regret that I have dwelt so long on this matter which is not connected with the subject of our discourses, but the fault is yours.

RUANO

I rejoice much to know this of which you think little, and I must take the blame on myself. The LASERPICIUM is not

[1] Wazir to a prince of Hamadan, afterwards at Ispahan.

ASSA-FETIDA, nor is it odoriferous, therefore that LASERPICIUM, of which Dioscorides and Pliny write, does not seem to be the ALTIHT of which Avicenna and other Arabs write.

ORTA

The Arabs who mention this drug, such as Rasis and Avenrrois, say little about it, and Serafio, in speaking of ALTIHT, says that all that Galen and Dioscorides write on the subject refers to LASERPICIUM.

RUANO

For many reasons your information goes to prove that ASSA-FETIDA and LASERPICIUM are different drugs, for LASERPICIUM is a medicine for preparations used for curing, and also for cooking, while ASSA-FETIDA is for medicine only by itself and to be used very seldom. To use it in cooking would damage all the food by reason of its horrible smell.

ORTA

I will leave you with this error and proceed, for if you want to know my intention it is necessary that you should turn your attention to these new writers, and be pleased to hear my truths without rhetorical colouring; for the truth must be depicted in its nakedness.

RUANO

I have told you many times that I desire nothing better than to get rid of my errors, and to sow my understanding with new seeds.

ORTA

Well, you must know that the thing most used throughout India, and in all parts of it, is that ASSA-FETIDA, as well for medicine as in cookery. A great quantity is used, for every Gentio who is able to get the means of buying it will buy it to flavour his food. The rich eat much of it, both Banyans and all the Gentios of Cambay, and he who imitates Pythagoras. These flavour the vegetables they eat with it; first rubbing the pan with it, and then using it as seasoning with everything they eat. All the other Gentios who can get it, eat it, and labourers who, having nothing more to eat than bread and onions, can only eat it when they feel a great need for it. The Moors all eat it, but in smaller quantity and only as a medicine. A Portuguese merchant highly praised the pot-herb[1] used by these Banyans who bring this ASSA-FETIDA, and I wished to try it and see whether it pleased my taste, but as I do not know our spinach very well, it did not seem so palatable to me as it did to the Portuguese who spoke to me about it. There is a respected and discreet man in these parts, holding an office under the king, who eats ASSA-FETIDA to give him an appetite for his dinner, and finds it very good, taking it in doses of two drachms. He says there is a slightly bitter taste, but that this is appetising like eating olives. This is before swallowing, and afterwards it gives the person who takes it much content. All the people in this country tell me that it is good to taste and to smell.

RUANO

And did you find a bad smell in the pot-herb that you tasted?

ORTA

The nastiest smell in the world for me is ASSA-FETIDA, but the pot-herb seasoned with it has not a bad smell. You need not be much surprised at that, for onions have a very bad smell, yet the dishes seasoned with them are very good. The truth is that there is a good deal of habit in the matter of smells. For instance the betel (that which these people are

[1] A score of species, or gardeners' varieties of Amaranths, *Amaranthus viridis*, etc., are used as pot-herbs in India, and seasoned with just a whiff of assafoetida they are most appetising.

continually chewing) has a very nice smell to those who use it, and to me a very nasty one. As a medicine ASSA-FETIDA is used by itself, and not, as you suppose, only with other ingredients. In this you are mistaken, as Sepulveda was mistaken, for Guarinero and many others use it by itself. Among the Indians it is good for the stomach. A Portuguese in Bisnaguer has a valuable horse who was broken winded, and for this reason the king did not wish to buy it. The Portuguese cured it by administering this YMGU with flour. The king gave a good price for it when it was cured, and asked how the cure was effected. The man replied that it was with YMGU. The king said that this should not cause surprise, for this food is the food of the gods which the poets call nectar. The man, but in a low voice and in Portuguese, observed that it should rather be called the food of devils.

RUANO

One doubt assails me. How can the Banyans eat it continually, when Matheus Silvatico says that it is poison, referring to Galen as his authority?

ORTA

Galen and the Greek druggists do not say this. On the contrary they say that it is good as an antidote to poison, and for the plague, lumbago, and diseases of the kidneys which are contrary effects. Matheus Silvatico may be pardoned for these mistakes like many others. The Indians apply it for toothache. Pliny says that it gave such pain to a person who applied it to his tooth that he jumped out of the window.

RUANO

Is this medicine very dear in India?

ORTA

Yes (for amongst us it has little value), the cause being that it soon deteriorates, and the men who supply it keep up the price. There is much in Mandou, Chitor, and Delhi, and besides it comes as merchandize from Ormuz to Pegu, Malacca, and Tenasserim and those parts, and when it fails to come the price is extremely high.

RUANO

Do they use the roots or leaves, because the roots, leaves, and branches are commended by the ancients?

ORTA

I have told you that I have not seen the tree nor can I describe it.[1] No people known to me use anything but the gum which is obtained by making cuts in the tree. This information was given to me by a man who had taken the medicine. He also told me that they had shown him the shape of the leaf, which was like our hazel leaves.[2] They added that, to preserve the gum, they kept it in bullocks' hides, first anointed with blood, and then mixed with wheat flour, so that when they find a thing that looks like bran they do not take it to be a falsity at once, as some say, but make sure. If I should ask an educated Banyan why he uses a medicine which comes mixed with bullock's blood, I should not be far wrong in assuming that his answer would be that it is necessary to preserve the drug in this way.

RUANO

The *Laserpicium*[3] of old had a somewhat translucent reddish-yellow colour, while what we use is turbid and dirty.

[1] It is a very large herbaceous plant with an erect stem 70 feet high. It is found round Samarcand and in the Persian province of Lar, where it was first described by Kempfer in 1687.

[2] Quite wrong.

[3] *Laserpitium*, and *Laser*, were originally one and the same drug; and *Silphium* the plant from which *Laserpitium* or *Laser* was obtained in the Cyrenaica; and judging from the *Silphium* plant represented on the coins of Cyrene, Laserpitium was assafoetida [see footnote, page 41], and Pliny, v. 5; xvi. 33 (61); xix. 3 (15); and xxii. 23 (48 and 49). Modern botanists have identified the *Silphium* plant with *Thapsia Silphium*, Umbelliferae. "Laser Syriacum" is a synonym of assafoetida; and "Laser Gallicum" of angelica, "the Holye Ghostes Roote."

SEVENTH COLLOQUY

ORTA

You must know that it comes to India in two forms, one clear and clean,[1] the other turbid and dirty. The Banyans clean the latter before eating it. The clean sort has a colour like shining brass, and this comes from Guzerat. The Guzeratis say that they get it from Chitor, Patane, and Dely. The other kind comes from the Estreito and Ormuz, and has a higher price. The merchants who find the bright kind do not buy the other, which is consumed in medicines and food. Some eat it like bread, when it is called APAS.

RUANO

Is the smell the same in all?

ORTA

That which comes from Guzerat is considered the best, being brighter and the smell stronger. The smell of that which comes from Ormuz is not so strong, but, to our noses, both smell very bad, and the worst of all is that which is brightest and is held to be the best. When I asked some Banyans which smelt the best they replied it was the kind that came from Guzerat, the smell of which is the worst and strongest. This is because they have become accustomed to it. To many persons the smell of liquid storax and of civet is bad because it is strong, though generally it is considered very good. For me I do not care for the smell of any of the gums, and I like that of our myrrh somewhat better. Avicenna divided smells into fetid and sweet smelling. People say that leeks have a fetid smell, and such like things. We consider after the manner of the ancients, for we do not call a thing odoriferous because it smells sweet, but because the smell is strong. In this way a reed is called odoriferous which, in the

[1] A dark brown, translucent, brittle mass.—HANBURY.

opinion of many, ought to be called fetid. Myrrh also has a bad smell, and aloes worse, and the ear of spikenard still worse. I have purged many persons with it who did not like taking rhubarb.

RUANO

This does not appear to me bad, but it would be better that ASSA-FETIDA should be what we use, with the sweet smelling BENJUY.[1] You have not given me a chapter on BENJUY.

ORTA

If it is a simple newly found for our use, why must we give it an ancient name?

RUANO

Because it is more reasonable to think that the root of the BENJUY tree should be good for seasoning food, and there is no reason to suppose that ASSA-FETIDA would be good. It is well known that the Banyans like it because they are accustomed tó eat vegetables and other unsavoury food as the people of our Europe eat them. According to what Antonio Musa says, those who navigated from these parts to seek for the BENJUY describe the tree as being like the tree of LASERPICIUM. They say further that the people of the same land, being obliged to speak the truth, call that gum LASERPICIUM.

ORTA

I know not who the Spaniard was who was so shameless as to tell Antonio Musa of Ferrara such a big lie. As I have told you, speaking of BENJUY, the tree is very different from that of the ASSA-FETIDA as it has been described. The BENJUY is only known in Sumatra and Siam, and in these countries it is only called COMINHAM and not LASERPICIUM. BENJUY is not

[1] This is certainly Gum Benjamin, *i.e.* "*luban-Jari*," "the Olibanum of Java," or Benzoin, the balsamic resin of *Styrax Benzoin*, Styraceae; the Asa-dulcis, and Laser Cyrenaicum of the older Pharmacists; see "The Nomenclator" of 1585, and compare p. 58.

found in Armenia, nor Syria, nor Africa, nor Cyrene, and there is no memory of it among the dwellers in those countries. The principal place whence the BENJUY is obtained that comes to these parts is Arabia, and I say this without denying that it also comes from other places, such as Dely, Mandou, and Chitor. For the Guzeratis and Deccanis who buy it from us say that it also comes from those places, though not in any great quantity. So that your Musa is wrong when he says that it comes from Africa, Armenia, Judaea, and Syria, for in all those parts it has yet to be sought for; for they would bring more profitable merchandise if they should have it.

RUANO

I really trust that you will not be annoyed by my questions. Ruelio, a man sufficiently learned and deserving of praise, who was the translator of Dioscorides, says, in his work on the nature of plants, that in France a large and thick root grows which is black outside and white within, going on to paint the leaves and general appearance. He says that as well the root as the seeds and the exudation have a very pleasant smell, and, the virtue of the medicine being well established, he gives it very grand names, such as the imperial root, the angelic root, the root of the Holy Spirit.[1] He says that it is useful in various ways, being warm and dry in the third degree. It is an antidote against poison, preserves from contagion of plague and other diseases. He says that by taking one grain of it in the mouth, in the winter with wine, and in the summer with rose water, it is a preservative against plague on the day that it is taken, and against other infirmities. It is there called LASERPICIUM GALLICO[2] by the veterinary surgeons. They say that the smell of the juice is like BENJUY, and the learned are of opinion that it is BENJUY, and the OPUS CIRINAICO which grew in Judaea and was introduced into

[1] See note p. 41.

[2] See note p. 47.

France. Hence it is said that the word was BEN JUDEO corrupted into BENJUY.

ORTA

You praise this root very highly, but the tree is very different from the BENJUY, as you will see when we speak of the latter, which is a great tree. The ASSA-FETIDA is not so large, nor is it the LASERPICIUM CIRINAICUM, some of which is said to be found in Judaea. I have enquired among drug merchants from that country, and they told me that there had never been such a plant there in the memory of man.

RUANO

So let it be, and now let us hear what sort of thing ANIL may be.

ORTA

ANIL is not medicinal, but commercial, and therefore we need not occupy ourselves with it. However, I may tell you that ANIL is so called by Arabs and Turks, and in other languages, especially Guzerati, where it was known as GALI, but is now called ANIL. It is a herb that is sown, and is like what we call MANGIRIQUAM. They gather it and leave it to dry for a time, and then pound it well, and put it in the sun for some days. When it is quite dry it has a green colour and afterwards has a blue tinge, and finally darkens. It is best when it is clear of earth. To prove it more surely, it is burnt with a lamp and should give a very fine ash. Others put it in water, and if it floats it is good. So that it must be light and of the right sort.[1] As it would be a serious thing to philosophize further it will be well that we should go to dinner and leave the ANIL to the contractors.

[1] See Yule's *Marco Polo*, ii. 370, and Duarte Barbosa. Indigo, *nilini* in Sanskrit [from *nila*, blue], *nil* in Hindi, *avuri* in Tamil, the 'Ινδικὸν βαφικόν, and 'Ινδικὸν μέλαν, of the Greeks, is prepared from *Indigofera tinctoria*.

SEVENTH COLLOQUY

Ruano

But first tell me what fruit that is about the size of a nut that smells so sweet.[1]

Orta

It is not a fruit that is used in medicine, but it serves to season food with a sharp taste, making it more appetising. The wood has a pleasant smell, and as the wood retains the appetising acid in itself, they call it Ambares.[2] It has a cartilaginous bark, is yellow when ripe, and of a yellowish colour when not ripe.

[1] Acosta writes nearly the same of Assafoetida in his lxii. chapter, but gives no plate. He has a very short chapter on Anil.

[2] *Spondias mangifera*, Willd., the "Hog Plum"; Sanskrit, *amarataka*, commonly called *ambara*. Acosta's xlv. chapter is on the Ambares tree, with a plate of the leaves and fruit.

EIGHTH COLLOQUY

BANGUE

RUANO

WHAT is the difference between that which they call BANGUE[1] and AMFIAM? It seems to me that they are one, for when you abuse your servants you sometimes call them BANGUE and sometimes AMFIAM. I, therefore, wish to know whether there is any difference between the two words.

ORTA

The AMFIAM we call opium, of which I will speak to you when we come to it. I will now satisfy you respecting the nature of BANGUE, its tree and seed. Antonia! give me what I told you to bring.

[1] BANGUE, that is *bhanga* [Sanskrit for "broken"], called by the Arabs *haschish* [whence the English word "assassins"], the gathered and dried leaves of *Cannabis indica*, var. *sativa*, the *kinab* of the Arabs. The whole dried plant is called *ganja* [="the plant"]. *Bhanga* is the larger leaves only, the preparations from them being called *sidhi*, and *sabza*; the preparations of the smaller leaves are called *savia*, and *sukhu*; all are highly aphrodisiac. *Charas* is the gum resin exuded by the plant, and is smoked. *Majum* is a confection of *charas*, *bhanga*, and *ganja*, with opium, and spices added; and *tadhal* a milder *majum*. The *haschish* in it gives its evil fame to opium smoking. *Haschish* means grass; and, so, *Aba-haschish*, "Father of all Grasses," "a botanist."

EIGHTH COLLOQUY

ANTONIA

Here is the tree of the small ones, and see here is the seed, and here is what they sell in the drug shop. For you told me to bring them altogether.

RUANO

This seed is like that of flax (ALCANAVE),[1] except that it is smaller and not so white, and the little tree is also like flax, so we need not discuss them because we already know all about it.

ORTA

It is not flax (ALCANAVE), for the seed is smaller and not white, and the Indians eat either the seeds or the pounded leaves to assist or quiet the women. They also take it for another purpose, to give an appetite; and our writers say that the branches have much inside and little rind, which is contrary to what the flax (ALCANAVE) has.

RUANO

Do they make cords of this bark?

ORTA

No.

RUANO

Is there anything else from which they do make cords?

ORTA

Yes. From the fruit of the palm, which I shall touch upon further on. Also in BALAGUATE they make cords from the roots of a very large tree, and, to confess the truth, they also make them from the flax (ALCANAVE) which is plentiful there, but not in the Deccan or Bengal. I saw there our flax from which we make our shirts, and all this flax is merchandize

[1] The old Portuguese name for flax, now called CANHAMO. The Arabic form is *kinab*, the Persian *kanab*, and the Greek *κάνναβις*.

to be met with in the above countries. They call it ALCI.[1] But there is very little of the flax (ALCANAVE) on the mainland. What is here is not the flax (ALCANAVE).

RUANO

Be it so; and now tell me how this BANGUE is made, and how and for what it is taken.

ORTA

They make the pressed leaves, sometimes with the seeds, into a powder. Some inject ARECA VERDE, and those who drink it become beside themselves. For the same purpose they mix nutmeg and mace with it, and there is the same effect in drinking it. Others inject cloves, others camphor of Borneo, others amber and ALMISQUE, others opium. These are the Moors, who are much addicted to it. The profit from its use is for the man to be beside himself, and to be raised above all cares and anxieties, and it makes some break into a foolish laugh. I hear that many women take it when they want to dally and flirt with men. It is also said, but it may not be true, that the great captains, in ancient times, used to drink it with wine or with opium, that they might rest from their work, be without care, and be able to sleep; for the long vigils of such became a torment to them. The great Sultan Bahadur[2] said to Martin Affonso de Souza, to whom he wished every good thing and to whom he told his secrets, that when, at night, he wanted to go to Portugal, Brazil, Turkey, Arabia, or Persia, he only had to take a little BANGUE.[3] This

[1] ALCI here, in the Sanskrit *atasi*, commonly called *alsi*, the flax plant, *Linum usitatissimum*, widely grown in India for its oil seed.

[2] In 1536, after the Portuguese had got possession of the fort of Diu, Bahadur Shah was in the town. One night he came and battered at the gate of the fort. When it was opened he came in shouting and nearly dying of laughter, reeling about until he fell down and went to sleep. He had been taking *bhang*.—*Lendas de India*, GASPAR CORREA, iii. 754.

[3] This passage is quoted by Yule in his *Glossary*.

was made up into an electuary with sugar and spices, and was called MAJU.

RUANO

Has it this pleasant effect on everybody?

ORTA

It may be that it has this effect when we have become accustomed to it. I myself saw a Portuguese jester, who was for a long time with me in Balaguate, eat a slice or two of the electuary, and at night he was pleasantly intoxicated, his utterance not intelligible. Then he became sad, began to shed tears, and was plunged in grief. In his case the effect was sadness and nausea. Those who saw or heard of it were provoked to laughter as if it was an ordinary drinking bout. Those of my servants who took it, unknown to me, said that it made them so as not to feel work, to be very happy, and to have a craving for food. I believe that it is so generally used and by such a number of people that there is no mystery about it. But I have not tried it, nor do I wish to do so. Many Portuguese have told me that they have taken it, and that they experienced the same symptoms, more especially the female partakers. However, this is not one of our medicines and we had better not waste any more time over it.[1]

[1] Chapter lxi. of Acosta is on *bhanga*, taken from Orta; two pages.

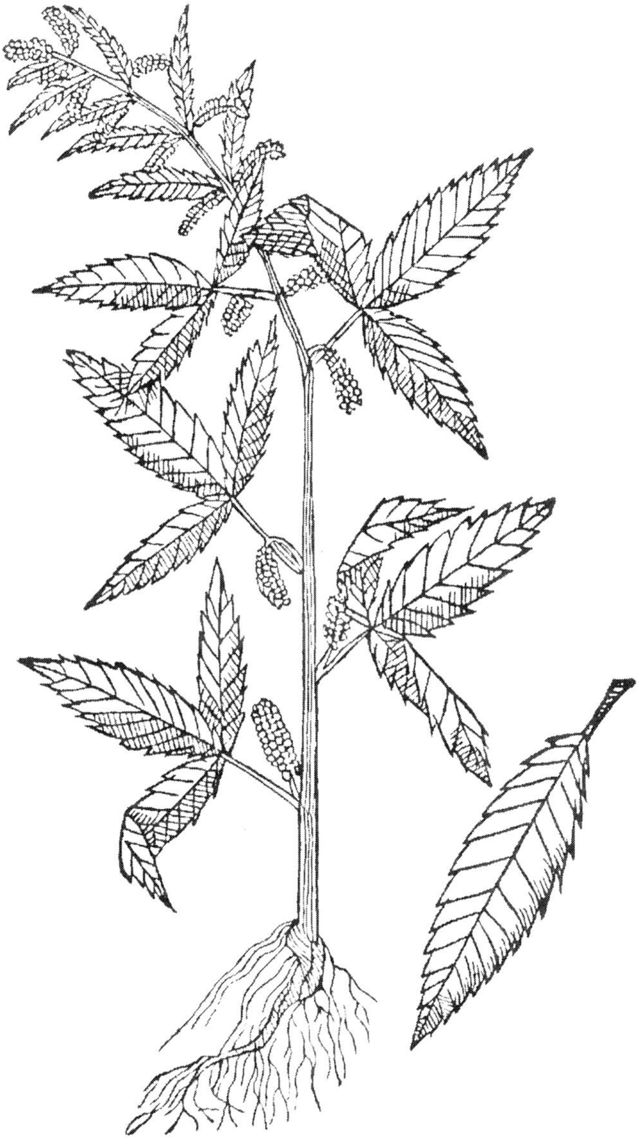

Plate III.—Bangue.

NINTH COLLOQUY

BENJUY

RUANO

TALKING of LASERPICIUM[1] you told me that ASSA ODORATA was not BENJUY as some learned people assert. Now we will discuss it, and with such suavity as will make the conversation agreeable. For me the smell of the kind they have here is better than that of Portugal. This may be from the great abundance of it here.

ORTA

You are quite right in judging the smell to be better, for this is not the BENJUY that is used in Portugal. That is called BENJUY DE BONINAS and costs much more.

RUANO

You spoke of both, and referring to ASSA-FETIDA you told me that they were not better for seasoning dishes than ASSA-FETIDA.

[1] *Styrax Benzoin.* Hanbury says that Garcia da Orta was the first to give an intelligent and lucid account of benzoin (*Pharmacographia*, p. 362). It has no medicinal properties, and is chiefly used for incense in the Greek Church. See foot-note, p. 49.

ORTA

What I then said to you I say to you again. No one uses BENJUY to season dishes, but ASSA-FETIDA is very generally used for that purpose. The things which smell bad themselves, such as leeks, garlic, and onions, do very well for seasoning dishes, and I further told you that the people of this country know very well how to prepare dishes made with those vegetables.

RUANO

Now I want to know the name of the BENJUY tree, what gum it yields, where is its native place, what it is called in Arabia, and whether any Arabian or Greek author has written about it.

ORTA

To answer your last question, I know of no Greek who wrote on the BENJUY. AVERROES calls it BELENIZAN, BOLIZAN, PETROZAN, and describes it as warm and dry in the second degree, aromatic, humid and weak as a stomachic, and comforting, has a pleasant taste in the mouth and strengthens the limbs. I do not, by these words briefly quoted, understand that they describe the BENJUY. If part of this text has to be withdrawn so much the better. Among the moderns Antonio Musa and Ruelio speak of BENJUY, and Antonio Musa says that it is the ASSA-DULCIS or ODORATA, and for this he gives the reasons I have already stated, namely, that the dwellers in the country where it grows, constrained to tell the truth, call it ASSA-DULCIS, and this is confirmed by Portuguese from Sumatra, or by persons who had heard from them. I told you that all the dwellers in the countries where it grows call it COMINHAM, and I also told you that the Portuguese, without any shame, said that this was not true.

RUANO

With regard to Antonio Musa I must tell you that he

says more, so that you may be supplied with all his remarks. He says that the BENJUY tree is a native of Africa and Armenia, and that it also grows in India; also that, according to Dioscorides, the root makes a flour like bran, that he has often found it, and he further says that the best of all grows in Cyrene and Judaea.

ORTA

Do not try to frighten me with Dioscorides or Galen, because I merely speak the truth and say what I know. Although they may call it OPUS CIRENIACUM (which means juice of Cyrene), I know that the principal source of it is in India, near the Ganges (which the Indians call Ganga), and is the BENJUY which is called AMENDOADO in Siam; and of all this BENJUY that is brought to India most of it comes from Arabia, Turkey, and Persia. If you are anxious that some small quantity should come from Judaea and Palestine, I tell you that I spoke to Moors and Jews who bought it here to take it to their country as merchandise. It is not to be believed that they would buy it to take to Palestine if they had better there, as you say.

RUANO

Answer me respecting what Ruelio says that a root grows in France which they call the angelic root, the root of the Holy Spirit, and the imperial root, which is warm and dry in the third degree, and is supposed to have the same virtues, or greater, than those you attribute to ASSA-FETIDA.

ORTA

I say that it may well be that there is such a root and exudation, and that it may have such and such virtues as are stated, for so learned a man may be supposed to tell the truth. Certainly in India it is good for many infirmities which he mentions. But the use of it being to repress the lusts of the flesh, which is what he says, no one would make any profit by bringing it here, because the Indians do not seek for medicine to repress, but to stimulate such lusts. It may be useful to repress, while the ASSA stimulates. So that it is clearly not the same, their effects being contrary. Nor is it found in Judaea as you say, according to the information I have obtained. It is clear that it was equally unknown in ancient times, for if not so there would have been some memory of it among the people, and it would have been praised by David and Solomon, who wrote with such laudation of perfumes. It may well be that Ruelio, who said that it was called BENJUDEUM, which means "son of Judaea," was deceived by the name. Certainly it is better to believe that it should be called BENJAOY, which means "son of John."[1]

RUANO

Well, you have replied to what these doctors say. I answer with what a Milanese states, that it comes from mount PAROPANISO, and that the people of Macedonia state that it comes from the Caucasus, where it has a strong scent and is better than our BENJUY. This author is Ludovico Vartomano,[2] and he says that the best of all comes from Malacca. Tell me whether this is true?

ORTA

You believe this Milanese to whom I do not give credence. As for the Macedonians, a number of Roumanians and Turks are here every day and take back BENJUY as merchandise. As for Ludovico Vartomano, I have spoken here and in Portugal with men who knew him in India, and said that he went about in the dress of a Moor, and that he returned to us and did penance for his sins. They told me that he never went beyond Calicut and Cochin, nor did we then frequent the seas which we now navigate. As for what he says about Sumatra, he never was there. It is true that the good kind is worth much

[1] Compare pp. 49 and 58.

[2] Varthema does not say this.

in its own land. That which now comes is that which we call BENJUY DE BONINAS. I did take this Ludovico, whom you quote, for a truthful man. But afterwards I looked into his book and found that he wrote in it what was false. For speaking of Ormuz, he said that it was an island and one of the richest cities there could be, and that it had the softest water in the world. Yet in Ormuz there is no water but what is brackish, and all the provisions come from beyond the island, nor is it very good water that does come from outside.[1] Also when this Ludovico speaks of Malacca he says that there is neither wood nor water. This is all false, for there is plenty of good water and abundance of wood.[2] From this you will see how little he is to be trusted as regards anything in India. Then turning to what this Milanese says about the BENJUY of Macedonia, I may say to you that it is possibly storax. May God grant you salvation; for I have laboured to obtain knowledge, yet I have never heard of storax coming from anywhere but Ethiopia, whence the myrrh also comes.

RUANO

So let it be, that God may be served. Now tell how many kinds there are, what the tree is like, and what it is called.

ORTA

There is one kind, the most saleable of all, which is called AMENDOADO, as it has some white almonds inside; and it is valued with reference to the number of almonds. This kind comes nearly entirely from Siam and Martaban. That which,

[1] Varthema did not say that the water was good; on the contrary, he said that no water was found on the island. Orta was misled by a bad Spanish translation. See *Hakluyt Society's* vol. for 1863, p. 95.

[2] Varthema does not say that there is no wood nor water in Malacca. But he does say that it is not very fertile (*ibid.* p. 225), a statement which is confirmed by Barros. The Spanish translation of Varthema was made by Christoral de Arcos in 1520, from the Latin version, reprinted 1523 and 1576. There is a copy of the first edition in the Grenville Library. The translator says that he could not procure the Italian version, and that the Latin is obscure.

according to Antonio Musa, came mixed with rice flour I believe to be a mistake, for the gum is the same, some thick and some clear, some almost black, which becomes whiter by being exposed to the sun for some time. This is sometimes brought in flour, which Antonio Musa calls rice flour, but it is of almonds, as you may see yourself by pounding some. There is another darker BENJUY in Java and Sumatra which is cheaper, and another black kind from young trees, also from Sumatra. That which we call BENJUY DE BONINAS fetches ten times the price of the others. That is the BENJUY which they sent me, the other day, as a present.

RUANO

I saw that BENJUY, but you have not told me about the value of the other called AMENDOADO.

ORTA

You saw it the other day. It smelt much better and, when rubbed in the hands, it emitted a delicious fragrance.

RUANO

Yes, I saw it, and you told me that, owing to the strong smell, it was called BENJUY DE BONINAS, or "of flowers." But I would not give so much money for it as is given here, perhaps that is because I am not so great a lord.

ORTA

I tell you what I have often thought, and that is that this BENJUY DE BONINAS is mixed with liquid storax, which is here called ROCAMALHA,[1] for there is certainly a smell like that of storax in the BENJUY DE BONINAS. To try an experiment, the BENJUY was mixed with liquid storax made in the country, and the smell of one was better than that of the other, but not so good as that of BONINAS.

[1] "Rose-malloes," both words being a corruption of the Sanskrit *rasa-mala*, "perfume-garland," the name given by the Malays to the "Liquid-storax," or "Liquid-copal," obtained from *Liquidambar orientalis*.

NINTH COLLOQUY

Ruano

I now remember that, the other day, you bought from a man who came in the same ship with me, ten quintals of liquid storax. You told me that you wanted it to send to Malacca, but it has not served for anything else but to mix with the Benjuy.

Orta

You are not mistaken in this, for I got it only because people are very fond of the smell. I intended to send most of it to China, while the rest would be used up in our country. That this is the truth will be manifest, for that which is taken to China, when they have a great quantity there of the kind called Rocamalha, is not sold because little is used in that country. The inhabitants of this land call all the different kinds of Benjuy by the name of Cominham. The Moors give them the name of Louanjaoy, almost like incense of Java, because the Arabs first heard of it in that form; Louan is their name for incense. The Deccanis and Guzeratis call it Udo.

Ruano

This derivation commends itself to me, for we call the incense Olibano, taking the word from the Greeks, and these also seem to have imitated the Greek word, corrupting it to Louan. I am satisfied on that point, so now tell me about the appearance of the tree, if you know it.

Orta

The tree of the Benjuy is tall and handsome, offering good shade with tufted leaves not very regularly arranged.[1] The trunk sends out high and thick branches. Some grow in the forests of Malacca in damp places. The small trees, as I said,

[1] The tree grows in Sumatra and Java. It is of moderate size, with a stem the size of a man's body, and a beautiful crown of foliage. See footnotes, pp. 49 and 58.

yield the Benjuy de boninas which are those of Bayros, better than those of Siam, while those of Siam are better than any others. Some blows are given to the trees to make the gum exude, which is the Benjuy, in greater quantity. I have seen the leaves preserved in vinegar and also the branches, which proves that my description is correct. While growing this leaf appears smaller than a lemon leaf and not so green, white underneath. From the branch it looks to me like a vine leaf, not so long but broader. I gained a knowledge of all these things by the payment of money, for the person I sent for these leaves and small branches was very well paid. Besides his labour in the forests he was in great danger from tigers which wander about. These tigers are called Reimoes in Malacca.

Ruano

I give you many thanks, and hope that you may see some new things from Malacca, contrary to what you have said and written to me, so do not be annoyed at your loss.

Orta

I promise you that if God spares my life I will not fail to write every year as to anything that may amend what I have said, or that I may hear of in the way of corrections, and if you should be living in Castille you will thus know about it; for him to whom I shall write will send on what I direct. As you said at first that the Amendoado is not so fragrant as the Preto from young trees, I know that old gum loses its scent in time like everything else. If two or three of the almonds are taken and put over a brazier, the scent will not be so sweet as that of the Preto, because the white is beautiful and the black has a better scent; mixed together as they are for sale, one with another, they remain more beautiful and more fragrant.

TENTH COLLOQUY

BER

Treats of the apples we use here and of a fruit we call BRINDOES, *also of the names and titles of the kings and lords of these lands. This is a colloquy which has nothing to do with medicine; but it is here inserted at the request of Dr Ruano as pastime for those who may read it in Spain. It also treats of chess and of its pieces.*

RUANO

THOSE small apples that we are eating for dessert seem so good that I should much like to know whether they are apples of the ANAFEGA (jujube tree) or a different fruit; I should also like to try that vermilion fruit that those girls are eating.

ORTA

The last fruit you have mentioned is one which you will not have much reason to wish to taste, much less to write about, as it is very acid.

RUANO

Never mind, I will taste by way of a medicine.

ORTA

In our land we call it BRINDOES.[1] It is more or less vermilion outside, but within it is such a bright vermilion that it looks like blood. Some of a darker colour are not so acid, and these have the blackness from being riper, but inside they are always vermilion. It pleases the taste of many people, but not for food, nor for medicine, as it is very sour. The tamarind is better. BRINDOES serve for a dye, and they are taken to sea to make vinegar. Some have taken them to Portugal, and have found a use for them.

RUANO

Now tell me about the other apples, their name, and about the tree and where it grows, and whether the apples are from the ANAFEGA.

ORTA

The name in Canarese is BOR and in the Deccan BER.[2] The Malays call it VIDARAS, and theirs is better than ours, but not so good as those of Balaguate, which are long[3] and very delicious. The tree is different from the jujube.[4] Those which you praise are brought to the Nizamoxa. They are small, and I am told that they are from a tree different to the jujube. They are in this land and are brought for sale by the apothecaries. These are for the most part sharp to the taste[5] or a little acid.

RUANO

Those we are eating are not without a sweet taste.

[1] *Garcinia indica*, Guttiferae; *G. hanburyi* being the source of Gamboge, and *G. Mangostana* of the "Mangosteen." *Brindas* is the Portuguese name of the fruit of *G. indica*.

[2] The BER or VIDARA is *Zizyphus Jujuba*, in Sanscrit *badari*, in Hindustani *bir*, and in Mahratti *bor*. It is a small thorny tree, and, both wild and cultivated, a favourite domicile of the *tussah* silkworm.

[3] This long kind was also noted by Dr. Wallich.

[4] *Zizyphus vulgaris* of the Concans and Goa territory, scarcely to be distinguished from *Z. Jujuba*.

[5] "Pontico." The word is obsolete. It means pointed or sharp when used to describe a particular taste.

ORTA

That is true, and others are even sweeter. But none ripen to such an extent as to enable them to pass by the name of ANAFEGA, for they always have a taste of sharpness, so that they cannot be pectoral like the jujubes from which we make syrup. We make festivity with this fruit, because we have not got the pears and pippins of Portugal. The tree is rather thorny and about the size of our apple trees, and the leaf is also like, but perhaps not quite so rounded.

RUANO

Let us to horse and ride into the country, and on the road you must tell me about this Nizamoxa,[1] for you have mentioned him several times.

ORTA

I have told you that there was a King of Balaguate whom I cured many times, and his son several times, for which I received, at various times, more than 10,000 *pardaus*.[2] He also offered me 40,000 *pardáos* of rent for visiting him during some months of each year, which I did not accept.

RUANO

Let us turn to a subject more acceptable to our ears. I would ask, sir, for the name of the king, and I should like to know what the name means, and those of other kings of this country. I do not want our colloquies to be all medicine, but to have some talk about other things to awaken our interest.

ORTA

I am disinclined to waste a colloquy on matters that do

[1] Bahram Nizam Shah.

[2] A pardau is 300 reis. It would be 60,000,000 reis of modern money.

not appertain to science, because readers will say that I have been engaged in writing a story-book.

RUANO

Let the blame fall upon me when you make this book, for I am sure that many in Spain will rejoice to hear the things you relate.

ORTA

Then I comply with your demand. Know that there was a powerful king of Dely[1] who, about three hundred years ago, conquered all this land and Balaguate. At that time Cambay was also tyrannically seized by Moors and Rajputs. They were Gentios who had been its lords. This King of Dely took Balaguate[2] from some very powerful tribes, whose descendants are now called Venezanos,[3] and others who inhabited that land were called Colles.[4] To this day the Colles, Kenezaras, and Reisbretos[5] live by plunder and robbery. The Kenezaras and Colles are in this land of the Deccan, and up to this time they cannot be subdued and made to submit to the kings.

RUANO

They must be a valiant race.

ORTA

Yes, they are. The kings also are grasping, for they pardon those who share the plunder with them. This kingdom of Dely is far in the interior of the country to the north, marching with the territories of Coraçone.[6] It is a very cold land, with snow and frost as in our country. The Mongols, whom we call Tartars, have been in possession of Dely for more than thirty years.[7] I was acquainted with a brother of

[1] Ala-ed-din Ghilzy, 1296 to 1316. [2] Deccan. [3] Brinjarries.
[4] Koles, a wild hill-tribe. [5] Rajpûts.
[6] Khorasan. He seems to include a vast area under that name.
[7] Baber took possession of Delhi in 1526. Orta's date is quite correct.

the King of Dely[1] at the court of Sultan Bahadur, King of Cambay, who honoured greatly this brother of the King of Dely. Afterwards this kingdom of Dely was taken from the Mongols by a soldier of fortune, outlawed by the King of Bengala for having killed his brother.[2] He rose against the King of Bengala and killed him, and afterwards subdued Dely and many other kings. For some time he was the greatest lord of all, and persons of credit told me that his dominions covered 800 square leagues. This King was the first Patane who held the mountains bordering on Bengala. He was called Xaholam, which means king of the world.[3] Of this potentate's history a greater chronicle might be made than that of the famous Tamirham (whom we corruptly call the great Taborlam), and some chroniclers have Tamirlangue, because Tamir was his proper name, and *langue* means lame, which he was. But he was of another time, and I return to the King of Dely. He conquered the Deccan and Cuncam,[4] and was lord of them for some time. But as he could not rule countries at such a distance he returned to his own dominions, and left them to his nephew who was crowned king. This ruler always favoured strangers from Turkey, Roum, Coraçone, and Arabia, dividing his kingdom into captaincies. Thus Adelham (called by us Idalcam) held the coast from Angediva to Cifardam, which is 60 leagues, bounded by other captaincies towards the interior. Nizamaluco had the coast from Cifardam to Negotana, which is 20 leagues, bounded in the interior by other lordships and by Cambay. These two held the Cuncam between them from the sea-coast to the summit of the mountain range called Guate,[5] which include much land, and are very high in many

[1] This was Muhammad Zeman Mirza, married to a daughter of Baber, and brother-in-law (not brother) to Baber's son Humayun. He was a fugitive at Cambay in 1534, and in 1535 Orta went to Diu with the expedition of Martin Affonso de Sousa, and visited Cambay.

[2] This was Shere Khan, an Afghan chief, who expelled Humayun and reigned at Delhi for some years, 1540-1545.

[3] Afghan.

[4] Konkan.

[5] Ghauts.

places. I have crossed them by some passes. There is something worth noting with respect to these mountains which contain many beautiful plains. As *bala* is the Persian for a summit, and *guate*[1] is a mountain, Balaguate is as much as to say "over the mountains." In Balaguate are the territories of Imademaluco, whom we call Madremaluco, and of Cotalmaluco and of Verido. All these captains were foreigners, Turks and Roums and people of Coraçone; except Nizamaluco, who is said to be a son of Decanim, son of the Tocha of the King of the Deccan. As the wife of this Tocha had slept with the King of the Deccan, Nizamaluco was looked upon as one of the royal caste, while the others were slaves bought with the king's money. As these captains were not inclined to obey the King of Deccan, they agreed among themselves that each should be master of their own lands, and that the King of the Deccan should be seized at Bedar, the principal city and capital of the Deccan. He was seized there and delivered to the care of one of them named Verido. He and the others, either themselves or by their representatives, made their salaams to the captive king on certain occasions every year.

RUANO

Is salaam the Arabic for peace? If so, I call this a false peace.

ORTA

Others rose in revolts in concert with those who imprisoned the king. These were Mohadum Coja,[2] and Veriche, who was a Gentio, and they seized very powerful territories and rich cities. Mohadum had Visapor, Solapor, and Paranda,[3] which were afterwards taken by Nizamaluco. Veriche was established in the territory which borders on

[1] Ghaut.

[2] Muhammad Kasim, styled Hindu Shah, whose literary title was "Ferishta."

[3] In Sholapur.

Cambay and on the land of Nizamaluco. The great-grandfather[1] of Adelham, who now reigns, was one of those captains who rebelled, and he was a Turk. He died in 1535. He was always very powerful and twice took this city of Goa from us. The grandfather of the present Nizamaluco,[2] who now is, father of my former friend, died in 1509. The Imademaluco (or Madremaluco as we corruptly call him) was Chergues by nation, and had originally been a Christian. He died in 1546. The Cotalmaluco who died in 1548 was also one of those who rebelled. He was a native of Coraçone. Verido, who died in 1510, was Ungaro by nation, and also originally a Christian, as I am informed on good authority.[3]

Ruano

Let us come to the names, and tell me who it was that took the cap, and do not pass on until he has passed.

Orta

He was Ambassador of Idalham, whose grandfather was lord of this island. These Moors gave distinctions at their pleasure. As among the Gentios *Rao*[4] means King, and *Naique*[5] means a captain, when these Kings take a Gentio into their service, and do not wish to give any very great title,

[1] This was Yusuf Adil Khan, who came to India as a slave. He became a powerful chief and established his capital at Bijapur, taking the title of Adil Shah. He died in 1510. His son and successor was Ismail Adil Shah, who died in 1534. Ismail's grandson Ali succeeded in 1557, and ruled in Orta's time.

[2] Nizam ul Mulk. His son fixed his capital at Ahmednagar, his name being Ahmed Nizam Shah. His son, named Buhram Nizam Shah, who succeeded in 1509, was the great friend of Orta. He died in 1550; Hoçein succeeded. The Portuguese called these kings Nizamaluco.

[3] Fath-Ullah, called Imad-ul-Mulk, was a Hindu, not a "Cherques" or Circassian. He was called Imademaluco by the Portuguese. Cotalmaluco was Sultan Kuli. His capital was at Golconda. Kasim Berid, called Verido by Orta, was a Georgian slave who rose to be prime-minister of Mahmud Shah, the Deccani king who was deposed. His son took the title of Berid Shah.

[4] Raja, Ray or Rao.

[5] Naik, a chief.

they added the title *Naique* to his name, and when they wanted to grant a very high distinction *Rao* was added to the name—as *Chita Rao*, whom I knew. It was a proud name, because *Chita* means a leopard, so that *Chita Rao* means a King as strong as a leopard. Among the Mongols or Tartars *Ham* means a King, and those they favour receive that title, which we have corrupted into *Cam.*[1] Rao by itself was the title of the King of Bisnaguer[2] who, in times past, was much harassed by Adelham, and in our time extends his power over all the lords of the Deccan, who all obey him, for all things befall in their turn. But to return to our subject. As *Adel* means justice in Persian, they give this title to the Adelham, the lord of this country, as much as to say King of Justice.

Ruano

The name is not suitable, for neither he nor others are accustomed to dispense justice. But tell me why in Spain they use the word Sabays.

Orta

Some people have told me that the name was given after a captain who was called by it, but afterwards I came to know the truth that Saibo in Arabic and Persian means Lord, and for this reason the name is given. Also because *Maluco* means a kingdom in Persian, and *Neza* a lance, they called my friend Nizamaluco, that is, "the lance of the kingdom." *Cota* is a fort in Arabic, and Cotalmaluco means "the kingdom's fortress."[3] *Imad* means summer, and Imadmaluco means the kingdom's summer. *Verido* means a guard, and

[1] Khan.

[2] Vijayanagar, called also Narsinga. In 1565 all the Musulman chiefs of the Deccan united against the Hindu King of Vijayanagar, and defeated him in the battle of Talicot.

[3] It was *Kutb*, pole star, not *Cota*, Ficalho.

Melique-verido means the King of the guard. Some only called these *malucos* by the name of *meliques*, which means little kingdoms.

RUANO

Does *maluco* properly mean a kingdom?

ORTA

No. Only a region or province.

RUANO

You have now satisfied me about everything except Xa. Why do you say Nizamoxa, and Adelxa?

ORTA

Xa Ismail[1] rose in Coraçone, and his son and successor is Xatamas,[2] now living. Being of low birth[3] Ismail made war against the false laws of the Grand Turk.[4] He rose to be one of the greatest sovereigns of the world. He ordered that his followers should take the side of the sect of Ali, which is opposed to Mafamede,[5] and against those who did not join that sect he made cruel war. His son, named Xatamas, ordered the denunciation of the Deccan Kings who called themselves Xa, which is the title of King. Thus they are called Adelxa, Nizamoxa, Cotumixa, and thus they assume the name of king when they are not able to coin anything more valuable than copper. Nizamoxa and the others at first accepted this law, receiving it from the ambassador, but afterwards they disregarded it.

[1] Shah Ismail.

[2] Shah Thamasp.

[3] He was not of low birth, being a lineal descendant of Ali and Fatima, the daughter of the Prophet.

[4] Not against the Turk, but against the Aku Koyunlu chief, then sovereign of Persia.

[5] Muhammad. On the contrary, Ismail was a zealous Musulman but of the Shiite sect.

RUANO

I always thought that he was called Xeque Ismael, and not Xa Ismael. I also thought that the man was called Sofy?

ORTA

It is true that Xeque is an ancient title used by the Xeques of Arabia, but Xa means King in Persia, and Xa Ismael means the King Ismael. The Turks and Roums called him Çufi, because there was a great captain who was named Çufo or Çufi, and hence they fixed the name of Çufi on Xa Ismael, in memory of their great captain.[1] As you play at chess, I will say something to you that you will be glad to know, and which also has nothing to do with medicine.

RUANO

Very many thanks for this.

ORTA

Xa means King, and when they say that the King makes a move in chess they do not say *Xaque*, but Xa.

RUANO

This is a very curious thing which I am glad to know about. Do they play well at chess?

ORTA

Yes, well, but differently from our game. That you may not be bored I will tell you the names of the pieces arranged for battle.

RUANO

I will take no excuse, so tell me.

[1] That was not the reason. There was no such captain. Sufism is a mystic philosophy of the Persian Shiites.

TENTH COLLOQUY

Orta

They call the King Xa. The Queen (Dama) they call Goazir, which means Constable. The Bishop (Delfim) they call Fil, the word for an elephant. The Knight (cavalo) is Guora. The Rook (Roque) is Roch há, meaning a tiger. The Pawn (Piam) is Piada, which means a man on foot; and thus you have this battle array marshalled.[1] You must pardon me if I have bored you with frivolous tales.

Ruano

On the contrary, I have enjoyed them much.

[1] Chess is said to have been invented in India, where it is called *chaturanga*, or the game of the four *angas*, *i.e.* "the four bodies" of an army, elephants, horses, chariots, and foot soldiers. The Indian tradition is that the game was invented by the wife of Ravana, the King of Ceylon, in the Hindu epic of the *Ramayana*, in the hope of weaning him from his immarital and marital misdemeanours. Passing into Persia, and simultaneously into the hands of the Arabs, the former changed the name of the king piece into *shah*, whence our word "chess," and the latter into *sheik*, whence our word "check," when the king is in danger, and phrase "check-mate," *i.e. sheik-mat*, "the King's dead." The Low Latin name of the game was *Ludus Seace-orum*. In India the Queen [*Farzin*] moves obliquely, one square at a time, and the Bishop [*kisses*] obliquely, three at a time, without hindrance from any other pieces in the way; and the pawn but one square at a time. Owing to these differences the people of India qualify our game as *Faringi Chaturanga*, the "Frankish Chess," and often, *Rumi-Chaturanga*, "Roman," *i.e. European*, or *Turkish* Chess.

In Spanish Rey (King), Dama (Queen), Alfil (Bishop), Cavallo (Knight), Torre (Castle), Peon (Pawn).

Plate IV.—Mançanas de la India.

ELEVENTH COLLOQUY

CALAMO AROMATICO AND DAS CACERAS

RUANO

NOW tell me the names of the CALAMO AROMATICO[1] among the nations you know, for several names come to my knowledge where I have been. Our modern doctors have grave doubts whether it is the ACORO. Some say that it is, while others say that GALANGA is the CALAMO. In this way a mistake has arisen, which it is necessary to have cleared away. I therefore come to you, who have a knowledge of these simples, to establish the truth.

ORTA

That which in Portugal is used by apothecaries for CALAMO AROMATICO, and which is also a medicine much in demand in India, as well for men and women, as for horses in their infirmities, is called VAZ in Guzerat, BACHE in the Deccan, VAZABU in Malabar, DARINGO by the Malays, HEGER in Persia, VAICAM in the Concan, which is near the sea, and CASSAB ALDIRIRA in Arabia.

[1] *Acorus Calamus*, Linn., Aroideae.

CALAMO AROMATICO

RUANO

Serapiam, who is an Arabian and an authority, calls it ASSABEL DIRIRI.

ORTA

Serapio's name is corrupt, and Avicenna has amended it. The Arabian apothecaries give it, for the word CASSAB means CALAMO and ALDIRIRA aromatic, for DIRIRE has the same meaning as aroma with us. This may be gathered from Avicenna. As the Malays receive this medicine from Moors who come from Coraçone, they call it corruptly DIRIMGUO. This medicine is much used in Goa and throughout India. Little is grown in the gardens here in Goa. The smell as well as the taste is bad. The greener the plant is, the stronger and more horrid is the smell, although Ruelio says the contrary. Some medicines smell better the drier they are, such as the sandal and the aloe. They sow much in Guzerat and Balaguate, and when it is growing the smell is not so strong as when it is pulled up. They do not raise it near the sea-coast, because that which grows in such a locality is withered and rots, that which comes from Balaguate is grown to the west. The women use it a great deal for nervous complaints. All the rest that is expended is for the horses in the cold season, who are given it mixed with aloes, and *ameos*, a kind of cummin, with some salt and butter sweetened.[1] They call this medicine *arata*.

RUANO

Does it grow in any place outside India? It would seem so, for Galen and Hipocras call it CALAMO YNGOENTARIO, Plutarch CALAMO ARABIO, and Cornelio Celso CALAMO ALEXANDRINO.

[1] In India it is a common practice to put ghee into horses' food. It is mentioned as allowed for horses in the Ain-i-Akbari, and Nikitin also mentions ghee as forming an ingredient in the food of horses.

ORTA

I have asked many Coraçones and Arabs who come to sell horses in India if they have the plant in their land, and they all told me that they had none but what came for sale from India. I asked them how they knew the use of it. Their reply was that they knew it very well, for that there was no medicine in their land or in this that they all knew so well. Those who say that it is common to Indians and Syrians are in contradiction to what these merchants told me, as well as the physicians of the Deccan kings. Thus those that call it a plant of India say well, and those who give it to Arabia say well, for it was first seen in India and Arabia. Those also say well who call it Alexandrine, for it is taken thence by the Venetians, and to Beyrout and Syrian Tripoli.

RUANO.

Menardo says that he saw it in Panonia, and that it was very fresh, from which it would seem that it came from the neighbourhood.

ORTA

We have faith in what we hear and see; but it may be that he was mistaken, and that what he saw was sown in some flower-pot or basket, as they sow ginger which grows; but what I told you is the truth, that it comes in the course of commerce.

RUANO

This that we use, is it from the root or the cane?

ORTA

It is the cane, for the root is small, but sometimes the root is mixed with the cane. They are wrong who say that it is the root only, and they say it only to base on it their opinion that ACORO is CALAMO AROMATICO or GALAMGA.

CALAMO AROMATICO

RUANO

Why do you call it AROMATICO, for you say that the name comes from the Arabic?

ORTA

I do not say AROMATICO because of the scent, but because it is a drug coming from these parts. Further, I would not say CALAMO ODORATO but JUNCO ODORATO. So now you can see the difference between cane (*cana*) and reed (*junco*). You must also know that what is within the CALAMO is not like a spider's web, but it is a porous substance of a yellowish colour. In this Avicenna and Serapio are mistaken, yet they had better opportunities of ascertaining the facts than the Greeks.

RUANO

Modern writers say that CALAMO AROMATICO is an ACORO, because the root of ACORO, which is administered as a cure, is not a cane but a root which we see in the druggists' shops.

ORTA

You need not trouble yourself about that, because it is only the reed that is sold and used and not the root, and if you want to see it, here it is green and dry.

RUANO

I do not doubt it, for I have it before my eyes. But tell me how ACORO can be ESPADANA, for they say that one is black and the other white, and that it is biting and warm in the third degree, and we do not find any asperity or warmth in it, and that not only in cold regions but in warm also. How much more does this show that it cannot be a medicine warm and dry in the third degree. Planted in another locality it would not be warm. For these qualities are according to the species, and cannot be drawn out at all points, as is seen in the

Acoro. Hence without doubt there is good reason for it not to be the Acoro, nor to sell it as that.

Orta

I confess to you that the Acoro is not the Espadana. But what we need is not to seek the localities where Galen, Pliny, or Dioscorides say that it is found, and this because those doctors were not very curious, and it will not be found from them whether it is Calamo aromatico. For Avicenna and Serapio have three chapters of information respecting Calamo aromatico,[1] Acoro and Galamga. Those who write of Calamo refer to that in India, and not in other parts, and they do not say that the Acoro is anywhere but in Europe, whence it came to be known to us; for we do not speculate as Menardo, Lyoniceno, and others speculate. None of the physicians among the Arabians, Turks, Coraçones, or in India know the Acoro. For when I cured the Nizamoxa of a tremor, I took great pains to investigate the matter, but I never succeeded in learning what Acoro was. But I was told that they had it in Turquia, using the Arabic name. The Calamo is warm and dry in the second degree, and Acoro in the third, so that they cannot be the same thing. If you do not find Acoro seek and look in the books what you may put in its place.

Ruano

Why should not the root of Galamga be Acoro, for all the Sinais take Acoro?

[1] We now know that Galangal [the Lesser] is the rhizome of *Alpinia officinarum*, and [the Greater] of *A. Galanga*, Gingerworts; and "Acorus Calamus," or "A. Calamus odoratus," of *Acorus Calamus*, the "Sweet Flag," an Aroid; while "Calamus aromaticus," the "Sweet Cane," is *Andropogon Schoenanthus*, a Grass, the source of "Palmarosa Oil": *A. Nardus* being the source of the "Citronelle Oil" of Ceylon; and *A. citratus* of "Indian Lemon Grass Oil." *A. muricatus* is the source of the fragrant "Khaskas Grass" of India.

ORTA

Here you see two methods, of Java and China. Planted here the leaves are not like gladiolo, and are very short and the shape of a spoon, as I will tell you when we come to speak of the GALAMGA, and I will show it to you both green and dry. The GALAMGA has another difference. It is warmer and has not the other properties of ACORO and CALAMO. For the two latter are for the nerves, and the GALAMGA for the stomach. These medicines, the GALAMGA and CALAMO, are articles of commerce in this land, where they began to be used, and to be sent to the west.

RUANO

So that you desire that we should share a drug so well known as ACORO?

ORTA

I do not wish to share, but I do not wish that in India we should mix up the two others, or one of them. I say that if there is confusion it is not the fault of the Indians but of the others. For Pliny says that the best is in Pontus, afterwards he says it is in Galatia, then in Crete.

RUANO

You say this, but what would you substitute for ACORO?

ORTA

I would use CALAMO AROMATICO in greater quantity, not being so warm and dry, a grain less. In this way I used it for the Nizamoxa and his father. You can try it if you like, but I know that it is not ACORO that we use for CALAMO AROMATICO; and what Marcello says, that it is cinnamon, is so false that it is unnecessary to refute it.[1]

[1] Chapter lxiii. of his book is devoted by Acosta to CALAMO AROMATICO. It is all

RUANO

It seems to me that it would be good to eat. Tell me what fruit that girl has got. It looks like JUNCA AVELANADA or JUNCO ODORATO.

ORTA

It is a fruit that sprouts in the mud underground, and afterwards comes up about the length of a finger, with leaves fastened to each other. These leaves are very green, like those of the ESPADANA. When the mud becomes dry it rises up like an earth tuber. When it is dried, it is like chestnuts, but it has not a good taste when it is not dry.

RUANO

I am glad to know this, and now tell me its name.

ORTA

It is called CACERAS,[1] and is not used as a medicine, but we eat it.

---

copied from Orta, but he gives a plate of the plant, with leaves and flowers. *Trupa bispinosa*, *T. bicornis*, and *T. natans*, the "Water Chestnuts" of China [*T. bicornis*], India, and Central and Southern Europe.

[1] *Scirpus Kysoor*, Roxb. The native name is *kachera*.

Plate V.—Calamo aromatico.

TWELFTH COLLOQUY

CAMPHOR

RUANO

THERE is very good reason why we should discuss the camphor, as it is so much valued and used in medicine;[1] but it was not mentioned by Galen nor by any Greek author except the modern writer Aecio.[2] No doubt we owe much to the Arabs in some things, but we do not get a complete notice even from them, it being unknown in those lands, so that there could not be a perfect account.

ORTA

This certainly is so, for I who have been so long in this country am only able to get to the exact truth about anything after much labour. The reason is that the Portuguese, who navigate over a great part of the world, only procure a knowledge of how best to dispose of their merchandize, of what they bring there, and what they shall take back. They are not desirous of knowing anything about the things in the countries

[1] *Laurus Camphora.*

[2] Aecio was a native of Amida in Mesopotamia, a student in Alexandria, and finally a physician in Constantinople.

they visit. If they know a product, they do not seek to learn from what tree it comes, and if they see it they do not compare it with one of our trees, nor ask about its fruit, nor what it is like. As I am unable to visit all those countries and should not be given leave, by those who govern them, to go beyond where they reside, because they want to make use of my old age and the learning it has acquired, not having learned physicians in those countries, for these reasons I am not to blame if I am obliged to speak to you of these medicines doubtfully, and with so much caution.

RUANO

It is well said that he who does not know does not doubt, and from this point of view you are not only worthy of pardon, but you deserve praise.

ORTA

Camphor is of two kinds, one called CAMFORA DE BORNEO, which I have never seen in our regions, at least since I have been here, and I am not surprised because it costs as much per pound as CAMFORA DE CHINA costs per *quintal.* What is brought here is made of rounded sticks of the diameter of a crossed hand: and being in that form it seems to be a compound and not simple.

RUANO

This that I have not yet seen describe to me, and first show it to me.

ORTA

Here is a small piece, but not of the best. Girl, bring that piece of CAMFORA DE BORNEO.

SERVANT GIRL

Here it is, Sir!

TWELFTH COLLOQUY

ORTA

Well, you must know that this piece you see, about the size of a grain or a little larger, is the worst kind. For among the Gentios, Banians, and Moors they make up four kinds for sale, which they call *Cabeça*, *Peito*, *Pernas*, and *Pé*.[1] A pound of *Cabeça* is worth 80 *pardáos*,[2] *Peito* 20, *Pernas* 12, *Pé* 3 or 4, if that. Some people sift this camphor in a sort of winnowing machines for sifting seed pearls,[3] which are made of copper with perforations of different sizes. The camphor which passes through the largest holes is sold for one price, and those passing through smaller holes for another. For these sifters are of four kinds, with different sized holes. The Banians are such expert tradesmen that they never mix one size with another, and make their account with such accuracy that no one can deceive them. This that you see here is the refuse, and is black. It is of little value. There is much of this camphor in Borneo, Bairros, Sumatra, and Paçem. The names given by Serapiam and Avicenna are all more or less corrupt. I know that this is an article of commerce much used in this country. What Serapiam calls PANÇOR is PAÇEM in Sumatra, and what Avicenna calls ALÇUZ may be ÇUMDA, islands or main land on the confines of Malacca. That which Serapiam says is a product of CALCA, must be a corrupt form of MALACCA, for it is found in BAIRROS, which is a port of that region.[4]

[1] *Cabeça*, head; *peito*, breast; *pernas*, legs; *pé*, foot.

[2] Acosta has *escudos*.

[3] Acosta, copying from *Orta*, has the word *tharnero*, a sieve.

[4] The Camphor of Borneo and Sumatra comes from a large tree, *Dryobalanops aromatica* [*D. Camphora*, Colebrook], of the family of Dipterocarpeae.

> Olha tambem Borneo onde não faltam
> Lagrimas, no licor coalhado e enxuto
> Das arvores, que camphora he chamado
> Com que da ilha o nome he celebrado.
>
> *Lusiadas.*

Pacem in Sumatra is the Portuguese name for a port, Pasei of the Malays, and Basma

CAMPHOR

RUANO

I rejoice much to know about this medicine, so noble and precious, and I wish to hear from you, first whether it is gum or pith as Avicenna and others held, and if it is at first with vermilion and black stains, made white by distillation or fire, or whether this is false.

ORTA

It is gum and not pith that exudes from the tree in the way I told you that I saw it come, and presently you will see the gum in the stick which comes out by some chinks, so that the camphor is seen to distil through them. This I saw very clearly on a table which an apothecary had. I also saw it on a stick the size of a man's thigh, which was presented to the governor Dom João de Crasto. Also, in this city, there is a merchant, who has a board a *palmo* long. All these sticks are shown to be from the CAMFORA tree. I will not deny that some of this gum falls into the hollow of the tree, as we often see in the trees in Portugal. But it is at first quite white, without any vermilion or black streaks, and it is not distilled, as some writers say, or boiled, to make it white. It is only that from China that is kneaded, and of this you need have no doubt, for Avicenna and Serapio were misinformed; long distances make long lies. I have been told by persons worthy of credit that if a man goes to fill a calabash with this camphor, and another man sees him with the calabash full, he kills him and takes the calabash without being punished, because they say that it was his luck.[1]

---

of Marco Polo. Bairros is a small port on the west coast of Sumatra in 1° 59′ 35″ N. It is the Kansur, T-'ansur, Kaisur of the Arabs, Pançor of Serapio and Orta.—Yule, *Marco Polo*, ii. 268, 285. The Camphor of Formosa is obtained from the *Camphora officinarum* of Linneus, a Laurel bloom.

[1] Acosta leaves out this remarkable rule, and says, in its place, that he had a set of very delicately carved chessmen in a box all made of camphor wood, judging from the smell and from what he was told. Camphor did not exude from the chessmen nor from the box. But by rubbing with the hands the smell of camphor gets stronger.

TWELFTH COLLOQUY

RUANO

Why do you say that the Greeks do not mention camphor? You may remember that Serapio, speaking of camphor, refers to Dioscorides, and you will not forget that you said that you would show me the mistake.

ORTA

You should not be surprised at this, for the assertion in Serapio is an error. As to the way in which camphor is adulterated, I know that the article from Borneo is often mixed with very small stones, or with a gum called *chamderros*, which looks like crude seed pearls, or else it is mixed with flour or sawdust. All these things must be prepared for, by those who wish to speculate in camphor. I know of no other way of adulterating. The black and vermilion streaks are said to be caused by the article having been knocked about, or exposed to damp. The Banians remove this blemish by washing secretly in a pan of warm water with soap and lemon juice. After it has been well washed they put it out to dry in the shade, when it becomes much whiter, and does not lose much weight. I saw this done, the Banian confiding in me secretly, because he was a great friend of mine.

RUANO

Have you found authors who mention these two kinds of camphor?

ORTA

Yes; Serapiam says so obscurely, mentioning that most of this camphor comes from Hariz, and less from China. From this it is to be understood that the largest quantity is from Chincheo, and that this is greater than the other from Borneo, whence a quantity not more than an *oitava* comes. All this is true, the text of Serapiam having been twisted, for the pieces of Chincheo (which we call China) are of 4 ounces or more.

RUANO

Tell me about the tree.

ORTA

A man worthy of credit told me that the tree was like a walnut tree, with whitish leaves, like those of a willow. He did not see the fruit, but it may have been that it was there and he did not see it. The wood is grey, and much like that of a beech tree, sometimes darker. It is not light or porous, as Avicenna says. It is massive, and it may be that what Avicenna saw was old. They say that the tree is tall, wide spreading, with a good crown, and pleasant to the sight. It exudes the camphor from itself, as you saw it ooze out, and as I saw it exude from a piece of the wood. I saw another pole the size of a man's thigh, which I have already mentioned. It may not have been camphor, though the smell was very like it. I have seen another plank a *palmo* in size, which gave out some camphor, and it was like the heart of beech wood.

RUANO

They tell me that it is true that a multitude of animals take refuge under the shade of this tree from fierce wild beasts.

ORTA

All this is a fable. In that land there are tigers (called *reimoes* by the Malays), and the shade of a tree would be no defence from them.[1]

[1] The Malay names are *ariman* and *rimau*. Acosta adds that he has seen tigers, and that the only way to keep them off is by making a circular fire, lighting it with the match of an arquebus. He continues to say that he has himself tried this plan with them, both in the forest and on the plain. Any flickering flame, as of a torch, will keep a tiger off; and so will the jangling of little bells fixed on the top of a stick.

TWELFTH COLLOQUY

Ruano

Is there more yield of this camphor in one year than in another ?—for they tell me that when there are many thunderstorms the yield is good, and when few—bad.

Orta

Avicenna, Serapiam, and Aecio were misinformed on this point. In the island of Sumatra and its vicinity there are always many thunderstorms; for it is on the line where there is more or less wind every day, whence there is always camphor every year. As the thunderstorms are not the reasons of the production of camphor, they cannot be called the cause, except through an accident, or what may happen occasionally. This sort of reasoning is called by philosophers a cause without the consequent effect.

Ruano

Give me an account of the camphor of China or Chincheo.

Orta

The camphor of China is supposed to be partly made up of the other kind from Borneo, the rest being of the kind from China of less value. It is mashed and they make loaves of it, as you have seen, not because it had vermilion and black streaks at first. I only say this conjecturally, and as the opinion of some people. For this camphor does not come from Canton, where all the people trading with that country go, but from Chincheo, whither few people find their way. A person worthy of credit told me that much of it is very cheap in China, others say the contrary, the deduction being that these sticks are compounded. For the camphor of Borneo is an article of commerce in Chincheo, and the people of that land say that it is wanted to mix with a kind that is not so good. The Banians of Cambay confirm this; for they say in secret that when the camphor of Borneo falls short, they mix a little of it with a quantity from China, and call it all falsely camphor of Borneo. These Banians say further that the camphor of China must be a composition, because it wastes with time, while that of Borneo never does.

RUANO

What is your own opinion?

ORTA

I say that there is camphor in Chincheo, but not so good as that of Borneo, and that the two pounded together make a good mixture, being included under one genus; but, being so compounded, it evaporates and wastes, while that of Borneo does not.

RUANO

Menardo says well that the camphor is a new thing, and that he believes it to be compounded and not simple.

ORTA

It is not my opinion that so much is compounded. There are two kinds of camphor, and that which evaporates is not very corruptible; but the things of which it is composed are more liable to corruption. Rhubarb scarcely lasts four months in this land of rain. So that there is much to corrupt the camphor of China, which is kept in India.

RUANO

Avenrrois says that there is another very different kind of camphor. He affirms that it is engendered in the sea, and that it is warm and dry in the second degree. What is more wonderful, he says that amber is a sort of camphor and that it comes into the sea by fountains. Have you by chance heard of that kind?

ORTA

I never heard of it; and the people of India think so much of this medicine that if they had heard of what you say, it would never have been forgotten. If the amber had been a kind of camphor it would not be in such esteem in China, whither it is brought, and fetches the high price I have already mentioned. Besides, amber is warm in the second degree, and camphor is cold in the third, so that they cannot be included under the same genus. For the qualities belong to the species, and never can lettuce be warm or pepper cold. On this, therefore, you may rely.

RUANO

Andreas Belunensis, of whom you do not speak ill but rather praise, says, in his dictionary, that the water of camphor, according to the Arabs, runs and flows from the camphor tree, and that the said tree and water are warm in the third degree. As camphor is usually spoken of as cold, it is necessary to know how this may be, and I would ask whether you have seen such water, or heard it mentioned.

ORTA

I have enquired of many people respecting this water, as well druggists as merchants, and have heard nothing. If it had been known I must have heard; for in Balaguate they have water of sugar cane, and sell it. The fault is not with the Belunense, but with the Arabic book he quotes from.

RUANO

Ruelio and Mateolo Senense say that the kind from China is best, and that the best of all the camphors is that which is purified for a barbarous king, whom they call the King of China.

ORTA

You can tell Ruelio and Mateolo Senense that though they may know Latin and Greek very well, they should not fill their mouths with the appellation of barbarous to those who are not of their generation, and their assertion is wrong. For the camphor of Borneo is sold by *cates,*[1] and that of China by *bores*:[2] the *cate* being 20 ounces, and the *bor* about 600 lbs. The King of China is not one who makes camphor. He is one of the greatest sovereigns that the world knows, and to write adequately of him and his kingdom it would be necessary to write a great volume. I know that the merchandize coming from there consists of silver bedsteads; richly worked services of silver; silk, woven and unwoven; gold, musk, pearls, copper, mercury, vermilion, and porcelain twice the value of silver. I am afraid to tell you what quantity of silk entered the cities of Goa and Cochin a year, during these last years.

RUANO

Tell me, for I am sure you will speak the truth.

ORTA

700 *bores,* each *bor* containing 3 cwts. and 16 lbs. By this you will see the riches and greatness of this land. In Goa, when the other monsoon has come, all this silk is disposed of.

RUANO

Tell me the names and qualities of camphor.

ORTA

CAPUR and CAFUR are the words used by the Arabs and

[1] "Un peso de veinte onças," according to Acosta.

[2] The *bor* is nearly 600 pounds.

all other people. For the letters O, F, E, O, P are letters much akin, indeed sister letters among the Arabs, so that all speak of it in the same way. If any writers speak of camphor with another name, they have been deceived, or else the books are badly printed. As for the qualities, Rasis gives cold and humid, Avicenna gives cold and dry in the third degree, and some or all other writers follow Avicenna.[1]

RUANO

It has appeared to many modern writers, from its smell and its quality of evaporating, that its quality is warm, and it seems to me that they have reason; because the scent of cold things is not so strong, as may be seen in the cases of sandal and roses.

ORTA

I thought this to be true for a long time. But when I saw great warmth in ophthalmia and in a burn treated with camphor, and how it has the same effect as snow, I saw that the contrary was the case. Moreover, the people of this land, Gentios as well as Moors, and in its native place, consider it to be cold, and the feeling of touch and taste are not likely to deceive and to make that appear warm which is cold and dry in the third degree. It is easy to reply to the argument of scent. For camphor has the property of evaporating and sending outwards all that it has within, while sandal and roses, being styptic, retain what they have within, and do

[1] Acosta here becomes enthusiastic about China:—"In the grandeur of the kingdom, in the number of people, in the excellence of its polity and government, and in its wealth, it exceeds every other country in the world. So many and so excellent are the things worthy of memory in it that I know not with what tongue any one would attempt to explain what there is in this grand kingdom. And though things far off generally sound grander than when seen near, it is the contrary in China. He who wishes to learn something of the number of notable things in China should read the book written by the Reverend Father Friar Gaspar de la Cruz of the order of St. Dominic."

not send it away. Many things that are cold and dry are also inflammable, such as hairs and tow.

RUANO

Avicenna says that it causes sleeplessness; so how can it be cold, for cold things have a tendency to cause sleep?

ORTA

It causes both insomnia and sleep. A little of it applied externally or internally makes for sleep. And much use of the smell of it, drying the brain, causes sleeplessness. There is nothing to be surprised at in these contrary effects. Now it really is time for dinner.

RUANO

These tarts certainly have a very pleasant savour. It seems to me that it comes from the roasted slices of some fruit that they contain.

ORTA

Antonia, pluck some CARAMBOLA,[1] for so we call them in Malabar, and we stick to the Malayalim names, because this was the first land we knew.

ANTONIA

Here they are.

RUANO

They are beautiful, sweet, and not very acid. They are the size of a small hen's egg, and a bright yellow colour. What seems still better in them, they are slit into four parts, making four small quadrants of a circle.

[1] *Averrhoa Carambola* and *A. Bilimbi*. The fruits of both contain a pleasantly acid watery pulp, and are very good candied, or stewed to a syrup.

ORTA

In Canarese and Deccani it is called CAMARIZ, and in Malay BALIMBA. I do not know any use for it in medicine, unless it might be given as part of the diet in fevers. With its juice and other things they make a lotion for the nerves of the eye, and it does good. Many people find in it a pleasant taste from what we call acid-sweet, for these are a little more acid. They also make an excellent preserve, which I ordered to be made instead of acid syrup. I will give you some to taste presently. Antonia, go and fetch some CARAMBOLA preserve.

ANTONIA

Here it is.

RUANO

Give me some acid syrup to take every morning, because it tastes very good.

THIRTEENTH COLLOQUY

CARDOMUM AND CARANDAS

RUANO

HERE is a tangled skein that we have to unravel, a formidable knot to untie, such an one as Alexander cut to save himself the trouble of unfastening it. For this reason it seems good to me that I should have the mistake, if there is one, explained by you. If you are able to untie the knot, well and good; and if not, it must be broken. The problem is how it can be explained that the use of *Cardamomo, mayor y menor*, as we now use it in Europe, is not in conformity with the teaching of Galen, Pliny, nor Dioscorides.

ORTA

I will let you know very well what it is that we call CARDAMOMO[1] *mayor y menor*, and you shall see it as clear as

[1] *Elettaria Cardamomum* (see page 28) yields the true Cardamoms of Malabar, Mangalore, Mysore, and Ceylon; those of Ceylon being especially qualified as "mayor." The Chinese include "Grains of Paradise," *i.e.* the seeds of *Amomum Meiagueta*, under their general denomination for Cardamoms. We distinguish the seeds of *Alpinia chinensis* as "Chinese Cardamoms." A synonym of the plant is *A. alba*, and the *A. Cardamomum* of Roxburgh may be another synonym of it. They are all Gingerworts. The Indian names of Cardamoms are: Sanskrit and Mahratti *ela*, Concani *alu*, throughout India *ilaichi* and *ilachi*, Arabic *kakilahisighar*, and Persian *kakilahikurd*.

the light of noon. For these medicines are much used as merchandize in this country, and also sent to Europe, Africa, and Asia; but whether the name CARDAMOMO is correctly placed or not, I am unable positively to affirm.

RUANO

Begin in a good hour, and tell me the names, Arabic, Latin, and Indian.

ORTA

Avicenna has a chapter on CACOLLÁ, dividing it into greater which he calls QUEBIR, and lesser, which he calls CEGUER. Thus he has CACOLLÁ QUEBIR and CACOLLÁ CEGUER, which is the same as CARDAMOMO, greater or less. By these two names the two kinds of CARDAMOMO are known to Arabian physicians and merchants. They have both kinds in India, in largest quantity from Calicut to Cananore, also in Malabar and Jaoa, but not in such quantity or so white in the shell. In Malabar it is called ETREMILLY, in Ceylon ENÇAL. Among the Bengalis, Guzeratis, and Deccanis it is called by some HIL and by others ELACHI. That is among the Moors, for among the Gentios of those parts it is called DORE. This is the reason that there is such confusion in the name in Arabic writings, for some call it by a name used in India, and others by the Arabic name. This confusion gives rise to mistakes.

RUANO

Serapiam calls one CACOLLÁ and the other HILBANE.

ORTA

The passage is corrupt; it should be CACOLLÁ and HIL, and the increase to it of BANE should be BARA, which means great in Deccani, just as CACOLLÁ, as all the Arabs call it, or

CACULLE, as Avicenna spells the word, or ELACHI, mean what we call CARDAMOMO.

RUANO

And what do we call it in Latin or Greek?

ORTA

Neither the ancient Latins nor the Greeks knew the CARDAMOMO, so why do you desire that I should tell you the name? You may believe that Galen never wrote of it, and this has been said by Avenrrois in his chapter on it. For Galen says that the CARDAMOMO is not so warm as the MASTURÇO, but more aromatic and pleasanter to the taste, with a little bitterness. Now all these qualities are not in our CARDAMOMO, which has not the taste of MASTURÇO, and is not bitter.

RUANO

Have not Pliny and Dioscorides written of it?

ORTA

Dioscorides says that the best comes from Comagena, Armenia, and the Bosphorus, and that it is also brought from Arabia and India. They had not got that which we call CARDAMOMO. What they had was that mentioned by Galen and Dioscorides, and not this in India, so that it follows that they are two different things and not one. If we desire to say that that which Avicenna and Serapiam call CORDUMENI is it, I will not contest the point, because it is not what Avicenna and Serapiam call CACULLÁ and HIL. Much less will it agree with the conditions laid down by Dioscorides, that it is hard to break, closed in a capsule, sour, and a little bitter, and that the scent flies to the head, all which conditions are not found in the CARDAMOMO but the reverse; for it is not hard to break, the scent does not go to the head, it is not bitter, nor is the taste like that of pepper or cloves, and though it causes water in the mouth it is more pleasant.

RUANO

Well, then, why should they call it CARDAMOMO if, as you say, it was not known to the Greeks?

ORTA

Because, as Terence says, "Davus[1] confuses everything." This Davus was Geraldo Cremonense, the translator. Not knowing this drug, by reason of his great distance from the country where it grows, never having navigated, and having no commercial relations with India, he put the name which seemed to him best. It would have been much better to have left the name in Arabic, as it was an unknown medicine; and not to have fallen into the error which this Geraldo has done.

RUANO

That of Pliny seems to be from this country.

ORTA

Pliny gives four species, one very green and thick (the best), one stubborn to rub, a third resplendent with the bright colour of gold, and a fourth smaller and blacker, another of various colours and which breaks easily. Now you see the CARDAMOMO here, that the shell is white, while the other is black and easily broken. To show you that it is not bitter, not black outside, much less green or of diverse colours, I will ask you to look at one. Boy, go and pick a black cardamom and bring it here. These black ones are much used because their scent is better for the mouth and for chewing and clearing the head.

[1] The slave of Simo, in the *Andria* of Terence. Davus was a most unscrupulous intriguing rascal.

Boy

Here it is.

Ruano

This is a very different thing. How says Valerio Cordo that the greater is almost like an acorn and the smaller one like filberts; but of these the largest is not bigger than the kernel of a nut, with its husk. Dioscorides says that these grains are sown with other larger ones. Is that so?

Orta

It is sown like our legumes, and the highest comes up to three *palmos*, and from it hang the capsules, and in the capsule there are ten to twenty small grains.

Ruano

We come to Ruelio and Laguna, who are more recent writers, and to what they say of this drug. Ruelio says that it has fruit like an Amomo, and that it is gathered like the Amomo in Arabia.

Orta

You can see here that what he refers to is not Cardamomo, for it is taken from India to Arabia and to the west, while the kings of these parts send for the Amomo from the west to make up their medicine, of which I am an eye-witness. That the Cardamomo or Cacollá does not come from the west is proved by its being an article of commerce to be taken there, as is known by all merchants.

Ruano

On the authority of Theophrastus it is said to be a neighbour of Nardo and Costo.

THIRTEENTH COLLOQUY

ORTA

We find this to be a mistake, for NARDO and COSTO are in Mandou and Chitor, and the CACOLLÁ in Malabar.

RUANO

They also say that the seeds are white, and that they raise blisters in the mouth.

ORTA

That is false as regards CACOLLÁ, for the bark is white and the seeds are black, and taken into the mouth with water it does not appear warm, which makes the Indians say that it is cold.

RUANO

Laguna, who translated Dioscorides into Spanish, says that three kinds of CARDAMOMO are met with in druggists' shops, *mayor*, *menor*, and another kind called NIGELA; and that all have a strong aromatic and biting taste. He says that the *mayor* is like Fenogreek or Alfoluas, but blacker and smaller; that the *menor* is smaller and inclined to a brownish colour; and that the third is NIGELA CITRINA, only differing in the black colour; so that my conclusion is that the first kind is MALAGUETA or "Grains of Paradise," and that this is the CARDAMOMO of which Dioscorides writes. The same Laguna says that a merchant showed him all three kinds at Venice in the year '48, and afterwards he says a thousand bad things of the Arabs who confuse everything.

ORTA

What you say of Laguna is a proof that what he says is false, as I have said. For Dioscorides did not see the CARDAMOMO with bark, for he says that it is MALAGUETA; nor did Dioscorides ever see it, nor did he know whence it came. As to the *mayor* being brownish he is wrong; and the NIGELA does not come from this region, nor has it the qualities of CARDAMOMO. The merchant who showed him the three kinds which he said that he brought from Armenia to Venice, did not speak the truth when he said that they were true CARDAMOMO. If it had been true he would have brought it from India, through Alexandria or some other port.

RUANO

From your reasoning I gather that you hold that the CARDAMOMO of the Greeks is not the same as what is called CARDAMOMO by the Arabs, and Menardo and other modern writers are right when they say that the CARDAMOMO of the Arabs is a new medicine, and that it ought not to be used, because Galen and Dioscorides, the princes of medicine, did not use it.

ORTA

From the first I confessed to you that the CARDAMOMO of which the Greeks write is not the CACOLLÁ of the Arabs; but I have never said that there is no use for it, for each day brings new diseases, such as the Neapolitan disease which we call Castilian itch; and God is so merciful that in each land He gives us medicines to cure us. He who causes the illness provides the medicine for it. But as Temistio[1] says, our knowledge is a very small part of what we are ignorant of. And as we do not know the medicines which cure all diseases, we bring the *rhubarb* from China, whence we get the root or stick to cure Castilian itch,[2] the *cana fistola* we get from India, *manna*[3] from Persia, *guaiacam* from the West Indies. Please

[1] Count Ficalho suggests that this is Themistius the Paphlagonian, a friend of Julian the Apostate and a commentator on Aristotle. His remark is what, in an Irishman, would be called a bull.

[2] Acosta adds Sarsaparilla.

[3] Manna, the Egyptian *menu*, and Arabic-Hebrew *man*; the "Ros Syriacus," "drosomeli," and "aier dew" of our old Pharmacists, is obtained in Sicily from the Olive blooms, *Fraxinus Ornus* and other species of Mountain Ash; also from *Tamarix gallica*,

God we will always continue to search for and enquire about medicines. The lovers of the Greeks, when they find that medicines are experimented with in the countries where they grow, are unreasonable in abusing Avicenna, Abenzoar, Rasis, Isaque, and others, who cannot be denied to be learned, for thus using them.

RUANO

You say well. But as I have used your CARDAMOMO to cure according to Galen, is not this what he speaks of?

ORTA

I say that in the receipts of the ancient Greeks and Latins the Arabs were not followed, and for CARDAMOMO they used that of Galen; and if you did not know it, do not blame the others because you had not got it. In the receipts of the Arabs and modern Latins the Cardamomo *major* is used, which is the large one you see, and the *minor* which is the other.

RUANO

There is another bone to pick. These two have both the same appearance, and only differ in size, and all your imitators of the Arabs call the smaller one CARDAMOMO *mayor*, while the big one has never been seen in Europe; and for the smaller one they use a seed called *grana paradisi*, which the Spaniards call MALAGUETA. It seems to me that you undo all medicine and all methods of curing. Tell me where you see this error.[1]

ORTA

I will make you see it clearly. I have often asked in Portugal, and of people who came from Portugal to MALA-

the "Myrica" of the Greeks and Romans, in Southern Europe and Western Asia; from *Alhagi maurorum* in Western Asia and Persia; and from *Quercus persica* in Persia; and *Larix europea* in Europe [" Briancon Manna"].

[1] The Portuguese editor says that all this passage is unintelligible.

GUETA, if they have CACOLLÁ, which we call CARDAMOMO, in MALAGUETA, and the answer was no; that they had never found it there. I then began to consider why so learned a person as Avicenna should have made a division into CARDAMOMO *mayor* and *menor*, when the *mayor* is found in India and the *menor* in Malagueta, 4000 leagues away. I also saw that Avicenna called Malagueta by the name of CONBAZBAGUE, and there seems to be good reason for this, for it is said that it comes from Sofala, of which Malagueta is a continuation. It may be that in Sofala there are adjacent lands unknown to us, for the people are barbarous and unaccustomed to converse as men. Do you want Avicenna to write two chapters on one thing? Walking in these cities of the coast, I saw a Jew merchant from Turkey who told me that he had CACOLLÁ QUEBIR in his list of medicines that he wanted to buy. As I understood that CACOLLÁ was the same as CARDAMOMO and QUEBIR as great, I asked many merchants if they had CARDAMOMO in other lands and what it was like, but none of them could tell me. I found that they have it in Ceylon, but much larger and not so aromatic. This was told to me by the agent of the King who resides there, who went to Ormuz and Arabia for merchandize and made good profit. At the same time I sent one of my ships to Ceylon and obtained a specimen. On another occasion I was curing a great lord named Hamjam, brother of the King of Balaguate named Verido, and I dispensed a receipt containing CARDAMOMO,[1] *mayor* and *menor*, in the Arabic language, and they presented me, to make it up, with these two medicines. This should be no less than sufficient to enable you to see that both are alike, only one bigger than the other.

RUANO

In Portugal I saw the lesser, and not the greater. Which do you consider the best to use?

[1] Masudi and Edrisi enumerate cardamoms as one of the drugs coming from India, but the first to fix Malabar as the locality was Duarte Barbosa.

ORTA

I say that both kinds that are sent to Portugal are good, and from there are spread over Europe. Perhaps the smaller is more aromatic and better, and it may be said of the larger that it is better in virtue and less in quantity. This I say, but deferring to better judgment.

RUANO

I am astounded how, seeing these two heads of seeds, I presently said this is CARDAMOMO *mayor* and this *menor*, and from this time forward this shall be my use and practice; and CONBAZBAGUE and MALAGUETA only when I am thinking of the medicines of Arabia.

ORTA

I know of nothing more than what I have said to druggists, physicians, and others, and it may well be that this has not been a good thing for me, for they say afterwards that they found out these things and obtain glory by my work; while I only wish to bring profit to all. God will be a witness to what I say. There was a Viceroy in India who was very anxious for knowledge, and as he did not know Latin, he read Pliny in Tuscan. He was desirous of acquiring a knowledge of some simple, and commissioned me to tell him when I should find it. I took this *cardamomo mayor* to him to show him, and also the *menor*, telling him what they were called. He listened and said that my account seemed to be correct, but that he had much faith in an old druggist, and ordered him to be sent for.

RUANO

Was this druggist learned? Did he know Latin and Greek or Arabic?

ORTA

No. He was an old man who had been a long time in India. He was well acquainted with the practice of a dispensary, but his knowledge of Greek, Latin, and Arabic was of the kind possessed in Spain by a man who has never heard those languages spoken or read. For all that, he was a very good man, and because he had made the drugs in the dispensary at Cambay, which were sent to Portugal by the Government overseer, the Viceroy said that no druggist in the country knew about medicines so well as he did. The Viceroy asked this man whether one of the specimens before him was *cardamomo mayor* and the other *menor*. He said no; but that the smallest was *cardamomo*, and the other neither *mayor* nor *menor*. Then I said that both had the same smell, one being large and the other small, and told him to try them, for that the two kinds were of this land, and that with his experience he ought to know that one was *cardamomo mayor* and the other *cardamomo menor*. On this he broke out with loud shouts in good Portuguese romance, *de presumitur*, that I might affirm or deny, but that he knew, which was an argument of mint and polypody. Then I asked him why this was not *cardamomo*, and whether he could produce any other in the land. He replied, "Why! would God allow that I should not know after so many years, if you know so suddenly!" I replied that we might know many things to-day which we would forget to-morrow; also that occasionally small people like me revealed things which great people like him could not explain. With all these flatteries I never could get him to confess, for he had recourse to *de presumitur*.

RUANO

And did you then have a good laugh?

ORTA

I might have, but there would have been trouble in so

dealing with such a person. But a learned jurist, who was sitting in a corner, laughed enough for himself and me too, and I can hear him laugh now when I think of it.

RUANO

Did not this Viceroy know what you knew?

ORTA

Yes; and he knew me well in Portugal; and when the King took leave of him to go to this land, he said that it was not necessary to take any medicines with him. He took this advice and departed this life under my care; but he was more ashamed of the obstinacy of the druggist than anything else.[1]

RUANO

I should like to know this druggist.

ORTA

He is dead; and may God pardon him, for, apart from some foibles, he was a very good man. We will not talk more about it, for I told you the story to amuse you for the moment, and not to remain in your memory.

RUANO

I would mention to you that Andreas Belunensis, an Arabic scholar, says that CACULLE is CARDAMOMO *mayor*, and that ALCAL or HALEIL or CAYRBUA and EILBUA are CARDAMOMO *menor*.

ORTA

All these names are corrupt in the Arabic books, and what I have said above is correct. I do not say this without

[1] This Viceroy must have been Dom Pedro Mascarenhas. He died at Goa on June 23, 1555. He knew Italian, having been Ambassador at Rome, and may have had the Italian translation of Pliny by Londino.

recognising that the author knew much, but as he had never been in this country, he could not have accurate information.

RUANO

Do the natives use *cardamomo* much as a medicine?

ORTA

Very much, for they mix it with their betel to give it a pleasant scent. They say that it draws out inflammations from the head or stomach. They take it in syrup. They are wrong in calling it cold; but this is a trifle, for they say the same of pepper.

RUANO

Do the Indian physicians give the root for fevers? Mateus Silvatico says so, and that it grows from great protuberances on trees. Does any chance to grow in that way?

ORTA

They do not use the root, except that they take it in fevers. It only grows from seeds in the ground which has first been burnt, and in no other way. What Mateus Silvatico says is quite wrong, and as no one else has made the statement the error rests with him.

RUANO

How is it that so much pepper is used in Europe and so little *malagueta*, which is better, especially with fish?

ORTA

This practice arises from the French and German merchants. They tell me that *malagueta* does not do for seasoning food when it is cooked, for it should not be boiled. It is only of use for raw food, or food that has already been cooked. This is the reason that there is little demand for it. Now let us drop this, and let us partake of the fish which is cooked ready for eating, for it also has *cardamomo.*

RUANO

Very well; but what is that fruit that has been roasted? It looks like small green apples.

ORTA

It is called *carandas*,[1] and comes from the mainland and from Balaguate. The trees are the size of an arbutus, and the leaf is like, and the flower has the scent of honeysuckle. When the fruit is ripe it is very pleasant to the taste and dark like grapes. I heard of a man who made wine from it, and it may be that it would be good wine after being kept. Now this fruit is green, and the size of a filbert with the rind. It is larger in Balaguate when it is ripe, and it then gives out some viscous juice like milk. Some people put salt with it when it is ripe enough to eat. These green ones are salted. and in this country they take salted fruit to excite the appetite. They also serve it up with vinegar and oil, when it is called ACHAR. In this way green plums, apples, vine stalks and brambles with their fruit come from Persia and Arabia; for these Indians seek all manner of ways of eating.[2]

RUANO

So will I; and I have already tasted this fruit, as well as green apples.

[1] *Carissa Carandas* (Linn.), a Dogbane. Used for pickles.

[2] Acosta also devotes his chapter lxiv. to *cardamomo.* It is all copied from Orta, but condensed, and the last three pages omitted.

FOURTEENTH COLLOQUY

CASSIA FISTOLA

Ruano

IT is very necessary to know about the Cana fistola,[1] for in it your Arabs have given such a good medicine for purging, without trouble or injury to the patient. I believe for certain that the Greeks would have written about it, highly praised it, and experimented with it.

Orta

We have little necessity to speak of a medicine so well known and tried. The only contradiction is in the name, which was wrongly given by Geraldo Cremonense who, as I have said to you many times, had much better have left it as it was given by Arabs, for they were the inventors of this medicine; nor ought Nicolas Leoniceno and Menardo and so many other modern writers to say so much evil of the Arabian physicians, as the faults are their own. To tell the truth, the Arabs are deserving of praise, not of vituperation.

[1] *Cassia Fistula*, Linn., *Cathartocarpus Fistula*, Pers., Leguminosae.

RUANO

The other writers do not say many things against Avicenna, but more or less follow him, without making any difference in the names. This being so, tell me the name in the countries where the tree grows.

ORTA

It is found in all these parts, but the best comes from the north. The Arabs call it HIARXAMBER,[1] a name of four syllables, but the most commonly used among them. Avicenna has CHIARSAMDAR, which is corrupt. COMDACA[2] is the form it takes with the Malabar people and Canarese, while the Gentios of our land of Goa have BAVA SIMGUA.[3] The Deccanis and Brahmasis have BAVA SIMGUA,[3] the Guzeratis and Deccani Moors GRAMALLA.[4] They call its tree BAHO[3] among the people of this Canarese country. It is the size of a pear, with leaves like those of a peach tree, as green and slightly narrower. The flowers are yellow, like those of broom, and the scent is like that of green cloves. As the flowers fall, there sprout from the stem buds like those of the chestnut; and the cane is very green on the tree, and when it ripens it is not vermilion as Laguna says. It is six *palmos* in length to two *palmos*, the shortest. It grows, as I said, in all these countries, and in Cairo, but the best is in Cambay, and the hardest. They may also have it in Malacca and Sofala, but the people who go to those places have very little curiosity.[5]

[1] Persian, *Khiyar-i-Chambar, i.e.* "Cucumber of necklaces" (from its long strings of golden flowers).

[2] *Konna, i.e.* "The King."

[3] These names signify "the good and strong medicine."

[4] *Girmala* of Gujerat.

[5] "Cana fistola" is here, I take it, the pulp of the pod of *Cathartocarpus Fistula*, which is a gentle aperient, and the "Pulpa Cassiae," *Cassé*, and "Cassiae atramentum," and, also, the "Siliqua Aegyptia" and "Cassia Fistularis" of our older Pharmacists. The root is considered to be a febrifuge. The word *cassia* is Syrian, meaning "peeled," *i.e.* bark.

CASSIA FISTOLA

RUANO

Is the tree planted or wild?

ORTA

I have never seen it except in a wild state. I am told that in the West Indies it was originally wild and gave out a large hollow cane. They sowed seeds in the convent of San Francisco de la Vega, in the island of Santo Domingo, and the plants grew and yielded very good *caña fistula*, raised from cuttings and seeds. In this way each person planted the trees on their properties until the quantity that was raised sufficed to supply all Castile. But I consider that the Portuguese are more fortunate for, without sowing, they have such a quantity that in Cathay the price of a *candil* (522 lbs.) is 360 reis, equal to one *pardão*,[1] and praise be to God that He confers such benefits on us daily.

RUANO

Of what composition do the Indians consider it to be?

ORTA

They do not have much faith in the graduations, but they call it cold. Avicenna considered it to be temperate, with qualities both cold and warm, and that it was damp. Serapiam makes it temperate. Mesuel says that it inclines a little to warm, and this would be from its sweetness. Antonio Musa finds it to be warm and humid. All these views may be maintained, for the physician judges from outward feelings.

RUANO

It is used as a medicine in India?

ORTA

Yes; as a purgative. They make mouthfuls, scraping the cane as we do.

[1] A *pardão* of gold was worth 360 reis, and a *kandi* varied from about 500 lbs. at different ports.

RUANO

Are the grains purgative also?

ORTA

No, they are thrown aside; and I marvel much at Menardo saying that the grains are purgative, being a thing which requires more art to open than to relax. If he was deceived it was by giving some mixture compounded, and as the seeds may have been found in it, they would lower the effect considerably. For these lubricating medicines do not draw more than the materials they encounter. For this reason one ounce of *caña fistula* purges more than thirty grains of *escamonea*. Also, it may be that the imagination of purging would purge more than that with which Menardo purged.

RUANO

Is it used to provoke menstruation, or to make conception easy, or to discharge the *placenta*?

ORTA

It is not used for any of these things.

RUANO

I did not ask this without a hidden reason, for we use the powders of the rind in a decoction of *artemisa* or in an egg, with 4 ounces of honey, and Sepulveda says that this was the result of experience.

ORTA

This Sepulveda was not an evangelist. It is much more by the decoction of *artemisa* that menstruation can be promoted, and not by this rind, which is very cold and dry. As for discharge of the *placenta* that is not of much consequence, for, without medicine, in the course of nature such things cease to have retentive virtue and loosen by themselves.

RUANO

Well! what will you say of Avicenna, who orders it to be given to facilitate conception?

ORTA

I should doubt very much whether this was the meaning of Avicenna. I look upon it as a rule that when he uses the word *cassia* in writing of purgative medicines he means *cassia fistula*, and on all other occasions when he uses the word *cassia* he means *cassia lignea*. I now turn to Andreas Belunensis, who says that the correct word is COGOMBRO SECO and not CANA FISTOLA, hence those who use *cana fistola* wrongly remain clear of reprehension, let the imitators of the Greeks say what they like.

RUANO

In Portugal they told me that dysentery was very prevalent in India because the cows eat *cana fistola*, causing the meat to be purgative. Tell me whether this is so or not.

ORTA

I was also told in Portugal by a former Governor, and by an actual Viceroy, that they did not like to take *cana fistola* for the same reason, and their physician agreed with them in this. I told them the truth, saying that it was not as was supposed in Portugal. For in India the trees were so high that the cows could not reach them, nor were they so numerous that the cows could be maintained by them, for the cows, in that our land, were innumerable. The reason is that the Gentios raise them, but do not eat them. Moreover, the *cana fistola* is hard in the rind when it is green, and would not be liked by the cows so much as green herbage, which comes up several times in the year. I also asked about it, and was told that the cows did not eat *cana fistola*, those of whom I enquired laughing at me.

FIFTEENTH COLLOQUY

CINNAMON

Ruano

ONE cannot eat any spice with pleasure except cinnamon. It is true that the Germans and Flemings eat pepper, and here our negresses eat cloves, but Spaniards do not eat any of the spices except cinnamon. I bear it in memory that the food smelt strongly of it, and not of any other spice. I asked the cook whether he used it, and he said no, but that many dishes were seasoned with cinnamon water. In place of what we call *cassia lignea*,[1] the word *canela* is often used. It will be as well that we should discuss it now.

Orta

Canela, and what we call Cassia lignea, are one and the

[1] "Canella," or "Wild Cinnamon," is in modern Pharmacy the bark of *Canella alba*, Murray, a Bixad of the West Indies. Here it is "Cassia" or "Chinese Cinnamon," the bark of *Cinnamomum Cassia*, a Laurel of Cochin China and Southern China. True Cinnamon is obtained from *C. zeylanicum* of Ceylon. "Cassiae Pulpa" is, as stated in a footnote, p. 114, the pulp of the pod of *Cathartocarpus Fistula*.

same thing; but the ancient writers saw this spice after it had come from such a distance that they could not have had a correct notice of it, and as the price was very high there arose a thousand fables which Pliny and Herodotus repeat. They relate them as true, when in reality they are quite fabulous. I do not propose to speak of them here, because every one now knows the truth. The price being so high, and the avarice of men still higher, the drug was often falsified. As the false kind could never be exactly like the real spice in every respect, they made two kinds, one the true cinnamon and the other the falsified one, both being usually of the same species.

Ruano

Tell me what you know, and at the end I will mention any doubts which occur to me, for I do not wish to remain with them. I wish to hear from you the names in all languages, in the lands where cinnamon grows, and in Arabia and Persia, for by these names we shall be able to obtain a knowledge of *cassia lignea* and of cinnamon. For my present view, with others who have written on the subject, is that the true cassia is not the true cinnamon.

Orta

I will satisfy you on all these points. Neither the Greeks nor the Arabs knew the cassia, and this was on account of the great distance and the little communication with the region where it grew. Those who brought it for sale to Ormuz and Arabia were Chinese, as I will explain to you presently. From Ormuz it was taken to Aleppo, a principal city of Syria. Those who brought it thence to the Greeks said that they had it in their country and in Ethiopia. It was taken with superstitious rights. The priest divided what remained into shares for the devil, for his worshippers, for the king, and for the priests.

FIFTEENTH COLLOQUY

RUANO

How is this? Do you mean to say that neither the cassia nor the cinnamon are in Ethiopia or Arabia?

ORTA

Yes! and I am astonished that you did not know it; for Ethiopia has been reached by our ships and great part of it explored by our people, and in it there is neither *canela* nor *cinamomo* nor *cassia lignea*; and the Arabs themselves come to buy it and take it to their country, and at the time when they come the price is very high.

RUANO

Is this that you allude to the real *canela*? For they may not get the true cinnamon or cassia, but that other, and being a rude people they would not know the difference.

ORTA

The physicians of Arabia, Turkey, and Coraçon are very learned men, and all call this thick *canela*, which they use, *cassia lignea.*

RUANO

What proof can you give me that it does not grow in Ethiopia?

ORTA

I say that both the Ethiopias are very well known to the Portuguese; for the coast of Guinea, which is the Ethiopia below Egypt, is known to us, not only the sea-shore, but also the interior; and as I have told you, a priest explored from the island of San Tomé to Sofala[1] and Mozambique, and came thence to Goa, and I knew him very well. From the Cape of Good Hope to Mozambique and Melinde many persons have

[1] That this priest went from San Tomé overland to Sofala is doubtful.

been along the coast, and these have come here, and none have ever seen the *canela*. So that we know that the *canela* does not grow in either of the Ethiopias, either above or below Egypt.

RUANO

Is this because they are few who have the curiosity to enquire?

ORTA

It is not so altogether. For the natives of the island of San Lourenço, who are very barbarous, showed some men who came to trade with a fruit of the size of a filbert without a head, and as it smelt like a clove[1] they wished to be shown the tree in case it should be cinnamon or *cassia lignea*. They were shown what appeared to be a medicine quite as odoriferous. The medicines were never better known than at present, especially by the Portuguese, so you must not suppose that such precious drugs are wanting with us, for the plants and fruits are now better known than ever. Of course the grafting makes diversity, as well as the transplanting them from one land to another.[2] For the love of me, do not suppose that we are deficient in cinnamon. We have as much as we require. I will now tell you the names.

RUANO

I say that I claim my right, as the lawyers have it.

ORTA

The *cassia lignea* is called by the Arabs SALIHACHA, and the Persians give it the same name. The Indians who do not know physic through the Arabian books, give it the same name as they give to cinnamon, because in all this country

[1] *Ravensara aromatica* or *Agathophyllum aromaticum*, Lauraceae, of Madagascar. *Ravensara* is its native name.

[2] Orta here recognises the variability of species.

they make no difference between the names of cinnamon and *cassia lignea*. In truth, no person can see any difference between them, whether physician or druggist.

RUANO

I rejoice much to know this.

ORTA

The Chinese have navigated to this land from a very remote period, and as the people were barbarous and unlearned they adopted laws and customs from them. The Chinese went in such large vessels and in such a way that, if it will not bore you, I will tell you many things which do not directly belong to the subject of our colloquy, but which may interest you.

RUANO

I shall be very much obliged if you will, and we have time to devote to it.

ORTA

I know the number of Chinese ships that navigated, having counted those which went to Ormuz as recorded in their books, and there were 400 junks which entered the port of the island Jeru, now called Ormuz. They also say that 200 junks have been lost on the rocks of Chilâm. Junks are long vessels, which have their bows and sterns alike. In Calicut they had a fortified factory, which still exists, and is called *China Cota*, or the fort of the Chinese. In Cochin they left a stone as a mark, in memory of their having been there. When the King of Calicut (called Zamorin or Emperor) besieged Cochin because the Portuguese held it who were then engaged in the discovery of India, he destroyed the place and carried off this stone as a trophy, which cost him very dear. On this stone the King of Repelin was crowned, the Zamorin placing the crown on his head, who received homage from him ; and the stone was left at Repelim by order of the Zamorim. Repelim is four leagues from Cochin, and there the stone remained until 1536. Then Martin Affonso de Sousa, the not less invincible than fortunate captain, burnt, sacked, and destroyed Repelim. The king fled and many with him, and many of those were slain who did not escape, as I can testify, being an eye-witness. The stone was taken back to Cochin, and the king ordered great festivities to be made over it, and thanks to be given to those who brought it. He remained deeply obliged to the Captain, Martin Affonso de Sousa, who twice drove the King of Calicut out of his territory, and sent him the *sombreiro*, which he took with the *praus* in Beadalla (there were 57), where 15,000 men were killed, remaining with him no more than 300. He also took 600 pieces of artillery, and more than 1000 muskets. As the deeds of this great captain are many, I will not tell you more. These that I have mentioned are not told in praise, for he has acquired more praise than any one else of his time, but I tell you of them to make clear what I said about the Chinese.

Ruano

But I want to know about the cinnamon, for, in discussing the plants, you always give me some history of them.

Orta

These merchants bring from their country gold and silk, porcelain, musk, copper, seed-pearls, alum, and many other things. They sell some in Malacca and bring thence sandal, nutmegs and mace, cloves, and aloes. Proceeding on their voyage they trade in Ceylon and Malabar. From Ceylon they bring a great deal of good cinnamon, which costs them very little money, while the sailors without any money collect wild and bad cinnamon in the woods. They also go to Java, and obtain pepper, cardamoms, and other drugs on this Malabar coast. They carry all to Ormuz or to the coast of Arabia, where merchants come to buy, and convey the merchandize to Alexandria, Aleppo, and Damascus. Asking these Chinese what sort of thing is this cinnamon which has such a smell and taste, they tell the fables which Herodotus and other ancient writers relate, to make a better sale for their goods. As the cinnamon from Ceylon was different from that of Malabar and Java they gave two names for the plant, but only one for the skin or bark. As a fruit is better in one country than in another, so the cinnamon of Ceylon is better than all others, all being cinnamon. They do not send any other cinnamon than that of Ceylon to Portugal. In Ormuz, because it is the bark brought for sale from China, they call it DARCHINI, which in Persian means Chinese skin. They send the same to Alexandria and the other places mentioned above, changing the name so as to get a better price from the Greeks, and calling it *cinamomo*, which means odoriferous skin, like *amomo* brought from China. To the bad kind from Malabar and Java they put another name, which is what it is known by in Java, CAISMANIS, which means sweet skin in Malay. Though it is one species they put two names, calling the good DARCHINI or Chinese skin, and CINAMOMO, which is *amomo* of China, and the other CAISMANIS or sweet skin.

RUANO

DARCHINI is not an Arabic word, then how is it that it is used by Avicenna and Rasis and all the Arabian writers?

ORTA

No, it is Persian; for many words in the *Kanun* of Avicenna are Persian. The name for cinnamon in Arabic is QUERFÁ. As it is said by Andreas Belunensis that it was the name for thick cinnamon, I communicated with the Arabs, who assured me that QUERFÁ or QUERFÉ was the name for any kind of cinnamon. The Greeks corrupted the word to CASSIA, and it was CAISMANIS that they called CASSIA. These are all the names that the Arabs have written to designate it, and those that are written differently are corrupt, such as DARSHAHAM and others. Well, this is the truth; and I call God to witness that the druggists cannot deny it, for *cassia lignea*, bad cinnamon, is mixed with very fine cinnamon, of which there is such abundance, and they excuse the doubling of the weight of the *cassia lignea* for cinnamon.

RUANO

This that you say about the weight of *cassia lignea* being doubled in place of cinnamon does not want for authority, for Dioscorides and all the others say the same.

ORTA

For me the testimony of an eye-witness is worth more than that of all the physicians, and all the fathers of medicine who wrote on false information. So that what the Greeks and Latins called cinnamon, the Arabs call QUERFÉ or QUERFÁ, the Persians DARCHINI, those of Ceylon CUURDO, the Malays CAISMAO, and the people of Malabar CAMEA. It is found that Serapio adopts DARCHINI, which is a tree of China, giving a derivation that for you is false, which was increased by the translator, and which to me is the truth.

RUANO

It may be remembered that you have said that the *cassia lignea* was first called *caismanis*, meaning sweet stick, and if this is so, cinnamon may be bitter stick, as Menardo explains the Greek verb, which signifies that at least it may be corrosive.

ORTA

This verb, interpreted by Menardo, means to furnish with

a bite soft and fragrant; and he further says that bitterness is outside things that are fragrant, but that it gives them a sweet smell and sharp taste. I also say of this, replying to Menardo, that the people of this country have not more than three kinds of tastes—sweet, acid and bitter, and what is known well and is not bitter they call sweet. So that the things they know well they call sweet, and so give it the name of sweet stick.

RUANO

A modern writer says that our *cassia lignea* is not that of the ancients, for he says that it is black and without smell, and that there was some cassia which is called *pseudo cassia* by Dioscorides, or false cassia.

ORTA

It is likely enough that in ancient times the cinnamon was falsified. Its great abundance makes it unnecessary to do that now; but one of the drugs that deteriorates most in the land is the cinnamon, owing to its being brought after such long voyages. It may be that good cinnamon is mixed with some that is damaged and is without smell and is not vermilion, and it may be held that it is not cinnamon at all, just as a dead man is not a man.

RUANO

What did those learned Persian and Arabian physicians use instead of cassia to cure that king who was your friend?

ORTA

The thick cinnamon of Malabar. I argued with them that nothing except fine cinnamon should be used, but they were obstinate and would not listen to reason; though the king was convinced and was on my side. Returning to the *cassia*, it is impossible to understand these modern writers, for some say that there is no true *cassia lignea*, and Menardo says that there is, and that it is sold under the name of *canela* and *cassia*, and this same Menardo says that it is not the true *canela.* Valerio Cordo says that he would not go so far as to say that we have not got the true *canela*, and that we must have some species of it. Laguna, quoting Galen, says that *cassia lignea* is converted into cinnamon, or it would be better to say that cinnamon is converted into *cassia lignea*, for one species cannot be turned into another more perfect species by lapse of time, but into another less perfect. So much for these authors. I say that one species can never turn into another, but that the good cinnamon may at times deteriorate and be called *cassia lignea*, not because the *cassia lignea* and cinnamon are different species, but one species grown in different countries. Afterwards, Amato Lusitano held that he had all the species, and he is followed by Mateolo Senense and some others. After these Laguna says that he who goes from India to Lisbon will find all the species of cinnamon. But, speaking the truth with you, I could never see more than two kinds or three, all of one species, the *canela* of Java, of Ceylon, and of Malabar. When Laguna says that in going from India to Lisbon a man will find all the species of cinnamon, I say that what he means is that he will find good and damaged cinnamon, some better, some much better, but not five distinct species as he says.

Ruano

Well, I know that he says further that, in the time of the Roman Emperors, if a stick of true cinnamon was obtained it was considered a great treasure, so no wonder if it was so hard to get. It is said that in the time of Pope Paul a piece was found which had been kept since the time of the Emperor Arcadius, and after 1400 years there was great rejoicing over it.

Orta

I will answer all this. I say that you can get more

knowledge now from the Portuguese in one day than was known to the Romans after a hundred years. The stick which was given may have been brought from Lisbon, and would not have deteriorated. The piece belonging to the Emperor Arcadius may have been preserved by the will of God, or it may have been an imposture.

RUANO

Ruelio says that the cinnamon smells like the *origanum*.

ORTA

The stick does not smell otherwise than as the bark smells, but not so strongly or intensely, and there is no *origanum* in the whole island of Ceylon, nor in Malabar, nor in any other part of India unless it is brought from Ormuz.

RUANO

Some say that we have cinnamon, but not that which was highly praised, called *mosselitico*. They say that the better cinnamon is, the longer it lasts. Others say that it lasts thirty days, and that it lasts longer when it is powdered. What is your answer to this?

ORTA

The first I will answer when I tell you where the *canela* comes from. As to the last, the drug of which we are treating lasts for a very short time before it begins to deteriorate. As to what you say that it lasts longer if it is powder, this is wrong, for the proper stick lasts longest. In houses where they eat cinnamon powdered over their dishes, they do not keep it more than from one day to another, because it deteriorates here in India. The bark as well as the stick can be conserved better in some countries than in others, lasting longer where the humidity is least. In other lands the physicians reckon with the climate and are guided by experience; so that they know very well that it will not keep for thirty years. As for the other *canela* which had been kept since the time of the Emperor Arcadius, my answer is that I should like to see it before I believe it.

RUANO

Another statement of Antonio Musa, given on the authority of Theophrastus, is that the ancient cinnamon had many knots, while this of our time has not.

ORTA

Theophrastus is wrong, for he was not a man of this country to know what the tree is like. When the bark is taken off neatly, you will clearly see the truth.

RUANO

I say that I shall see at the end how my doubts have been solved.

ORTA

The trees are about the size of olives or rather smaller, the branches are numerous and not crooked, but somewhat straight. The flowers are white, the fruit black and round, larger than a myrtle, or between that and a nut.[1] The *canela* is the second bark[2] of the tree: for it has two barks like the cork tree, which has bark and shell. The *canela* is the same, except that the two layers are not so thick and distinct as in the cork tree. First, they take off the outer bark and clean the other. The outer bark, cut in squares, is then thrown

[1] It is a small evergreen tree richly clothed with beautiful shining leaves. The panicles of greenish-white flowers are not sweet.

[2] The Canella of modern pharmacy is the bark of *Canella alba*, Murray, "the Wild Cinnamon-tree."

on the ground. When on the ground it rolls itself up in a round form, so as to look like the bark of a stick which it is not. For the poles or sticks are the size of a man's thigh. The thickest of the bark is the thickness of a finger. It takes a vermilion colour, or that which is given when burnt by the sun; or more like ashes mixed with red wine, very little of the cinder and a great deal of the wine. The trees are not so small as is stated by Dioscorides and Pliny,[1] and they are numerous, so that the price is very low in Ceylon. For more than thirty years it has not been possible to buy any, except from the agent of the king. This year's bark is taken, and leaving the tree for three years it renews its bark.[2] There are many trees, the leaves like a laurel. The trees that yield bad *canela* in Malabar and Goa are much smaller than those of Ceylon, and are all wild, growing of themselves. The root gives a water the colour of camphor, and is considered to be cold. The king forbids the roots to be pulled up, so as not to destroy the trees.

RUANO

Is this *canela* white or vermilion or black?

ORTA

That which has not been well dried remains white or grey, when very dry it is black. When well dried it is vermilion, as I said before. The root is almost without taste, and smells like camphor. The fruit is not pleasant to the taste. They also distil the flower, but it has not so good a smell as the water distilled from the bark. Laguna says that they only

[1] The drug name "Canella" does not occur in Pliny or Dioscorides. It is the French *cannelle*, diminutive of *canne*, a "cane," from the Latin *canna*, a reed.

[2] Ficalho characterises this as an unpardonable error, which is also adopted by Gaspar Correa; a result, he supposes, of some vague reminiscence of what occurs with the cork tree. But in the cinnamon tree to strip the bark causes the death of the branch.—LENDAS, i. 652.

distil from the flowers, but it is true that they distil more from the bark before it is dried. It is a very mild medicine for the stomach, or to remove the pain of the colic, from its cooling quality. For it draws off the pain as I have often seen. It improves the complexion, removes a bad smell from the mouth; and for Portugal it is certainly a very good article of commerce. It is sent there in sufficient quantity, for besides being a good medicine, it is useful in seasoning dishes, as is the practice in India.

RUANO

Is Ceylon the only place for the best cinnamon?

ORTA

I have not heard of any other place.

RUANO

Well, Francisco de Tamara,[1] in his book on Customs, says that in the strait of the Red Sea cinnamons and laurels cover the water when the sea rises. Also our Castilians, who write about the West Indies, say that there is cinnamon in many parts of those Indies, especially in a land called Zumaco. They also say, speaking of China, that there is much cinnamon and spices there. Answer me all this.

ORTA

As for what Francisco de Tamara says you can reply that he repeats what others have falsely written, and that the Portuguese who navigate the Red Sea have never seen anything of the sort, sailing over it every year. The other chronicler who says that there is *canela* in the West Indies also does not tell the truth, for they say that the fruit is like

[1] A professor in Cadiz. He wrote *Juan Bohemo de las costumbres de todas las gentes* (Antwerp, 1556).

acorns of the cork tree, whereas the fruit of the *canela* in Ceylon is like an olive, but smaller and more round. It would be well if some of this *canela* could be seen in Spain, when it would become known that it is another tree that yields this fruit and bark, and that the trees are different,[1] as the pear tree of *engoava* differs from other pear trees. What is said about China is well known to be false, for this drug is brought there from Malacca, and it is known that they have it not in China.

RUANO

What use do they make of the fruit of the *canela*?

ORTA

They extract oil from it, as we do from the olive. It looks like French soap, and has no smell unless it is warmed, when it gives out an odour something like cinnamon. It is used to warm the stomach and nerves.

RUANO

Is all the cinnamon of Ceylon very fine?

ORTA

No! some is very bad, which has not rolled up well or was very thick from not being of that year, and being old is not good. I speak of Ceylon, for that of Malabar and other countries is all very bad. The quintal of Ceylon *canela* is worth ten *cruzados*, and that of Malabar one *bor*, which would make four quintals to the *cruzado*. The Malabar people take their *canela* to Cambaya, Chaul, and Dabul for sale, and thence it is taken to Balaguate.

[1] The South American Cinnamon tree, of which Gonzalo Pizarro went in search, is the *Nectandra cinnamomoides*, a Laurel bloom of British Guiana. In modern pharmacy the "Bark" [*Bibiru* bark] is referred to *N. Rodiaei*, Hooker.

CANELA

Ruano

Tell me the names of the kinds given by Pliny, to see if we can identify them in any parts of India.

Orta

We must do what we can, for what I said is true, and the names give themselves to it. As for *zegir* it may be that all the land of the Cingalese is so called, that is the people of Ceylon. For the Persians and Arabs call the blacks *zangues*, and all the people of Malabar and Ceylon are of that colour. Also, those rocks which are between Ceylon and India are called *Chilam*, whence we may derive the name *zegir*.

Ruano

And the *cinamomo musilitico*, so highly praised, where is that?

Orta

The island of Ceylon, a mountainous country, which is opposite Mount Cory, which is the Cape of Comorin, is where the seed is, on the scent of which Dioscorides puts his faith. And Pliny says that he traces that *canela* to the port of Genalabitas, as he calls Ceylon.[1] See how clearly he wants to speak of the port of Chingualas, which is Ceylon. For he says that by a direct road from the promontory of Cory one comes to a port of Genalabitas called Ocila. If you are not satisfied with these derivations I can give you no better.

Ruano

These derivations appear better, but those who say that the *canela* leaf is like *lyrio espadanal* say well.

[1] "Taprobane" and its productions are noticed by Pliny, vi. 22 (24), vii. 1 (2), ix. 35 (54), and xxxii. 11 (53).

ORTA

No: because the leaf of the *canela* is like an orange or a laurel leaf, that is, the shape is that of the orange and the colour of the laurel.

RUANO

Is oil made from the *canela* also?

ORTA

I have told you that it is only made from the fruit of the tree, and that it is made, as we make the olive oil, and that is the truth.

RUANO

I found in the receipts of a doctor of credit "*take cinamomo allipitinio.*" Does that place happen to be anywhere in Ceylon, or where is it?

ORTA

It is Aleppo, a principal city of Syria, whither the *canela* comes by Ormuz and Gida, where they sell it. They take horses and many kinds of silks and brocades to Ormuz. As that *canela* was good and new they kept that name for it; and that is the reason.

RUANO

I am satisfied, and it seems to me well that we have the true cinnamon and the true *cassia lignea*,[1] and that it does not fail us. When *cassia lignea* is found in the receipts, or cinnamon, we shall always know that it is the best cinnamon, for all are one, and the things written by the doctors touching their qualities apply to one as well as to the other. If God spares me to return to Spain, I will explain this error to many physicians and apothecaries. I will also tell that famous

[1] "Cassia lignea," or Chinese Cinnamon, is the bark of *Cinnamomum Cassia*, Blume, and *Cinnamomum Loureirii*, Nees, of Cochin China, and both also yield the "Cassia-buds" of modern pharmacy.

doctor Thomas Rodriguez[1] how the exhortation which Mateolo made to the physicians of the King of Portugal is all answered, and that you present it to him from your heart, complying with the request that he had made to you. Now tell me what you know of that celebrated island of Ceylon.

ORTA

The island of Ceylon is rather more than 80 leagues in circumference, and its length is 30 leagues by six to eight broad. It is the most fruit bearing and best island in the world. Some have said that it is Trapobana or Sumatra.[2] In front of its coast is the promontory they call Cape Comorin. It is very populous, though mountainous in many parts. The people are called Cingalese. It belongs to the King our Lord, and the native kings are subject to him. It is certain that this island is the most noble in the world. It belonged to one king who was killed by his grandsons, and they divided it amongst themselves. When the Portuguese came to this land they took counsel to cut and sterilize many trees, such as nutmegs, cloves, and pepper. In this island there are all kinds of precious stones, including diamonds; and many pearls, as we shall state farther on. They have gold and silver, and do not wish to bring it to the kings, but to keep it for treasure. They say that they combine sometimes to withdraw it secretly. The woods are full of all the birds in the world, many peacocks, fowls, and pigeons of many kinds, stags and deer, and pigs in great quantities. There are many fruits and orange groves in this land, which is all mountainous, and the

[1] Wrote Commentaries on Dioscorides (Venice, 1553 and 1557). Rodriguez was a famous Portuguese doctor of medicine. He seems to have been piqued by the exhortation of Matthioli to the Portuguese doctors and to have written to Orta on the subject; and Orta now redeems the obligation that was imposed upon him, to publish the results of his observations in India.

[2] No one ever said that Ceylon was Sumatra. It has been contended whether Taprobane was Ceylon or Sumatra.

oranges are the best fruit in the world for taste and sweetness. The land also yields all our fruits, such as figs and grapes. Certainly very good profit might be made of the oranges, for they are the best fruit in the world. They have flax and iron. The natives say that it is the terrestrial paradise. They have a fable that Adam stopped on the top of a very high mountain which they call Adam's Peak. They have other fables much stronger. There are many palm groves, and the elephants are the best in the world and very intelligent, and they say that the others which they have are obedient.[1]

[1] Orta visited Ceylon at least twice. He was with Martin Affonso de Sousa at the taking of Repelin, and when that captain landed at Colombo soon afterwards, Orta probably accompanied him. This was early in 1537. On the 15th of February 1538 the battle of Beadala was fought, and Orta seems to have been present. Beadala was in the Gulf of Manaar.

Plate VI.—Canela.

SIXTEENTH COLLOQUY

COCOA PALM

RUANO

I HAVE always heard it said that the coquos (cocoa-nut palm),[1] so called by the Portuguese, is a tree which yields many things necessary for human life.

ORTA

It yields so many necessaries, that I know no other tree that yields a sixth part.[2] It is well that you should know that we call it *palmeira*. The ancient Greeks wrote nothing about it that I have seen, and the Arabs have written little. This will be a good thing to tell them in Spain, though this much is known from those who come, as a notable thing. Coming to the names I say that it is called MARO and the fruit NAREL, and this word NAREL[3] is common to all, for it is used by Persians and Arabs.

[1] *Cocos nucifera.*

[2] The India nut alone,
Is clothing, meat, and trencher, drink and can,
Boat, cable, sail, mast, needle all in one.
G. HERBERT.

[3] *Narikela*, Sanskrit [the 'Αργέλλιον of Cosmas Indicopleustes], in Persian *nargila*, a name given to the "Hubble Bubble," or Indian smoking stand-pipe, from the shell of a

Avicenna calls it JAUZIALINDI,[1] which means "nut of India." Serapio and Rasis call the tree JARALNARE,[2] which means "the tree that yields coco." The Malabar people call the tree TENGAMARAM,[3] and the fruit, when it is ripe, TENGA.[4] The Malays call the tree TRICAM,[5] and the cocoa-nut NIHOR, and we, the Portuguese, with reference to those three holes gave it the name of Coco,[6] because it seems like the face of an ape or other animal. The tree is of considerable girth, and has the leaves on the highest part, like the leaves of our palms, though those of our palms are smaller. The flower is like that of the chestnut. The pith is very spongy. It likes sandy places near the sea, but inland it does not thrive.[7] They sow the cocoa-nuts themselves, and from them sprout small palms, which are transplanted, and in a few years if they are well treated they bear fruit. They should have water in summer, and manure in winter. Those which are near inhabited houses grow to be grand and beautiful, so that it seems that they do well near people. This may be caused by the dirt from the houses, and because they thrive well with plenty of refuse.

RUANO

Now begin to relate the uses of this tree.

ORTA

The wood, though not very good, is useful for many purposes. In the Maldive Islands they build a ship in which

cocoa-nut [*huka*, "casket"] being generally used to contain the water through which the smoke from the tobacco bowl is sucked through a long flexible serpentine pipe; and *nari-kela* simply means "water spring" or "squirt."

[1] *Jauzulhindi*, the "Indian nut."

[2] *Jaralnari*, the "Naril tree."

[3] The "Southern tree," *i.e.* introduced from Ceylon.

[4] The "Southerner," *i.e.* the nut originally coming from Ceylon.

[5] *Trinarajah*, "King of Grasses," is a Sanskrit name for the Cocoa-nut Palm.

[6] Spanish *macoco*, for "monkey faced."

[7] It thrives best in low sandy situations within the influence of the sea-breeze.—DRURY.

as well the vessel itself as the fastenings, the sails, and cordage are all made of the cocoa-nut palm. Two branches, which we call OLLA in Malabar, cover the houses and the ships. They have two kinds of palm plantations, one for fruit, the other for *çura*,[1] which is a rough wine, and when cooked is called *ORRAQUA*. When the tree is wanted for *çura* they cut some ends and fasten small pots under the place whence the *çura* is drawn, and climb up the tree, fastening their feet to notches that have been made in the stem. They distil this *çura* into a kind of aquardiente, and it gives a wine like strong water, and they put a wet cloth into it, as they do with spirits. This is called *fula*,[2] which means flower. The other kind is called *orraqua*, a small quantity of the other being mixed with it. Of the *çura* they make vinegar, putting it in the sun until it becomes acid; it becomes sometimes very strong. After they have drawn off this vessel of *çura* which makes a quantity, they draw off another to make sugar thickened in the sun or before a fire, which they call *jagra*.[3] The best of all is that from the Maldive Islands, which is not so black as that from other lands. The fruit, when it is new, has a very soft rind with a taste like artichoke, moistened with salt or without it. Inside, the water is very sweet and soft, but not so sweet as to cause loathing. This water lasts a long time, and softens the rind. It is what we call COQUO, and in Malayalim TENGA. There is in it some water not so sweet as the first, and at times it becomes a little sour. When this COQUO is green it is called ELEVI in Malayalim, and at Goa LANHA. The cocoa-nut has two rinds, very thick before coming to the pulp in the nut. When the

[1] *Çura* is meant for the Sanskrit *asara*, and *sara*, "essence." ORRAQUA is the Arabic *arak*, "exudation," the Rack of Anglo-Indians.

[2] *Fula* is the Hindustani *phul*, from Sanskrit *pushpa*, "a flower," literally a "pusher." The word occurs in our phrase "Gooseberry Fool," a stew of gooseberries, introduced into this country by Anglo-Indians.

[3] A corruption of the Malabari word *chakkara*, which is a corruption of the Sanskrit *sharkara*; from which is also derived the Greek *σάκχαρον*, the Latin *saccharum*, and our "sugar."

pulp is ripe and fit to eat, it is well to reach it by scraping the rind from the top, as Avicenna and Serapiam recommend. The first outside rind is very woolly, and of this they make the CAIRO,[1] a name used in Malayalim and by us. Of it they make the rigging and cordage for all the ships. It is very serviceable for us, for it is very flexible, and does not rot in salt water. All the ships are caulked with it, so that it serves as linen, as oakum, and as matting. These qualities make it good merchandize for Portugal, and the space it requires is small, which is the reason for so much being used. Yet it always appears that there are many palm groves in India, and that much CAIRO comes from the Maldive Islands. All the ships being caulked with it explains much, and the CAIRO swells out when put into salt water.

RUANO

This tree is a good thing, it has so much to give in itself. Laguna also says that it yields carpets or painted mats.

ORTA

He is not correct, nor had he good information. The half shell serves as a drinking-vessel for common people. When burnt it makes very good fuel for goldsmith's work.

RUANO

And does it not yield a drink which is good for paralytics, as Sepulveda says?

ORTA

I have often heard that said ever since I was a boy, but I doubt the authority, for I have never seen it written. I therefore think it must be a mistake, the more as it is not used here in that way. The fruit is not valued for the nerves, but the oil which is so separated from the bark, so contrary to its nature.

[1] Our "çoir," from Malabar *kayiru*, the fibrous rind of the nut.

SIXTEENTH COLLOQUY

RUANO

I have now tasted the fruit many times.

ORTA

Yet I must tell you that when it is young we call it COQUO, and the milk is most agreeable to the taste. It should have the rind taken in the middle where it is hard, for lower down it is not so well formed, and the milk tastes like green almonds. Some people eat it with the JAGRA already mentioned, a kind of sugar. If it was not for the great abundance of this fruit it would be more highly valued, as it is in Balaguate. With this COQUO pounded they make a sort of milk, and cook rice with it, and it is like rice boiled in goat's milk. They make dishes with it of birds and meat, which they call CARIL. They all dry the COQUO after removing the rind, and make it into dried pieces called COPRA, sending it to Ormuz and Balaguate and other lands, where there is less of it and not enough to dry, or wherever they need it. The fruit is wholesome, and is used like dried chestnuts in our country, for it tastes better than the COQUOS that they bring to Lisbon.

RUANO

How do they make the oil?

ORTA

They make it out of this same COPRA in great quantity, and it is very clear, like water. It lights up very well and is costly, being well refined. The common people eat it with rice, and they say it has a good taste.

RUANO

So say Avicenna and Serapio, that it is better than butter, and that it does not soften the stomach as butter does.

ORTA

There are two kinds of the oil, one made of fresh COQUOS, and the other from what we call COPRA, which is the dry COQUOS. That which is made from fresh COQUOS is extracted by pressure and applying warm water. It draws it out from the bulk which is at the bottom, and is expressed from the top, when the oil floats on the warm water. This is a purgative medicine which lubricates and softens. We give it to many to clear the stomach and bowels only. It does it very well, without any harm or danger. Many combine it with extract of tamarind, and I have found that very good. Avicenna, in holding this oil to be nourishing, is quite right, but not in saying that it does not mollify the stomach or loosen. The other oil made from the COPRA is a very good medicine for the nerves, and we find it very effective in spasms and for pains in old joints. Placing the patient in a small canoe, a little larger than a man's body, or in a large trough, and leaving him to sleep in the warmth the benefit is miraculous.

RUANO

They say that the oil kills the worms, and that the COQUO when eaten has the same effect, at least so say Avicenna and Serapiam.

ORTA

It is not within my experience that the oil kills the worms, but it appears very conformable to reason.

RUANO

Serapio stated to Mansarunge (said to be the ancient Mesue) that COQUO stops the diarrhoea.

ORTA

It is not contrary to reason that it should staunch the full stomach while the oil relaxes it, for the oil belongs to the air and the COQUO to the earth.

RUANO

Laguna says that some have the *oleo mel* of which Dioscorides treats in his first book, a very sweet oil taken from this palm. Tell me what you think of this.

ORTA

I say that this palm does not yield any oil except that which is extracted from the COQUO, hence I believe that he was deceived.

RUANO

Respecting the COQUO they sent to Portugal, said to be from the Maldives, both are to be counted as one species. For I saw a shell without any pith at all, of which they said much that was good, and to the pith which I did not see they gave still greater praise.

ORTA

I will give you an answer to this. But first I want you to say what you think of a savoury food from this palm, though it is not of much profit. The oil of this palm is bitter, and leaves joined together, the taste of the more delicate ones (which we call *palmitos*) being better than our *palmitos*, more like white and very tender chestnuts before the shell has fallen; but the *palmito* tastes better than that. And he who eats a *palmito* eats a palm, for it presently dries; and the older the palm the better the *palmito*. Returning to the COQUO of the Maldive Islands, it is much praised by the people of those islands and by the Malabar people, who maintain those islands.

RUANO

Is the COQUO esteemed by those other kings who have been cured by you, and by their people?

ORTA

No, I have not heard them speak of it; for it does not receive much credit from them. I have not known a case in which any one has been cured by it; and I have only heard many persons worthy of credit say that it is very good as an antidote to poison. They state also that it is good for many infirmities such as colic, paralysis, *gota coral,* and nervous ailments. They tell me that it is a preservative against other illnesses, by drinking water mixed with the COQUO and a little honey, the effect lasting for many days.

RUANO

You have been very negligent in not having tried the experiments yourself.

ORTA

I have not done so because there was no occasion, and I do not use it because there are other better medicines, such as bezoar stone, *triaga,*[1] *páo de cobra, páo de Malaca de contra erba, terra segillata.* As I find these efficacious I do not care to try others. It may be said that many men drink of this COQUO and find it very good; well, this may be the effect of imagination. I therefore do not care to say whether it is good or bad. I will not tell you anything is good without having been an eye-witness or ——[2] having the testimony of persons worthy of belief.

RUANO

I have been told that the Queen our Lady[3] sent for this COQUO every year, and that it was brought to her; so you will not deny to me that it is good as an antidote for poison, for it would seem that it was tried by some good physicians.

ORTA

When they tell me this I may believe and affirm, but not

[1] The modern meaning of this word is *treacle.* See footnote, p. 28.

[2] The editor, Count Ficalho, thinks that the words "sabendo o por" ought to be intercalated.

[3] Catalina, wife of Joao III. and sister of the Emperor Charles V., Regent in the minority of Sebastian.

now, as I have not seen it; until then I shall deny it, without feeling any shame.

RUANO

Then I must take it to Portugal, and find whether it has that effect. However, tell me or show me what they make of it.

ORTA

The shell of this COQUO is black, and brighter than the other COQUOS. It has an oval shape for the most part, and is not round like the others. The kernel within is very hard and white, inclining a little to yellow, and at the end of the pith there are rifts and it is very porous. It has not any strong taste. They take up to the weight of ten grains of wheat of this pith, in rose water.

RUANO

Is it the same species? It appears not to be, for the COQUOS we eat are much larger, and of a different shape.

ORTA

The COQUOS of the Maldive Islands are much larger. I have had one which would fit into seven *quartillos*. There are also two COQUOS in those islands which are antidotes against poison, but they are small and round.

RUANO

Well tell me your opinion, and what you know of this.

ORTA

The common fame is that these islands were once part of the mainland; and from being low they enlarged and produced these cocoa-nut groves. These became very old, and were strongly fixed in the earth, which is now covered by the sea. These COQUOS have neither trunk nor leaves, so that it cannot be decided whether they are the same species or not; but it would seem that they are not the same, for they have different effects. When I find anything contrary to this, I will write to you in Portugal, if God should spare me so long. For I hope to know it all well when I go to Malabar, God willing. Afterwards I came to know that the Coquos came fastened two in one, like the arch of a cross-bow; eventually they separated, and sometimes some of them come separated. The sea throws them on the beach. The Coquo is not so hard as that we see, nor so small and soft as the Coquos of the palms which we eat.

Ruano

A modern doctor has said many things in praise of the palm of these Coquos. In all he says the most notable is that the wine is made by pressing the Coquo, as I heard you say. For you told me that it is made by boiling and distilling as we make boiling water. Tell me whether this is true.

Orta

He is wrong as regards the wine, and in the way he says that the honey is made, and in some other things that do not matter. To conclude with this Coquo of the islands, I say that they draw the pith out of the Coquos, and put it to dry in the same way as the other product they call Copra, and it becomes very hard, as you see. The heart of it is like cheese from sheep's milk and very good. That Portuguese, who is well acquainted with these islands, told me that no person had ever seen the tree on which these Coquos grow, but that the sea throws them up. Any person who finds one on the beach and does not bring it to the king incurs the penalty of death. This gives the Coquo of the islands greater value. We will now drop this, and speak of the Costo, which is more used in medicine.

SEVENTEENTH COLLOQUY

COSTO AND THE COLERICA PASSIO

RUANO

THE COSTO was much esteemed in former times and it still has its admirers. I shall receive great benefit by having the road opened for me which leads to the truth about this medicine; and I shall feel neither affection nor hatred for any one, of whatever quality he may be.

ORTA

I have no hatred except for errors, and no love except for the truth, and with this proposition I may tell you that I have not any doubt about this medicine.

RUANO

Well we have all; for Galen with all the Greeks, Pliny with all the ancient Latins, and all the Arabs give many kinds of COSTO, yet the druggists in Spain and Italy, and in all other nations know of no kind but this of India. If there are others we have them not, either owing to want of care or avarice.

COSTO

ORTA

I believe that there is no other; and of this one I will give you the names, and the uses that are made of it.[1]

RUANO

I spoke, as a protest, that I may come with my opposition, when necessary.

ORTA

I say that in Arabic COSTO is called COST or CAST. In Guzerati it is UPLOT, and in Malay, where there is a great trade in it and it is much used, it is called PUCHO. I give you the name in Arabic, because it was known and used by both Greeks and Latins; and in Guzerati because that is the land nearest to the native region of the plant, and in Malay because the greatest quantity is needed there, and for transmission thence to China.

RUANO

And does not the Indian COSTO grow in Guzerat?

ORTA

It grows in regions which have often been subject to Guzerat, on the confines of Bengal, of Delhi, and of Cambay, that is, the land of Mandore[2] and Chitor.[3] Thence come many carts laden with this and UPLOT, ESPIQUE, and TINCAR and many other articles of merchandize, which are taken to the principal city of the kingdom called Amedabar,[4] which is in the interior, and also to the city of Cambayete which is on a bay of the sea. Thence it is distributed over the greater part of Asia, over all Europe and some parts of Africa.

[1] COSTO, "Costus Arabicus," is the root of *Aucklandia Costus*, Falconer, *Aplotaxis auriculata*, DC., *Saussurea lappa*, C. B. Clarke, Compositae. Its Eastern names are *kushtha*, Sanskrit; *kut*, Hindustani; *pachak*, Mahratti; *kost*, Persian; and *kust*, Arabic.

[2] Malwa.

[3] Rajput state of Mewar.

[4] Ahmedabad.

RUANO

How can they raise so many trees, for the part we use is the root?

ORTA

The smallest quantity is from the root, all the rest from the stem, and that from the stem is worth more than that from the root. The tree from which it comes has been compared to the elder. The flowers have a sweet smell. The appearance of it is white inside, and grey outside. Some of them are dark green and yellow outside. It gives out a very fragrant smell, and some people put it to their noses. Its strength causes headaches. The smell is not bitter, but rather sweet, and a little bitter when the flower is old. For when it is fresh the smell is sharp, as with other spices. They make a good deal into powder, when the smell is less and bitter; and this is the truth. The Indian physicians use it in many prescriptions. The merchants take it to Ormuz, whence it is carried into Persia and Coraçone. It is also taken into Persia and Arabia, by way of Aden, nor is much of this COSTO falsified there. They only bring a small quantity to Portugal, and it is believed that what they use in distant parts of Portugal is false, or something else is given for it.

RUANO

Serapio calls it CHOST.

ORTA

The word is corrupt, and in some books it is written CAST or COSTUS. The Arabs, as I have said, call it CAST, others COSTO, others COSTI; of this you need have no doubt.

RUANO

All mention three species: (1) Arabian which is white,

light, and aromatic; (2) Indian black, light, and bitter; and (3) "Land of Syria," the colour of a stick of box-wood and with a styptic smell: Also Costo sweet and Costo bitter. Though I have not seen Costo (sweet) it must be in existence, seeing that doctors of such authority write concerning it.

Orta

I asked many merchants of Arabia, Turkey, and Persia to tell me where the Costo that came from India was used, showing it to them in my hand. They all answered that the greater part was used in Turkey and Syria. The Arabs and Persians also told me that they made a profit by taking it to their respective countries. I then asked them whether they had it in their countries and they all answered in the negative. I asked the physicians of Nizamaluco, and they said that they had never seen any other Costo than that of India, and one of them was a physician of Xatamaz[1] and had been for a long time in Constantinople and Cairo.

Ruano

And what have you to say about the sweet and bitter Costo?

Orta

You know very well that things when they become powder also become much more bitter, and the colour, which at first was white, turns black as the substance becomes corrupt, and in the meantime is yellow. As this Costo comes to us by a long land journey, some of it has not begun to decay, but some has decayed and has lost its white colour. This is called the bitter Costo, and the other the sweet Costo. As the merchants who take it for sale come from different parts,

[1] Shah Tahmasp of Persia.

they take occasion to say that one sort is from Arabia, another from Syria, another from India; the truth being that all comes from India, the country where it grows.

RUANO

Laguna, a diligent writer, says that the druggists are worthy of blame who, through avarice or carelessness, do not procure the COSTO from Venice whence it comes from Alexandria, but use a medicine which is no more like COSTO than a quince is like a calabash. Others use the roots of Roman mint, and many herbalists in Spain told me that they had seen them so used. One showed me a shrub 5 *palmos* high and, reading from the book, we found what convinced us from indications in the same book.

ORTA

I say that Laguna is right. They bring the COSTO which has come from India to Venice, not adulterating it in the powdered state. For greater security and certainty it would be better to take it direct to Lisbon; which place it would reach in a more reliable way. For I have sent a quantity to the King in the year that the drugs were made; but little arrived there because the demand was not received nor was so much asked for. As for what you say about the herbalist in Spain who showed you the shrub of COSTO, neither you, nor the herbalist, nor the author of the book have seen the tree of COSTO at any time, so that you were all deceived. Forgive me for saying that a blind man, which was the Pandetario,[1] guided the herbalist and yourself. I say this because the tree of the COSTO is the size of a large arbutus, or an elder tree, or a juniper. And as for the bush, was the branch soft or slender or thick, and did the bark peel off well or not?

[1] Mattheus Sylvaticus, author of *Liber Pandectorum*, also cited in the Colloquy on Aloes.

RUANO

It was soft, and the bark came off well.

ORTA

Well, this other is different, the branch is hard and the bark will not peel off.

RUANO

Might not this sweet COSTO be lost owing to the great distances and the time taken in traversing them?

ORTA

No, because the lands whence it comes are now discovered and better known, while the errors of the past are rectified, as well as the deceit of men who seek to get a higher price for their wares. They give different names to them, and say that they came from great distances. It is enough for us that there is no other COSTO than this, and that discreet and learned people use it and require much.

RUANO

Do you allege this of people very wild and barbarous like the Scitas Asianos?

ORTA

The Chinamen are very subtle in buying and selling, and in mechanical trades; while in letters they do not give the advantage to any others. They have written laws in conformity with common right, and others that are very just; as may be clearly seen from a book on the subject which is in this India. One of these laws is that a man may not marry a woman who is known to have another husband. The men who go to China learn to practise just ways. The Chinese give degrees and many honours to literary men, and it is such men who govern the land under the king. On the pictures that they design are painted the chairs and the men reading, and the audience listening to them.

RUANO

That is true, for when they invented that art I was in Hungary, or in the more northern parts, which they say are on the confines of China.

SERVANT GIRL

A boy has come with a message.

ORTA

Tell him to come in. [*Enter a Page.*

PAGE

Don Geronimo sends me to request that you will come and visit his brother, and it must be soon, without waiting for the regular visiting time, for there is danger in delay; and my master will be greatly obliged to you.

ORTA

What illness has he, and how long has he been ill?

PAGE

He has the MORXI, and it is two hours since he was taken ill.

ORTA

I will come after you. [*Exit Page.*

RUANO

Is this the disease that kills many people quickly, and from which few escape? Tell me what it is called with us, and by them, and the symptoms, and the remedy that is used.

ORTA

Among us it is called *Colerica Passio.* The Indians call it

MORXI,[1] and we corrupt the word into MORDEXI. The Arabs call it HACHAIZA, a word which Rasis corrupted into SAIDA. It is more acute here than in our country, for it generally kills in twenty-four hours. I have known persons who have not lasted more than ten hours, and the longest endurance of it is four days. As there is no rule without an exception I have seen a man, with the gift of much endurance, who lived for twenty days, always vomiting *colora curginosa*.[2] Finally he died. Let us go and see this sick man, and you will see the symptoms, and have ocular testimony of what the disease is.

RUANO

Let us go. [*Exeunt.*

Road to the patient's house. Enter ORTA *and* RUANO.

ORTA (*describing the symptoms*)

The pulse is very low, and can scarcely be felt. Very cold, with some sweat, also cold. Complains of great heat, and a burning thirst. The eyes much sunken. Vomits much, but so weak that he is unable to discharge anything. No sleep. Cramp in the legs. Follow after me, I will show you the way.

Patient's room. PATIENT *and* DON GERONIMO *discovered; to these enter* ORTA *and* RUANO.

ORTA

Much health from God for this house. How long is it since you were attacked (*to the Patient*)?

[1] MORXI is the Mahratti *mordachi*, meaning "collapsed," and Gujerati *morchi*, corrupted by the Portuguese into *mordexine*, and by the French into *mort-de-chien*! See Yule's glorious *Glossary*, new edition, edited by William Crooke. Murray, 1903. *Ilyza* is also a general Indian synonym of *mordachi*.

[2] The Portuguese editor is very doubtful about the meaning of these words, and suggests a typographical error. Query—*cholera sanguinosa*, "red-bile," as distinguished from "black bile," two of the four "natural humours" of ancient and of Indian, or surviving ancient, medicine.

SEVENTEENTH COLLOQUY

Patient

It must be two hours that I felt myself attacked, and I was in great consternation. I have vomited nothing but water, without any bitter taste nor acid smell.

Orta

Have you any cramp in your legs?

Patient

I have been taken with it three or four times, and I treated it with strong rubbing, the hands being dipped in warm grease of cocoa-nut. When the cramp returned it was treated again in the same way.

Orta

What have you eaten to-day?

Patient

I have eaten fish, rice with milk, and some cucumbers, so that what I threw up smelt of cucumbers.

Orta

This case will brook no delay. Put chafing dishes and hot poultices to the body, and rub it with rough cloths. He must drink no water in any form. If it is necessary to quench thirst a little that has been boiled may be given. Cauterize the feet with warm iron; and administer an emetic. Put on a purgative clyster. Send to the apothecary for these things. Anoint him with warm oil from the nape of the neck down the spine, and on the legs. When the emetic causes him to vomit, and the clyster has made him purge, send and tell me what passed. Tell me whether the vomit was in quantity, or what, whether the patient becomes warm, whether he has any cramps, whether the pulse gets stronger; because we have to work in conformity with these symptoms. For in this disease the constant care of the physicians is necessary, and of the servants of the patient.

DON GERONIMO

All shall be done with all possible despatch. Here comes the druggist. [*Enter Druggist.*

ORTA

Make, with all speed, an emetic of boiled water with barley, cummin seed, and sugar, for that has been found to be very good for this disease. The clyster should be of boiled barley, oil, and honey strained. The oil for anointing should be of castor and *ruda*, for all should be mixed, to draw out the poison. For food take a fat fowl, first taking out the fat, and give it between slices of quince; if there are no fresh ones use conserves, first washed in white wine, with a little cinnamon or rose water. The doctor who is now present knows better than any one what ought to be done, and he gives me his hand in this, as a man having experience in this country. As he is present I say that a partridge, either of Ormuz or this land, or a cock or hen of the woods would be better. In all this you can do as I said.

RUANO

Of all this you can speak, for it is a long time since we have known each other.

ORTA

May God grant health to this house; and do not omit to let me know what happens. [*Exeunt* ORTA *and* RUANO.

RUANO

I am astounded at this disease, for though I have seen many patients with the plague, their cases have never been so severe and, for the most part, so serious. Why did he say that he had eaten cucumbers? I remember that the doctors say of some kinds of food that they become corrupted into a sort of poison. These are melons, cucumbers, and apricots, and that many get this disease after eating cucumbers. Yet this patient has done so frequently.

ORTA

Do you know in what way this happens? A gentleman, of great integrity, who had been suffering from this disease for thirty hours said to me, "Now I am not sick, I have no cramp in my legs, but I am unable to take breath, and this will kill me." I heard in what state he was, and that he was unable to take his breath.

RUANO

What men are most liable to take this disease, and at what time of the year is it most prevalent?

ORTA

Those who eat most, and those who consume most food. I knew a young priest here who died of eating cucumbers. Also those who have much intercourse with women. The disease is most prevalent in June and July, which is the winter in this country. As it is brought on by over-eating, the Indians call it MORXI, which means, according to them, a disease caused by much eating.

RUANO

How do the native physicians cure this disease?

ORTA

They give the patients rice water to drink, with pepper and cummin seed, which they call CANJE.[1] They cauterize the

[1] "Conjee" of Anglo-Indians, from Tamil *kanji*, "boilings," "rice gruel," "rice water"; the "ptisanarium oryzae" of Horace, *Satires*, ii. 3. 155; and Pliny, xvii. 7 (13).

feet in the way I ordered to be done to that gentleman. They also throw long pepper into the eyes to try its strength. For the cramp they fasten up the head, arms, and legs with very strong bands down to the knees, and from the knees to the feet. They give him their betel to eat. All these things are not wanting in reason, though they are done roughly.

RUANO

And you, the Portuguese, what do you do?

ORTA

We give them partridges and the best chickens to eat, or most of them; and we also give them wine with cinnamon. Administering these things to eat, I do not usually give much food, but rely upon exterior applications, such as anointing the stomach with warm oil of gum mastich and spikenard. I work with much care to clean the stomach with lavatory medicines only and with clysters. The treatment is varied according to the nature of the patient and the course which the disease may take.

RUANO

Is it not necessary to assist nature which is blind and may turn towards poisonous humours?

ORTA

Although this humour is poisonous, it is not infectious, yet it is good to throw it off soon, by evacuation and then by comforting the stomach with oil of cummin seed and powdered cinnamon; but first discharging the greater part of the humour.

RUANO

Have you any specially tried medicine?

ORTA

There are some. *Triaga*, given with wine or rose-water or cinnamon, according to the requirements of the case. Also *Pao de cobra*, of which we have already spoken; or *Unicornio*[1] or powdered *contra erva* of Malacca, which is good also for wounds from poisoned arrows. But the best medicine of all is three grains of *bezar*[2] stone, which the Persians call *pazar*. It is of such use that it almost miraculously dilates the powers of the heart. I have had many patients who said to me after taking it, not knowing what it was, that the medicine they had eaten had given them renewed force, and made the soul return to the body. I did much good to the Bishop of Malacca by giving him *bezar* stone with treacle. After evacuating great part of the material, much treacle should be applied in clysters, increasing the quantity.

RUANO

I never saw treacle in clysters given in these diseases.

ORTA

It is conformable to reason to give them in poisonous diseases, as happened to me in curing an overseer of the King's revenue of a poisonous diarrhoea, to which my brother physicians did not wish to give their consent. But when they saw the good result they were pleased with it, and tried it on many persons afterwards.

RUANO

Are there any other diseases like this in India that destroy the vital force in the way this does, and what medicines are used for them?

[1] Horn of Unicorn, or "Asinus Indicus," was a drug of the older Pharmacists of Europe.

[2] Bezoar, from Persian *pad-zahar*, *i.e.* "antidote to poison," was at one time a famous alexipharmic.

COSTO

Orta

Many men die with the vital force destroyed, owing to attacks of dysentery (the Indians call such infirmities Mordexi Seco), and in these cases they use external remedies, such as wine with cummin seed, and over that oil of spikenard or castor. For food I give something warm, with more care for its sustaining power than for its quality. I do not like to use yolk of egg, because it will subvert and corrupt, and as we have already spoken of the *bezar* stone we need not revert to it.

Returning to the Costo, I say that Mateolo Sinense alleges that some hold the angelica root to be a kind of Costo, but that he has not given it or tried it. Those who use it instead of Costo have more reason than those who use Roman mint which, I say, is not Costo. It may be a better medicine.

EIGHTEENTH COLLOQUY

CRISOCOLA

RUANO

THEY will question me and examine me as to where TINCAL [1] comes from, and why it is called CRISOCOLA,[1] so it will be well if we make mention of it here, and of whence it comes.

ORTA

Yes. But these are prohibited drugs, and for little you will lose much.

RUANO

I do not want to take it, but I want to know whence it comes and its name.

ORTA

It is called BORAX and CRISOCOLA, and in Arabia TINCAR, as well as in Guzerat. It is very little used in Indian medicine,

[1] Chrysocolla, or "golde solder," or "terra viridis," or "greene earth" of our older Pharmacists is a bi-borate of soda, found in its crude form in Cashmere, Ladakh, and Thibet. Its Sanskrit name is *tankana*, in Hindustani *tangar*, in Arabic *tinkal*, and in modern European commerce "tincal"; taken, it is thought, from the people [the Τάγγανοι of Ptolemy], who import it into India on their little pack-ponies called *tangun*.

or for skin diseases or surgery. Nor do we use it much, except in unguents, or for the teeth. As merchandise there is a demand in all parts, for gold and other metals to be well conglutinated. As to whence it comes, there is a mine or a mountain distant from the city of Cambayete about a hundred of our leagues, and it is brought for sale to Amadabar, and to the borders of Chitor and Mandou in great quantity, for in all those lands a great deal is used.

RUANO

About this there seems to be nothing more to be said. We will talk now about what you call AÇAFRAM DA TERRA.

ORTA

There is something to say of this medicine, as it is used by native physicians. It is a medicine and article of commerce which is taken much to Arabia and Persia. In this city there is little of it, but much in the Malabar ports of Cananore and Calicut. The Canarese call this plant ALAD,[1] and it is the same in Malayalim, more correctly MANJALE. The name in Malay is CUNHET, in Persian DARZARD, which means a yellow stick. In Arabic it is HABET. One and all declare that this AÇAFRAO does not grow in Persia, Arabia, or Turkey, but that all comes from India.

RUANO

This seems reasonable, for this medicine has a name in Arabic; and has not some Arabian author written about it?

ORTA

You have reason, but it is not my way to affirm anything

[1] Turmeric, the rhizome of *Curcuma longa*, called in Sanskrit *haridra*, *i.e.* "yellow-wood"; in Hindustani, *haldi* and *haled*; in Tamil, *manjal*; in Malay, *kunhet*; in Persian and Hebrew, *karkum*. The word "Turmeric" is a corruption of the French *Terre mérite*.

without first seeing well to it. I hold it for certain that Avicenna writes of this Açafram da Terra in Chapter 200, calling it Calidunium or Caletfium. This Avicenna was a man who, when he did not know a thing well owing to its not belonging to his country, quoted the statements of others. Nor is it very inconvenient to find the Arabic word corrupted by him. The Arabs, like the Indians, call it Aled, and the name is corrupted by Avicenna into Caletfium. There is a further proof that this view is correct in the chapter on Fece de Curcuma or Curcumani, which is in agreement with it, and where you will see that what I say is true; for Avicenna, when he was doubtful about anything, made two chapters of it.

Ruano

It does not seem to me that this is right, for he says that what we know to be Cilidonia[1] is Meimiram.[1]

Orta

I do not hold it to be quite certain. For in these two chapters he makes the medicine yellow, and says that it is very beneficial in eye diseases. As these qualities agree with Cilidonia it was said that this medicine was Cilidonia; but there is much greater reason for the simples treated of in these two chapters being Açafram da Terra.

Ruano

But for what do they use it in those countries?

Orta

To tinge and season their food, as well here as in Arabia

[1] "Mamiran," the *mamira* of the Indian bazaars, is said to be the root of *Coptis Teeta*, and also of *Thalictrum foliosum* of the Himalayas, and it has always been in repute throughout Southern and Western Asia as a sovereign remedy. Our old Pharmacists distinguished between *Chelidonium majus*, or "Greater Swallow Herbe," and *Chelidonium minus*, "Celondine the Lesse," or "Pilewort." The Arabs confounded the former with Turmeric.

and Persia, like garlic or our AÇAFRAM; they use this as being cheaper. They also used AÇAFRAM as medicine for all sorts of things, for the eyes, and for the itch, mixed with orange juice and cocoa-nut oil. In these chapters Avicenna commends it for these properties, so it must needs be that he used it for them. Avicenna speaks doubtfully on the subject because it did not belong to his country, and he did not know it well. You may, therefore, look upon it as a good medicine to take to Portugal.

SERVANT GIRL

The *curcas* have come from Cochin. Does your worship wish that they should be dished up in tamarind juice with the fowl, or thrown into the mutton?

ORTA

In both cases you can use it, and meanwhile bring me a little green AÇAFRAM DA TERRA.

RUANO

What sort of things are these *curcas* of Malabar.

ORTA

They are white grains larger than filberts with a rind, and not so round. They are white, and are known to be cooked earth tubers. They have them in Malabar, where they are called CHIVIQUILENGAS, which means small yams. I also met with them in Surat and Cambay. Coje Çofar, a native of Apulha who had become a Moor, told me that there were many in Cairo, where they were also called CURCAS. In Cambaia, where he was, he told me that they were called CARPATA. They sow them in Malabar, where I saw them first, and they come up in branches. But as they do not concern the question of medicine let us pass on without saying more about them, and you will do well to return to that subject.

SERVANT GIRL

Here is the green and the dry AÇAFRAM and the root.

RUANO

First I would ask whether any writer, especially any Arabian writer, has written on the subject of this drug.

ORTA

I do not much wish to have a whole chapter on this medicine, but speaking generally, I find that Serapio mentions it, calling it ABELCULCUT, which is corrupted to HAB ALCULCUL, which is probably a corruption of CURCAS. I say this because HAB means great sowing in Arabic, and AL is the article in the genitive. I am also moved to think this because Serapio says there is much use for it in the *colerico passio*, and that the cultivation increases. The people of Malabar themselves also say these things, from which we gather that all these are one. Rasis also speaks of these CURCAS and calls them QUILQUIL, perhaps corruptly. So now look at the AÇAFRAM root, green and dry.

RUANO

Inside it is yellow and outside it looks like ginger. The leaf resembles that of millet but larger, and the branch is made of leaves. The root neither burns nor is very bitter when green, yet it burns, though its great humidity prevents one from feeling it.

ORTA

Try it when dry. The root burns, though not so much as

ginger, whence I think it would not be bad to take it inwardly, and from this there can be no doubt that it is CURCUMA.

RUANO

I would make a request that you will take great care of this, and ascertain from the physicians what they know of it. I also see it to-day, but to-morrow we can return to the subject. This will be well, for what we do not know to-day we may know to-morrow.

ORTA

The more I hear of their discourses the more what I say seems to me to be true. For some say that CURCUMANI and MEIMIRAM is RUIVA DE TINGIR, and both roots seem to be one and the same.

NINETEENTH COLLOQUY

CUBEBAS

RUANO

We will talk of the *cubebas*,[1] although, as Sepulveda says, we very seldom use them by themselves, but only in compound medicines.

ORTA

It is not so in India. They are much used by the Moors in wine, and in their native land, which is Java, they are used much against chill of the stomach. You may believe that it is considered a very important medicine.

RUANO

I am surprised to hear you say so, for things that are most plentiful are the least esteemed.

ORTA

That rule is not without exceptions, for in Malabar there is much pepper which is taken all over the world, yet it costs as much in Malabar as in Europe.

[1] *Piper Cubeba* Linn., *Cubeba officinalis* Nug., of the Malay Archipelago.

CUBEBAS

RUANO

Tell me the names of this medicine.

ORTA

In Arabic it is *cubebe* or *quabeb* in the works on the subject, and hence QUABEBECHINI.[1] In Java it is called CUMUCOS or in the singular CUMUC, among all other Eastern people, except those who talk Malay, it is CUBABCHINI.

RUANO

Not only in Malacca is it known but also in China, for it is there called "of China."

ORTA

It is not grown in China, but it is brought there from Java and Sunda. As I have told you, the Chinese navigate the Indian Ocean and bring merchandise from where they find it in their voyages. Those of Goa and Calicut, the Guzeratis and Arabs hear it called CUMUC, which they corrupt into CUBABCHINI because the Chinese bring it, and this is the true origin of the name.

RUANO

I wanted to hear about the shape and form of the tree, and you only tell me of its origin. Do you not consider that there is more than one kind, for I have heard that there are many species?

ORTA

The tree is the size of an apple tree, and the leaves climb upwards like the pepper leaves, or, to make myself clearer, they creep up the tree like ivy leaves. The tree is not like a myrtle, nor are the leaves of that shape, but they are like the leaves of the pepper, the leaves of the *cubeba* being

[1] *Kababchini*—"Chinese Cubebs"—is the general Indian name of this drug, and *kababa* the Arabic, and *kumunkus* the Javanese. It is the "Carpesium" or "Cubebes" of our older Pharmacists.

narrower. They grow in clusters, not collected together in one cluster like grapes, but each one hanging from its stem. These *cubebas* are so highly valued in their own country that they are gathered there first, and sent away, and this because they do not come up when sown in other lands; and this may be why they decay in Europe and in India. I received this information from Portuguese worthy of credit, who had resided for a long time in the island of Java.

RUANO

May it not be another kind of pepper?

ORTA

No, it is not, for in Sunda the principal article of commerce is the pepper, which does not compare to a disadvantage with that of Java scarcely in anything. But this tree and its fruit are different. In Sunda itself, although they take it to China, it is in very small quantity for medicine, and not to eat as is the case with pepper which occupies twenty ships at least in the China trade. And these trees have flowers which smell sweet.

RUANO

Mateus Silvatico says, on the authority of Serapiao, that what the Mauritanians call CUBEBAS is the *mirtus silvaticus* of Dioscorides, and that the description of Galen touching *cubebas* shows them to be the *mirto agreste* of Dioscorides. How is it that none of them speak of the *cubebas* except that Galen treats of *cubebas* in *carpessio*,[1] and Dioscorides in the chapter on *mirto agreste*?[2]

ORTA

Does it not appear to you that Galen and Dioscorides may

[1] See footnote on previous page.

[2] The Greek and Latin writers had no knowledge of this drug. It was first made known by the Arab.

not have exhausted the subject, that they left many things unwritten about because they had not come under their notice? Serapio and the other Arabians speak of the medicines of India, and how does it seem that they profited by some medicine referred to by the Greeks? They would say that it is a medicine which is used by the Indians, and which the Greeks called by such a name. They might be mistaken from not knowing the Greek language very well. For this reason Serapio errs in what he says, and also the *Pandetario*. To this cause much may be attributed, for otherwise faults might be fixed on Galen and Dioscorides. For they have left many things to write about, as Avenrrois says in 5 of *Coliget*. It is clear that CUBEBAS is not *mirto agreste*, because the *mirtus silvestris* is called *ruscus*, and those who are not very good at Latin call it *bruscus*, which is a known plant whose root enters into the syrup of roots. The same view was taken by Ruelio, a diligent new writer. Besides this, *mirtus agrestis* has no smell while the CUBEBAS have a very good aromatic smell. The CUBEBAS do not have grains inside and the *mirto agreste* has them and are sweet, while the CUBEBAS have a sharp smell. I can also prove that the *carpessio* is not the same as CUBEBAS. It does not follow that it is more inconvenient that Galen should not write of CUBEBAS; it is not inconvenient, for the CUBEBAS grow on islands very distant from where he lived.

RUANO

I give the reasons for it. For Ruelio who was so learned, and the Italian Friars[1] who wrote the book on drugs, and who were so curious and such good druggists, did not hold *carpessio* to be anything else than the CUBEBAS of Serapio and Avicenna. For in the prescriptions where Galen put *carpessio*

[1] The Minorites, Fra Bartolomeo and Fra Angelo Palla, who published at Venice in 1543 some commentaries on drugs.

Serapio and Avicenna put CUBEBAS, therefore the intention of all was the same.[1]

ORTA

I do not say to you that Serapio erred in this, and it would not be much blame if he did, for he was a mortal. For the reason I have already given there were errors. Galen and Dioscorides had to write of all, and not to leave out anything. Now you say that it is no marvel that Avicenna should be mistaken also. So that Avicenna and Serapio might know this medicine, and not well understand Galen or Dioscorides. Ruelio says that there is better *carpessio* in Pontus, and that there is much in Syria, referring to Autuario. Tell me, for the love which we have for each other, when were CUBEBAS ever heard of in Pontus, Esclavonia, or Syria! It is taken there from India, as a merchandize yielding much profit. The Turks and Persians use a good deal of it, but very little reaches Portugal.

RUANO

May the *carpessio* be the *mirto silvestre* of Dioscorides?

ORTA

Neither one nor the other, for Galen says in the book of antidotes that there are some *festucas*; and you know that CUBEBAS and *mirto agreste* have fruit like nuts, so how can they be all one? For I assure you that only the fruit comes from Java without *festucas*, nor are there many species, but only one, nor is the tree cultivated but wild. I should not look upon it as inconvenient if it was planted in other lands with the same climate.

RUANO

The Friars say that the CUBEBAS are of different kinds,

[1] "Carpesium" is with all the older Pharmacists a synonym of "Cubebes." It is now the name of a genus of the Compositae. See footnote, p. 169.

some without taste, others bitter, and that they have others which are much better in their dispensary.

ORTA

I say that those without taste or bitter are corrupt, and that the others will have been more recently gathered and better preserved. There are many fault-finders saying that there are other species. I reply that this may be so, but that I have never seen, up to this day, more than one species, nor have I seen any one who has seen another.

RUANO

There are not wanting those who say CUBEBAS are *semente de vitiçe.*

ORTA

This is another new doubt. They say this because one species of *semente de vitiçe* has the taste of pepper, and these CUBEBAS have almost the same taste. But this is false because the *vitex* is *agnus castus.* As for what Antonio Musa and Serapiam say, that we want CUBEBAS, it would be better to say that they are mistaken in identifying it with *carpessio* and *mirto agreste.* What the Pandetario says, that CUBEBAS are called *cauli* by Galen, is also false, for that is a species of wild *dauco.*

TWENTIETH COLLOQUY

DATURA AND THE DORIÕES

Enter a Servant Girl

SERVANT GIRL

PLEASE, Sir, a negress of the house gave my mistress DATURA[1] to drink, then took her keys, and the jewels she had on her neck and in a box, and ran away with another negro. Will you please come and succour my mistress?

ORTA

How do you know this?

SERVANT GIRL

Because they have just caught the negress in the Passo-Seco, and found half the jewels on her, and she confessed that she gave the other half to her friend who went in the direction of Agaçaim. It may be that he too has been caught.

ORTA

Let us go and see this woman, who is an unmarried half-

[1] The drug DATURA (*dhatura* is its native Indian name) consists of the leaves of *Datura fastuosa*, Linn. vars. *alba*, and *D. Metel*, Solanaceae. The leaves of *D. Stramonium* and *D. Tatula* are used as cigarettes in cases of asthma.

caste. You will be interested. Those who take this medicine lose their heads. They always laugh and are very liberal, for they let people take such of their jewels as they choose, and only laugh or speak very little, and that not to the purpose. So that a robber has only to give this medicine in the food and the effect lasts twenty-four hours.

[*Exeunt* ORTA and RUANO.

House of PAULA DE ANDRADE, *who is discovered seated.*

Enter ORTA and RUANO

ORTA

God preserve you, lady.

PAULA DE ANDRADE

Im Im Im.

ORTA

Have you no answer to give me? How is this?

PAULA DE ANDRADE

Im Im Im.

ORTA

Rub her legs very hard and continuously with some coir and apply a clyster, also administering an emetic for which I will write prescriptions. If she is not better after two hours she must be bled.

The natives of this country do not consider the taking of this to be dangerous, and many take it as a joke. I saw two men, the younger of them being over fifty, to whom the sons of Nizamoxa gave it for fun. One was a huntsman and the other was a master for making bows and arrows; and both were cured without afterwards feeling any the worse in head or brain.

RUANO

Let us give it to one of your servant boys or girls.

ORTA

No. It would be against my conscience to do such a thing.

RUANO

Send me to seek for this herb.

ORTA

I will show it to you in the country as we ride. Meanwhile you must know that it is a tall plant with leaves the shape of *Bronca ursina*. The leaves are not very large and pointed at the end, making a prick like a lance, and round the leaf there are similar points. The leaf is set on a thick stalk, and has many nerves spread over its surface. The flower, which comes from the branches, is like the rosemary in colour. They are rounder and not so much the shape of a cube. They use this flower much, or the seed which is enclosed in it. The taste of the leaves and stalks is rather insipid with much humidity, and a little bitter. The smell is like raddish, but not so strong; whence I should think that it was vaporous, with some poisonous qualities. Girl! take this prescription to the apothecary, and tell him to make it up with all possible despatch. You others! mind you come and tell me what happens. Now let us go to dinner.

RUANO

Talking to a man who had been a long time in Malacca he told me that the best fruit there was in the world was one they called DORIÕES, and I remember that you have some practical knowledge of it.

ORTA

I have not tasted it, and of the men who have tasted it some say that it is good and others say the contrary, namely that the taste is not so good as that of melons. They say that at first it smells like onions, but that afterwards one begins to like it, until at last the taste seems very good. They say that a merchant came to Malacca and met one laden with merchandize and that he sold only DORIÕES.[1] This was told to me, but I do not know whether it is true or false. But in Malacca there are very good fruits such as grapes and mangos, which are not esteemed so highly as DORIÕES. But we should not waste our time over it, but dismiss the DORIÃO in few words, as it is not an article of medicine.

RUANO

They praise this fruit to me so much that I felt obliged to speak to you about it.

ORTA

The DORIÃO is a fruit about the size of a melon, and it has a very thick rind covered with small excrescences, so that in Goa we call it JACA. This fruit is green outside and has chambers within like rooms, and in each room a separate fruit. In colour and taste it is like white food. Its smell is universally praised. The fruits are the size of a small hen's eggs, when they are divided. Some are not white but a clear yellow. The flower is white rather inclined to yellow, and the leaf is the size of the palm of my hand, sharp pointed, green outside and a darker green underneath. A hidalgo of this country told me that he remembered reading in a Tuscan translation of Pliny "*nobiles duriones.*" Afterwards he asked me to find the words in Latin, but up to the present I have not found them. If I should do so I will write.

[1] The Durian, the *du-yen-yaing* of Burmah, *turien* of Siam, and *dourian* of the Malay Archipelago, is the fruit of *Durio zibethinus*, Linn., a Sterculiad. *Artocarpus integrifolia*, Linn., the *jaka* of India, bears the "Jack fruit" of Anglo-Indians.

PLATE VII.—DELA DATURA.

TWENTY-FIRST COLLOQUY

IVORY AND THE ELEPHANT

It is a colloquy which does not refer to medicine, but to pass the time.

RUANO

AS the bones of elephants are used in medicine it will be well if we discuss them and the elephant.

ORTA

Much has been written about the elephant; still there is a great deal to say, and something to wonder at, so that one ought not to want material for conversation on the subject. Beginning with the ivory, I must tell you that no elephant bone is ever used in medicine or in industry, only the teeth. Do not be deceived by those who write of *espodio*, saying that it is the burnt bones of the elephant, before you have made certain that it is not so, if God grants the time for this and for other things. I note this because, as regards the elephants which die here, the people make no use of the bones, but take the flesh to eat, and the teeth for objects of art.

RUANO

Do some have horns?

TWENTY-FIRST COLLOQUY

ORTA

No, for what we see is all from the teeth and pieces of them, and each elephant has not more than two tusks. The nails are not made use of, although Paulo Egineto says that they are. The elephant wants only the power of speech, to be a rational animal. More than that, in Cochin there is an instrument with which it says two words.[1] When the elephant had to eat, its master (called NAIRE in Malabar, and in the Deccan PILUANE) said that he had not a good caldron to boil the rice, and that the elephant should take the caldron to the *almoxarife*,[2] and that he would order it to be mended. So the elephant came with the caldron on his trunk, and the *almoxarife* told the NAIRE to take it to the caldron-maker, who would mend the bottom only, where it was damaged. The elephant took it to the house, but in cooking the rice, the water leaked out because the bottom was not properly soldered. Then the *Naire* gave it to the elephant, and the elephant took it back to the caldron-maker, who took it and mended it. But the workman left it worse than it was before, giving it a few blows with a hammer. Then the elephant took it to the sea, and put it in the water, looking whether water came out of the bottom. And as he saw that it did, he took it back to the caldron-maker, making many trumpetings at the door, by way of remonstrance. Then the caldron-maker mended it and soldered it well. The elephant returned to the sea to try it, and found it to be all right. Then he took it to the house, and made it ready to cook with. No man could show more sense. This happened as I say, and to-day there are witnesses who saw it, and many others who have heard it at first hand.

[1] The words were probably rice and betel. For Damian de Goes related that when Diogo Reveira was at the city of Bijayanagar he saw an elephant writing with the end of its trunk. He afterwards asked it what it would eat, and it replied in a clear voice, "Rice and betel."

[2] In Portugal the office of the *almoxarife* was to collect the royal dues. From المشرز *eminent, honoured.*

IVORY

RUANO

What is the word for an elephant in Arabic and in India?

ORTA

In Arabic it is called FIL and the tusk CENALFIL. In Guzerat and the Deccan ATI, in Malabar ANI, and in Canarese AÇETE. In the language of the Kafirs of Ethiopia YTEMBO. No one calls it BARO, as Simão Genoes says, for he tells stories with many lies. The Indians do not use it in any form for medicine, only the Turkish and Arabian physicians, who cure by following Avicenna.

RUANO

In this country they use, for the fine arts, a quantity of ivory which comes from Sofala, and they tell me that it also goes direct to Portugal as merchandize ordered by the King.

ORTA

You must know that from Ethiopia, that is from Sofala and Melinde, there comes to India every year 6000 quintals besides what goes to Portugal, which is very little in comparison. Besides this there are elephants in Malabar, but few and not tamed. In Ceylon there are many and very well trained, indeed the most valued of any in India. They are in great numbers in Orissa in Bengala and Patani, in the Deccan, and in the country of Cotamaluquo[1] which borders on Bengala. They also have them in Martaban, Pegu, and Siam, and they say that the King of Siam has a white elephant, and has the honourable title of King of the White Elephant. I know not whether this is true.

RUANO

You have not solved my doubt as to where so much ivory is used up.

[1] Golconda.

ORTA

Some ivory is used in China, where the demand is increasing. In Ceylon it is used largely for many things and is highly polished, such as boxes, combs, and many other things, also in Pegu. Of the 6000 quintals that come from Sofala, all is used in Cambaia, except the small quantity that goes to China. This is used every year.

RUANO

The supply is such that if you had not stated it I could not have believed it.

ORTA

You must know that the Devil puts a certain superstition in the minds of the wives and daughters of the Banians, being those who live according to the pitagorico custom. It is that, when any relation dies, the women break all the bracelets on their arms, numbering twenty or more, and presently have as many new ones made; and these bracelets are of ivory, except a few of tortoise shell. The Devil has arranged this in order that the ivory which comes annually from Ethiopia may be wasted, and always will be wasted while this superstition lasts. This ivory is valued according to the size of the tusks, for small tusks are worth little, and large ones have a high value, weight for weight. They also make other works of art from the tusks, but in smaller quantity.

RUANO

I am astonished at this superstition. I am told that the tusks of elephants grow again after they have fallen out; because I do not know how there can be so many elephants in the world.

ORTA

You have much reason in that. The elephants are long lived, but none have more than two tusks and do not change them; and the females do not have tusks. In Ethiopia the Cafres kill the elephants to eat their flesh raw, and sell the tusks to us. The elephants are caught by traps made with trees and in other ways. It may be presumed that there are more elephants in Ethiopia than cows in Europe.

RUANO

Of what illnesses do the elephants die, and what use is made of them in these lands?

ORTA

They are very melancholy and in much fear, more so at night than in the day-time. When they sleep at night they seem to see dreadful things. The cure for this is that their Naires should sleep on them, and always be asking them why they do not sleep. They are subject to fits of fury, when they break their chains and do much harm along the road they pass. The Naires cure this by taking them into the country, calling them many bad names and abusing them for their folly. For other ailments they take medicines peculiar to this land. As for their uses, they work at drawing and moving artillery from one side to another. They serve the kings in their wars; some kings have a thousand elephants, others less, others more. They go to war armed, especially the head and breast. Small bells hang along their sides, and iron points are fixed upon their tusks. Castles are fixed on their backs in which the Naires sit who guide them, and in which are placed battle axes, hooks, and other muniments. I have seen them in battle, and the mischief they did was to throw the enemy into disorder and sometimes to put them to flight. They tell me that they often take to flight themselves, when they do more harm to their own people than to the enemy; but this I have not seen.

TWENTY-FIRST COLLOQUY

RUANO

Have they any other way of fighting?

ORTA

Yes. They fight each other with their Naires, who teach them and guide them from above. It is a very cruel battle, fencing and wounding each other with their tusks. They fight with great bravery, and they often deliver such tremendous blows, one on the other, on their foreheads that they fall dead on the ground. A Portuguese of credit told me that he saw a very powerful elephant fall dead in an encounter with another. They also fight when inebriated, and have taken men in their trunks and torn them to pieces, which I have seen on several occasions.

RUANO

Pliny says that their blood is useful for several things as well as the liver and scrapings from the ivory.

ORTA

This may well be true, but they are not used.

RUANO

They say that the elephant sleeps with the female elephant, like a man with his wife, different from other quadrupeds.

ORTA

The truth is that their habits are the same as those of other quadrupeds. The only difference is that the male rests on a higher bank and the female on a lower one. I have been told this by Portuguese worthy of credit. I have seen elephants, but I have not seen them with their females; only this is what I have heard.

RUANO

Pliny also heard that the souls of elephants drew serpents out of their places.

ORTA

I cannot say how that may be, for I have neither seen nor heard anything of it.

RUANO

Pliny also says that the elephant, when he eats poison, seeks for an antidote to cure himself.

ORTA

I have not seen this and have no knowledge of it, nor have I ever heard of it in Ethiopia where the elephants most abound.

RUANO

Pliny also writes that the best elephants and the most warlike are in Taprobana, not in India.

ORTA

If by Taprobana Ceylon is to be understood, as some maintain, the elephants there are the best and the best trained. If by Taprobana Sumatra is intended, elephants are there but not so good as those of Ceylon. Very often men think that a thing comes from one land, when it comes from another more distant. For instance many believe that the best *lacre* comes from Sumatra, and hence the name *locsumatri*, but the best *lacre* does not come from there, but from Pegu ; and so it may be with the elephants of Sumatra.

RUANO

Are they capable of understanding the language of their country, as Pliny relates ?

TWENTY-FIRST COLLOQUY

Orta

Not only of their own country, but they can also be taught that of another land. The elephants brought from Ceylon to Guzerat and the Deccan can easily be made to understand their masters, and some that are taken to Portugal soon learn Portuguese. I will show you some in India that understand. They are so covetous of glory that if one is told that he is the king of Portugal he is much pleased. They are ashamed of anything wrong that they do. They are grateful for kindness shown to them. They are vindictive if injuries are done to them. I remember what happened at Cochin. An elephant threw some cocoanuts to a man, and the man broke them on the elephant's head. The good elephant kept the rind of the cocoanut in his mouth, holding it in one cheek. Seeing the man who had done him the injury, he drew out the cocoanut husk with his trunk, and afterwards he carried the cocoanut husk in his cheek, as much as to say, I remember the injury you did to me. By this you can see that the elephants have memories.

Ruano

Pliny says many other things about them, for instance that they make war with the rhinoceros over their pastures.

Orta

These rhinoceroses are in Cambaia where it borders on Bengala and Patani, and they are called Ganda. They are not so good or so gentle as the elephants, and for this reason they are not so well known, for it stands to reason that animals so large and fierce should not be liked. When I write of the *licies* I will revert again to this animal and relate all I know. Pliny also says that a headache is cured by putting a poultice of barley on it. But there is no barley in Ethiopia, whence the greatest numbers come. Of the other places there are some from Bengal, and a small quantity from Cambaia. So that I do not know how he can have tried it, although I know that the tame ones might make use of it.

RUANO

How are the elephants tamed and taught?

ORTA

The newly caught are tamed with whips, and with shameful words, and with hunger; then with good treatment as reward for good behaviour. They tell me that to tame the large ones in Pegu, they put them into large houses with many small doors, and prod them through the doors with swords and spears. Then they go in to them when they are tired, wounded, and starving, and when the elephants, badly wounded, lie down on the ground, they make friends and become masters. They wash the wounds and dress them with oil, and give them food. First with wounds and starvation, then with kindness, they succeed in taming them. I wish to tell you these things about the elephants because they are the most certain. Pliny relates many more.

SERVANT GIRL

Please, sir, Micer André Milanes, the lapidary, has called.

ORTA

Ask him to come in.

Enter ANDRÉ MILANES

I kiss the hands of your worship.

ORTA

And we kiss yours.

TWENTY-FIRST COLLOQUY

ANDRÉ

Do you wish to sell your great emerald or the smaller one? for I am ready to buy both. The small one is the finest.

ORTA

I will sell all, and I will show you both if you will tell me who is the buyer. This I will confide to your faith, that you will only show them to the buyer or to his agent, and that you will return them to me at once if he does not buy. Now tell me, did you see elephants caught and tamed when you were in Pegu?

ANDRÉ

Twice. The first time when the King and all the kingdom hunted, upwards of 200,000 people. They formed a circle, making it smaller and smaller, catching a great number of deer, pigs, and tigers, many alive, and others wounded or dead.

ORTA

I have seen Nizamoxa hunt in the same way, and take a great number.

ANDRÉ

Then 4000 elephants were surrounded—females, males and young. They were all let go, except 200, so as not to denude the forests of them. I saw this, and the 200 were tamed by encircling them with beams, and making the circle smaller and smaller, and the barrier stronger, until the space was only large enough for the elephants to stand in. Then cords were passed between the beams, and round the legs and tusks, so that they could not move. Then two cords were passed round each, and they were wounded until they wept tears, which I saw. On each elephant his master then mounted, taking the ropes from its feet. If they gave trouble they were wounded and starved. When they submitted their wounds were dressed and they were given food. Taking them out, each one was placed between two tame elephants to give it counsel, and in this way all were tamed.

ORTA

I had heard of this way of taming, but I did not know that there were so many to hunt in Pegu and Ceylon; and I have heard that there is another way of hunting. Do you know it?

ANDRÉ

The King received news of a very great elephant that ranged the woods, and he sent to it some of his elephants which were very tame and domesticated. He told their masters that he did not want them to go with the elephants, but that they were to keep near. The elephants came on, the females in front, and the wild one followed, feeding as it went, until it came into Pegu, which is a great city, and there it was enticed to a place where it was enclosed and the females left it; and it was tamed in the way which I have already described.

RUANO

This is very well, but Pliny says that when they meet pigs in the forest the elephants turn back and are frightened.

ORTA

I already know the contrary; for in the close places of elephants there are pigs and the elephants take no notice of them. In the woods of Malabar there are many pigs, where there are few elephants, and it is not said that they have any fear. It is true to my knowledge, what Pliny says, that they detest rats, and when there are rats where the elephants sleep, they sleep with their trunks coiled up, that the rats may not bite them. For the same reason they dislike ants. I request that you will see to the sale of my emeralds, and now let us go to dinner. Do not think I am frivolous in talking so much on this subject, for Mateolo Sinense, a learned man, talked much of elephants, and not with so many truths as I have related.

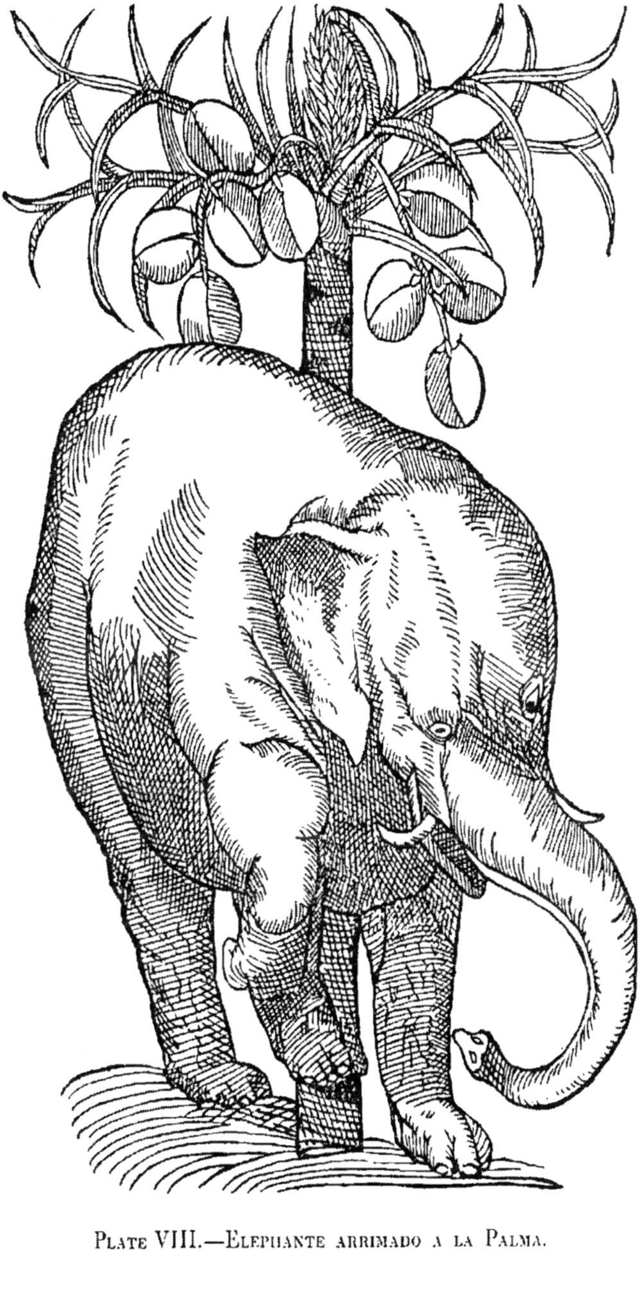

Plate VIII.—Elephante arrimado a la Palma.

TWENTY-SECOND COLLOQUY

FAUFEL[1]

RUANO

WE speak in Portugal of what is called "nuts of India." You tell me that the *betre* is much used by everybody here. We use it very little. Speaking the truth with you, I have never seen it, for we put in its place the vermilion sandal.

ORTA

Here it is a common thing to mix the food with the *betre*, and in countries where they have no *betre* they also use it for chewing with *cravo*. What you say about using vermilion sandal in its place does not appear right, for in its place they have a medicine which is often falsified, and they give a vermilion stick for it; for as the vermilion sandal wants the smell, and is not in Timor whence the other comes, as I will tell you in speaking of it, there is difficulty in knowing one from the other. This ARECA is more valuable and is less perishable. The reason it is not sent

[1] *Areca Catechu*, the *supari*, or Betel-nut Palm of India, *faufal* of the Arabs; the nut of which is chewed with the leaf of *pan* or *Piper Betel*, and a little finely prepared lime, as a masticatory, throughout India.

to Portugal is that the apothecaries do not ask for it, for neither they nor the physicians are sufficiently curious to trouble about it. I will now tell you the names it has in the countries where it grows. Among the Arabs it is FAUFEL. Avicenna calls it corruptly FILFEL. It has the same name in Dofar and Xael, Arabian lands. The FAUFEL[1] is very good. In Malabar they call it PAC, and the word for it among the Naires, who are the knights,[2] is ARECA, whence the Portuguese have taken the name, being the land first known to us, and where it abounds. In Guzerat and the Deccan they call it ÇUPARI, but they have very little, and only on the skirts of the sea. There is a better supply at Chaul because of the trade with Ormuz, and still better at Mombaim, land and island, where the King our Lord has made me a grant, a long lease (*emfatiota*). In all that land of Baçaim they are very good, and they are taken thence to the Deccan; and also to Cochin they take a small kind called CHACANI, which are very hard after they are dried. In Malacca there are not so many, and they are called PINAM. In Ceylon they are in greater quantity, and they are sent to parts of the Deccan, namely to Golconda and Bisnaga, also to Ormuz, Cambaya, and the Maldive Islands. The name in Ceylon is POAZ.

RUANO

Serapio says that this *areca* is wanting in Arabia.

ORTA

That is true to a great extent, for Arabia is a vast region, and there is *areca* only at Xael and Dofar seaports. For this tree loves the sea and will not thrive at a distance from it. Where it will grow they do not fail to plant it, for the Moors and Gentios do not let a day pass without eating it. The Moors and Moalis (who are those that follow the law against

[1] *Faufal* is the Arabic name; *paku* the Tamil; *poka* the Telugu; *adikai*, whence our "Areca," the Canarese; and *puvak* the Cingalese. [2] Cavaliers.

Mafamede[1]) keep a feast or fast of ten days, when they say that the sons of Ali, son-in-law of Mafamede, were besieged in a fortress and died. During the ten days that they were besieged, they sleep on the ground, and do not partake of *betre.*[2] In these days they chew cardamom and areca, which is much used to chew, as it clears the stomach and the brain.

RUANO

Now tell me how the *betre* is used, how it is administered, whether to help or to rectify.

ORTA

The *betre* is warm, and the *areca* is cold and temperate. The lime they use with the *betre* is much warmer. They do not use our lime from stone, but a lime made from oyster shells which is not so strong. With the *areca* they mix the medicines, you see, because they are cold and dry, and much drier when not dried in the sun. Then they add the *cate,* which is a medicine I have mentioned before; because with the *cate* it is a good medicine to open the gums, fortify the teeth, and compose the stomach, as well as an emetic, and a cure for diarrhœa. The tree from which it is collected is straight and very spongy, and the leaves like those of our palm trees. Its fruit is like that of the nutmeg, but not so large, and very hard inside, with veins white and vermilion. It is the size of the small round nuts with which the boys play. It is not exactly round, for it has a band round it, though this is not the case with every kind of *catechu,*[3] for I must not deceive you. This

[1] Muhammad. They did not follow any law *against* Muhammad, but were of the Shiah sect. [2] See footnote, p. 192 *supra.*

[3] The drug "Catechu" is prepared (1) from the leaves of *Uncaria Gambir* (Rubiaceae), of the Indian Archipelago, this variety being known also by the names of "Terra Japonica," "Catechu pallidum," and "Gambir"; and (2) from the heartwood of *Acacia Catechu,* of India, this variety being known by the names of "Catechu nigra" and "Cutch," these names being corruptives of Indian name of the tree, *katha-kikkal,* the "Catechu Acacia." *Katha* means "strong."

fruit is covered with a very woolly husk, yellow outside, so that it is very like the fruit of the date palm when it is ripe and before it becomes dry. When this *areca* is green it is stupefying and intoxicating, for those who eat it feel tipsy, and they eat it to deaden any great pain they have.

RUANO

How do these Indians eat it, and how do they prepare the medicine?

ORTA

It is usual to cut the *areca* into small pieces with some large scissors they have for the purpose, and then they chew them, jointly with the *cate*. Presently they take the leaves of the *betre*, first pulling out the veins with their thumb nails, which for this are cut to a fine point, and they do this that it may be more tender, and then they chew it all together. They spit out the first, after the first chewing, and then take more *betre* leaf and begin another chewing, expectorating what looks like blood. In this way the head and stomach are cleared, and the gums and teeth strengthened. They are always chewing this *bétre*, and the women worse than the men. The lords make small pills of the *areca*, mixing it with *cate*, camphor powder of *linaloes*,[1] and some amber, and this is made for the *areca* of the lords. Serapio says that in the taste with the warmth there is some bitterness. I tried this and found it with scarcely any taste. Serapio did not know this *areca* and could not ascertain the taste.

RUANO

Silvatico says that he has seen it, and that it was mixed with the cinnamon of Calicut.

ORTA

It may be that the Moors of Calicut take it to the Strait, and

[1] Lign-Aloes, Agallochum, "Eaglewood," or Calambac, the fragrant wood of *Aquilaria Agallocha*, Roxb. (Thymelaeaceae), of Assam, Bhutan, and Burmah.

that it may come mixed with cinnamon, but it was not the cinnamon of Ceylon. That of Calicut is much more black, and is called *checani*. That of Ceylon is whiter, and once seen is easily known.

RUANO

Do you know whether it is used for anything else, besides what you have described?

ORTA

I order this water to be distilled, and in secret I use it to cure choleraic diarrhœas, and it works well.

RUANO

This profits me little, for in Spain we cannot have it green for distilling. It must now be dinner-time.

ORTA

Be it so. We will go and wash our hands.

RUANO

One thing surprises me. It is that we always eat the figs[1] on the table and always know them well; but it is unknown to me whether they come by sea, seeing that you have such abundance on your table. It seems to me to be a very good fruit, and I do not get tired of it. It will be well, talking and eating, if you would tell me their names in all languages, what kinds there are, whether they do any harm, and your own opinion about them. For I know well that neither Dioscorides, Galen, Paulo, nor the Arabs mention them.

ORTA

I beg your pardon, Avicenna, Serapiam, and Rasis all

[1] He means Plantains, the fruit of *Musa paradisiaca* of India, the *kela* and *keli* of Western India, and *pisang* of Malaya. Its generic name is formed from the Arabic *mauza*, "taste." The West Indian Plaintain [Banana] is obtained from *Musa sapientum*. "Paradise" is, ultimately, the Sanskrit *para-desha*, a "far-country." Compare the "parvis" in front of churches.

write concerning them, and others have written on the subject whose works I have not seen.

RUANO

You tell me much, and I shall rejoice to hear about them.

ORTA

I sought for knowledge on this subject, and found it. In Canarese, Deccani, and Guzerati, and in Bengal they are called QUELLI, in Malayalim PALAM, and in Malay PIÇAM. For they are in all those countries, and have names, as well as in other lands. The Arabs call them MUSA, and both Avicenna and Serapiam devote a chapter to them, calling them by the same name: as does Rasis. They also have these figs in Guinea, where they call them BANANAS.

RUANO

What does each of these writers say of the figs, and do the natives say that they are good or that they do harm?

ORTA

Avicenna says that the nutriment in this fruit is small, that it promotes inflammation, that it is useful for chest complaints, that it irritates the stomach, and that it is well to take it after eating. Rasis says that it does harm to the stomach. Serapio says that it is warm and humid in the first degree, that it is good for burning in the chest or lungs, and that many use it who feel a weight in the stomach. So that it is clear that these writers were acquainted with the fruit. If this is not enough, ask any Arab what he calls AMUSA or MUSAY whether he is from Cairo or Damascus or Jerusalem.

RUANO

I rejoice much to hear this from you.

TWENTY-SECOND COLLOQUY

ORTA

You must know further that a Franciscan friar who was in Jerusalem, and writes of the mysteries of the Holy Land, praises this fruit. He says that it is called MUSA because it was the fruit of the Muses, and that they fed upon it. He says further that it was this fruit which caused Adam to sin, that the leaves are more than a *braca* in length and 2½ *palmos* in width, with a thick and green midrif in the centre. It sends out some red flowers where the fruit is to grow afterwards, which form an egg the size of a hand whence comes a bunch with a hundred, and sometimes two hundred figs.

RUANO

I do not know whether it was the tree of the terrestrial paradise, and in this I hold to the views of the sacred fathers. But I must confess that it is a very good fruit. I want to know whether it is useful for anything besides what is mentioned by these Arabs, and where the best are grown and in what way they are eaten.

ORTA

In Martaban and Pegu they say that they are very good, for in Bengal, where they are plentiful, those from Martaban are preferred. Those which have the strongest scent and the best taste are CENORINS, which have a bright colour. The CHINCAPALOES of Malabar are good, being green with a very good taste. Those of Sofala, which I have tried, are much praised, and I found them to have a pleasant taste. But as I knew them all when I first came from Portugal, I am not a good judge. The Cafres call them ININGA, as well as those on the coast of Abexim and in the Cape Verdes. As I said, these are as large as a *palmo* in Malabar, and Baçaim, and other parts. They are very good roasted, or dipped in wine with cinnamon.

RUANO

I have tried them three or four times, and liked them much better than quince roasted.

ORTA

They also cut them down the middle, and fry them with sugar until they are well toasted, when they taste very good with cinnamon on the top.

RUANO

I have also tasted them here on fish days, done in that way, and they tasted very good, but I did not know what they were.

ORTA

They take them to Portugal for ships' provisions, and eat them with sugar. It is very good food for the sea. The physicians of this country say that they are very good as diet in fevers and other illnesses. I know well that all these things that I tell you are of little importance, but I relate them that you may not say, when you return to Spain, that you have not been told of the things of the land, and not because they are all appertaining to medicine.

RUANO

Ruelio has a chapter on the figs of India, referring to Strabo and Theophrastus, and he makes out several species. In another place he speaks of foreign trees, and these men appear to have known about the figs of India.

ORTA

I have read the same author, and if he is right in one thing he errs in many others,—one on the nail and four on the horse shoe, as the saying is. However, he puts the last species as the one in which the tree bears the most fruit, and he says that they grow of themselves. This is true, for the tree is only planted once, and gives out a branch bearing sometimes two hundred figs more or less. Presently there grows from the foot another tree from the same branches or trunk; for the trunk has a gathering together of bark, and the figs grow in the eye of the tree, fastened to the stem.

RUANO

Perhaps the fruit which in Italy is called MUSA is the same.

ORTA

As I have not been in Italy, I am not sure. But there are some Venetians dwelling here who say that the fruit is in Venice, and that it is like a plum. It may be that the same kind of plum is in Spain, for they say it is very sweet.

RUANO

Mateolo Senense writes of a kind of Indian palm, and the description is not like this tree. He says this in the chapter on palms. But when he writes of Egypt he does not write well, and therefore I do not quote what he says.

ORTA

I know well that they have figs in New Spain and in Peru, and we have them in Brazil, in the Concan, and between Chaul and Goa, namely in Carapatam. They are also planted in some places in Portugal, as at the country house of Dom Francisco de Castelo Branco,[1] and for this reason it was not well to tell you about things that are known to all.

RUANO

These notices of the figs had not reached me in Spain, and you tell me so many good and interesting things that I considered it needful to ask for all you know. You have also told me the names of the authors who have written on the subject, information which I value highly.

[1] Camareiro Mor to Joaõ III. in 1527, son of the first Count of Villa Nova.

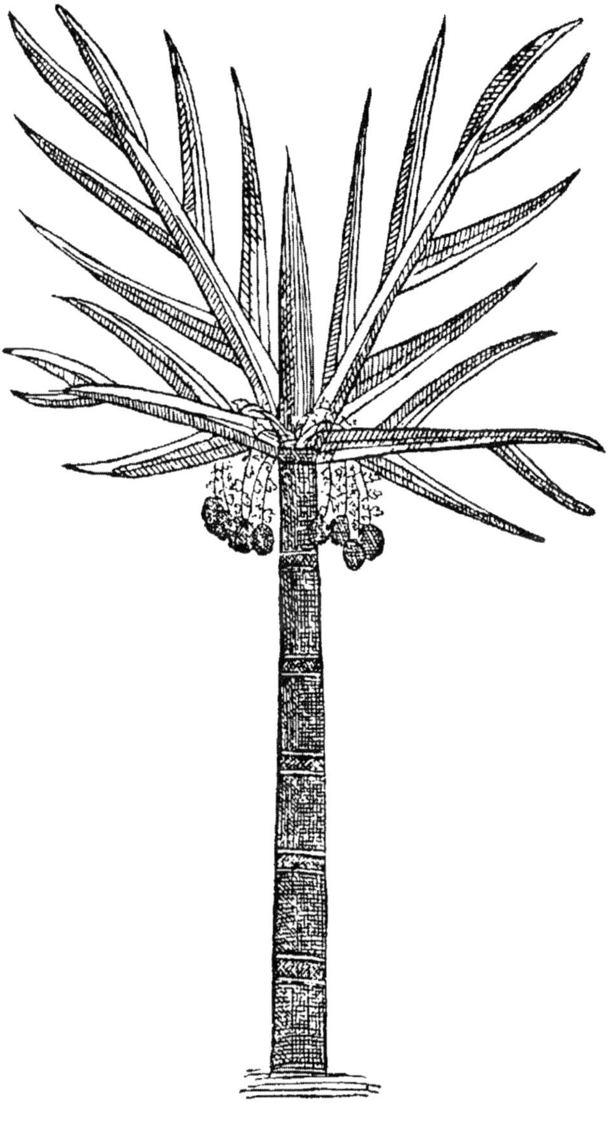

PLATE IX.—AVELLANA INDICA.

TWENTY-THIRD COLLOQUY

LEAF OF INDIA

RUANO

I AM reminded that you told me, speaking of the *betre*, that it was not *Folio Indo*. This was important to me, because the physicians who pretend much to a knowledge of what there is in these parts, and Laguna, with other modern writers, call it, in their writings, TEMBUL, and they say that it is the name given to it by the Mauritanos. Now you must tell me what the *Folio Indo*[1] really is, and show me that it is a different thing.

ORTA

It is clear that they are different things, as I told you. Avicenna has two chapters, the 259 on the *Folio Indo*, and 707 which is on the TAMBUL. This needs no argument, for

[1] The *μαλάβαθρον* of Dioscorides, i. 11, and *malobathrum* of Pliny, xii. 26 (59), the Sanskrit *tamalapatra* (*i.e.* "Cassia leaf"), and *tejapatra* (*i.e.* "aromatic leaf"), was from the earliest historical ages a renowned article of commerce between India and the countries on the Mediterranean Sea, and has now been clearly determined to be the leaf of various species of Cinnamomum, of which the principal are: *C. Tamala*, Nees, *C. Cassia*, Blume, and *C. zeylanicum*, Breyn. The Arabic *kadigi* is a corruption of the Indian word *tejpat*, a vulgar contraction of *tejapatra*.

the *Folio Indo* is called CADEGI INDI, and the *betre* TAMBUL. I have already told you the names for *betre*. The *Folio Indo* is called by natives of India TAMALAPATRA, which the Greeks and Latins corrupted into MALABATRUM. In Arabic CADEGI INDI, which means "leaf of India." Avicenna was properly translated, for it is in Arabic *lingoa de vaca* or *lingoa de passaro*, and *melam da India*, for these names all mean the same thing as *Folio Indo*. If you like I will show you. Girl, bring those leaves that I brought in my pocket, from the druggist's shop.

SERVANT GIRL.

Here they are.

ORTA

Look at them.

RUANO

They look like leaves from an orange tree except that they are sharper, and the colour is dark green. They have a nerve in the middle and two others which unite at the point, which will enable me to know them well, when I see them again.

ORTA

The smell is very pleasant, and is not so strong as that of the *espiquenardo*, nor as the apple. It smells like a clove, but it has not so strong a smell as cinnamon.

RUANO

Tell me the appearance of the tree, for these leaves look as if they hung over water, like the plants they call water lentils, as they are described by Dioscorides; for Dioscorides says they are like lentils.

ORTA

Wrong information was given to Dioscorides and Pliny,

for these leaves come off a large tree and not in marshy ground. The tree that yields this *Folio Indo* is grown in other places, including Cambaya, and the venders of medicine, called *gandis*, will understand you if you ask for TAMALAPATRA, because that is what they are called in that part.

RUANO

We have been deceived about this medicine as in many other things. In the land of Prester John, a Franciscan friar, who wrote *Modus faciendi*, says that he has *Folio Indo* in his hands, and that it is called *follias do arvore da canella*, that the leaves are not like these grown in water but on trees.

ORTA

Such leaves may well be those of the cinnamon, and the *Folio Indo* is not very different, but the cinnamon leaf is narrower and not so pointed, and has not the nerves like *Folio Indo*; but neither the cinnamon nor the *Folio Indo* are in the country of Prester John.

RUANO

Dioscorides says that some, from the smell, said it was a leaf of the spikenard, that they are collected by passing a thread round them, that the leaves are kept for sale, and that the dried marshes where they grow are burnt, because unless they are burnt the plant will not grow there again, and that the best is the newest, and that from white they turn to black, and that the scent goes to the head and often remains there, and that it is like spikenard, and has not the taste of salt.

ORTA

The smell is certainly not so strong as that of nard; and the nard is not a tree, and the way of gathering is not like that. They collect the leaves and make them into bundles, which they carry away for sale. They do not grow in marshes, nor is a fire necessary to make others grow. They burn all the land where there is to be sowing, but not the land which is left for the plants to grow. The colour is clear green, and the things they keep do not become so clear, but are nearer black than dark green, and some of them have not the smell of *salva*. It is true that the inner part is better, because its virtue is better preserved, and the scent comes to the head like other scents. Autuario says that the Moors call it TEMBUL, but he is mistaken like the others.

RUANO

Pliny says that there are curved leaves in Syria whence oil is obtained for unguents, and that in Egypt they are more abundant, but that the best come from India, and that they grow over water, and that they smell stronger than the *açafram*, and more like the *salva*, and that it is like the nard; and that when put into wine it exceeds all the scents, and that its price was something marvellous, 300 *livras* and the oil 60 *livras*. Pliny says this. Will you answer it, and satisfy me?

ORTA

Avêlo in Syria and in Egypt I do not know, but I have friendly relations with physicians of Cairo and Damascus and Aleppo, and all assure me that it does not exist in Syria nor in Egypt. The smell is not so strong as *açafram*, nor is it like nard. The nard comes from a place 200 leagues distant, and is a thing that is sown, while this is a large tree. As to other things I have already answered, confuting Dioscorides. The scent seemed so strong in *Folio Indo*, because then they had not *beijoim*[1] *de boninas*, nor *amber*,[2] nor *almisere*,[3] nor *calambuco*[4] as we have them now. For things go on increasing,

[1] See footnote, pp. 49, 58, 63 *supra*.

[2] See footnote, p. 20 *supra*.

[3] See footnote, p. 25 *supra*.

[4] See footnote, p. 195 *supra*.

though their virtues may not increase so much. But never believe that things producing scents will be lost. Like the cinnamon, that which was profitable in times past may give place to something else, as the world is more completely discovered.

RUANO

Neither Galen nor Rasis say anything new about the *espique*. Avicenna says that he has ascertained the same virtue and that the leaves are those of the *sasifrâo*, that it grows in water on swampy land without taking root like water lentils, whence some hold that it has the leaf of *golfam*, and that its oil has the virtue of *laserpicium*[1] but is stronger.

ORTA

I have proved what is in Dioscorides and Pliny to be false, so it is not necessary to answer further. For Avicenna, Serapio, and Rasis did not know more than the Greeks about this medicine. They only knew that, according to the Greeks, MALABATRUM was *Folio Indo*, and they translated what the Greeks said, merely adding some things about its use. All say that it is useful to provoke urine and for foul breath, and finally they say that it is good for all these things, like the *espiquenardo*.[2]

RUANO

These modern writers confess their ignorance, and in my

[1] See footnote, p. 149, on "Costus Arabicus."

[2] Spikenard is the root of the Nepalese Valerian, *Nardostachys Jatamansi*, the true νάρδος of Dioscorides, etc.; the adjective πιστική, "precious," applied to it, Mark xiv. 3, and John xii. 3, being a twist given to the Sanskrit *pisita*, meaning "fresh." The Sanskrit names are *nalada* and *jatamansi*, the latter name meaning "hairy-root." The older Pharmacists called it "Nardus Indica," "Nardus Assyria," and *Aspic d'outremer*, to distinguish it from "Nardus rustica," the root of *V. officinalis*, and *V. celtica*. The Arabic name is *sumbal-hindi*, the spikenard of India. Blane attributes "Nardus Indica" to certain aromatic Indian grasses, *Andropogon* sp., and undoubtedly in the bazaars of peninsular India almost any aromatic root is sold for *jatamansi* or *sumbal-hindi*.

judgment they do well. Others say that in its place they give leaves of clove or cinnamon, for the author of *Luminare majus* says that salesmen will sell leaves of clove and say they are *Folio Indo*. The other Franciscan, already alluded to, says that they substitute cinnamon leaves for it. Antonio Musa says that he saw it in Venice, and that they showed him *Folio Indo* of Syria, and *Folio Indo* of India, but that he did not know them. We must put something in its place in Spain, for the *Folio Indo* is wanting to us, and we must do without it.

ORTA

He who says that he substitutes leaves of clove does not, as it seems to me, say well, for between the place where the clove grows and that where the *Folio Indo* grows is a journey of two years. He who tells us of the cinnamon leaf is ignorant, and if these were sent there would be enough to supply all Europe. They could easily raise enough, and those who cannot get *Folio Indo* use cinnamon leaves in its place, though it is not so like it as other medicines. Avicenna orders THALISAFAR in its place, according to André Belunensis, but I do not know that medicine, nor do I believe that it is like *Folio Indo*. Of this opinion is Mateolo Sinense, against a modern writer.

TWENTY-FOURTH COLLOQUY

GALANGA

Ruano

GALANGA is a very necessary medicine, and I believe that the Greeks did not know it, at least under that name. It is a requisite in all dispensaries, so let us talk a little about it.

Orta

The name in Arabic is Calvegiam, and you will find it written by all the Moors Chamligiam or Galungem,[1] but Serapio has corruptly written it. All the Moors call it as I have said. There are two kinds of what we call Galanga, one small with a strong scent, which is brought from China to these parts, and thence to Portugal and other Western countries. This is called Lavandou in China. There is another larger which is found in Java and there called Lancuaz. It is larger and not with so strong an aromatic smell as the first. In India we call both Lancuaz. The

[1] The rhizomes of *Alpinia officinarum*, the "Lesser Galangale" of south-east China, and *Alpinia Galanga*, the "Greater Galangale" of the Indian Archipelago. The Arabs, Persians, and Hindus give to the rhizomes of both plants the name of *kulanjan*, from the Chinese *kauliangkiang*. The correct Malayan name is *langkwas*. See note, page 82 *supra*.

first is a small shrub of two *palmos*, with leaves like a myrtle. The Chinese say that it grows without being planted. The larger one in Java grows to 5 *palmos*. It makes large roots, and has knots like a cane, and the one from China has the same. The Java one has leaves the shape of a lance, and it has a white flower. It has seeds, but they are not sown, though in that land they are sown in gardens in small quantities, and used by the people as salads and in medicine, chiefly those who come from Java, who are midwives (called *daias*) and work as doctors. They propagate the plants from rhizomes, like ginger, and not in any other way. If you see anything written to the contrary, do not believe it; for Avicenna, Serapio. and other Arabs only had confused accounts of it. There are two kinds; that from China is the most praised, but they do not speak in this way from having a real knowledge, being, so to say, only gropers. This is the reason that Avicenna writes of it in two chapters. In one (321) he calls it CALUNGIAM and in the other (196) CASERHENDAR. Which is the one from China, which we use most, and which is the one from Java I do not know, because they do not write without doubts. It will be best for you to see, as well dried as green, and I will show them to you to-day.

RUANO

The Belunense, in his Dictionary, says that Avicenna writes of both, yet that there is not more than one, and the reason he makes two chapters is that he is doubtful; so that what he leaves out of one he can put into the other.

ORTA

Rather he does that where he is doubtful. To me it seems that he saw the two kinds of GALANGA, and therefore made two chapters. As we are certain about the medicine, we need not trouble so much about the names.

TWENTY-FOURTH COLLOQUY

RUANO

Dioscorides does not mention this simple, nor any of the Greeks, and the Arabs write little and doubtfully, which is a reason that we should follow the moderns so far as they write correctly. Antonio Musa, who was curious and intelligent, says that to Leoniçeño it appeared that what we, in the dispensaries, call GALANGA is the ACORO which has a root of *espadana*. It does not appear to be a root without scent, and has a warm and sharp taste (conditions necessary for the *acoro* which we falsely call *espadana*). And he says that the same appears to him, considering the GALANGA with its scent and taste.

ORTA

I said to you, speaking of the *calamo aromatico*, that the ACORO was not *calamo aromatico*. I will state the reasons on which I found my opinion. The *Acoro* is bitter in taste, the *calamo aromatico* is sharp to the taste. The *Acoro* has a white root and the *calamo aromatico* a yellow root. Now I will tell you that the GALANGA is much less like it than the ACORO. For the GALANGA is less warm and has a pleasanter smell. The things for which the GALANGA should be used are not those for which the ACORO is useful, according to the Arabs who have written on the subject. The use of GALANGA is for the stomach and for foul breath. The use of ACORO is for the head and nerves. I remember curing Nizamoxa of a tremor, and the physicians never made mention of GALANGA. Antonio Musa did not know the plant of GALANGA.

RUANO

The Italian friars who have written on the subject say that the GALANGA we use is the root of *esquinanto*.

ORTA

You can easily see that there is no reason in this, for the

esquinanto grows chiefly in Arabia, that is in Mascata and Calaiate, and China and Java are sufficiently distant from those parts; besides, the root of *esquinanto* is very much smaller.

RUANO

Menardo, and the friars who wrote on Mesue, say that the *calamo aromatico* is *acoro* and what we call *acoro* is not it. For love of me, tell me if you found what we call *acoro* in Spain would you use it, for it is there, and if you would not use it what would you put in its place?

ORTA

If I found the true *acoro* in Galicia, and it possessed the qualities ascribed to it by authors, I would use it. But if I saw what they call *espadana* in Portugal I would not use it, for it is not GALANGA. I say this without any doubt, *calamo aromatico* is more like ACORO than GALANGA.

RUANO

I will take your advice, if God is pleased that I should return to Spain.

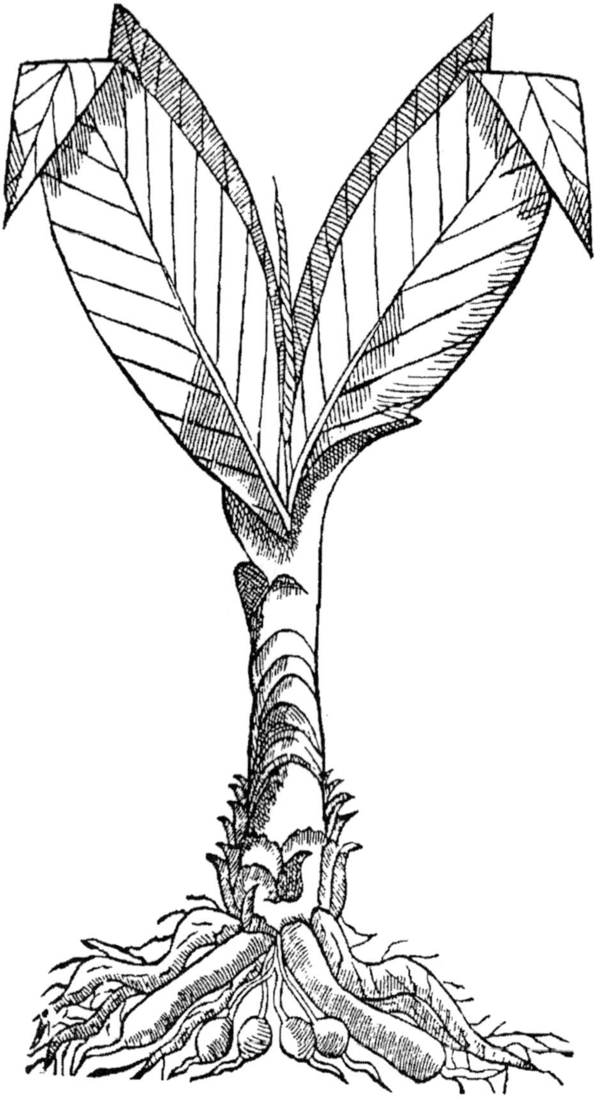

Plate X.—Galanga.

TWENTY-FIFTH COLLOQUY

CLOVES[1]

RUANO

WE will talk of GARIOFILO, for it comes from the same region as GALANGA.

ORTA

You make a mistake in not using the letter C, because in good Latin it is CARIOFILO, and in bad Latin GARIOFILO, as you may see in the modern authors who have written on it.

RUANO

I cannot agree, for so I have been taught all my life.

ORTA

If I show you in Pliny that it is called so, what will you say?

RUANO

I confess it to be better Latin, but usage excuses me.

ORTA

The Greeks did not speak of this GARIOFILO, only Paulo

[1] *Eugenia Caryophyllata*, Myrtaceae.

Egineta, who says it is the leaf of a nut, because GARIOFILO is supposed to have that meaning, but this does not appear to be known. And so Serapio says that in the Greek definitions the name is not to be found. Afterwards he refers to Galen and Paolo, and says that he translated it literally. I do not find it in Dioscorides.

RUANO

Well, I can give you the place where Galen speaks of it.

ORTA

In books which are properly of Galen you will not find it.

RUANO

In the second book of Dinamedis mention is made of GARIOFILO, and also in the third, and many Arabian writers say that Galen said it. Perhaps they translated some books of Galen which are now wanting, having been lost through the lapse of time.

ORTA

These books you refer to, in which Galen speaks of CARIOFILO, are not coveted by Galen. It is enough for me that Ruelio, such a diligent and laborious writer, says that it is not to be found in Galen.

RUANO

Well, this that you say is supported by Paulo, and Aeçio, and Pliny. It is said that in India it is very like pepper, except that it is larger and longer, and this is called CARIOFILO.

ORTA

I do not deny that these men talk of it, but I deny that Galen does. I further tell you that this medicine was found quite recently, first as medicine and for the scent, and then for culinary purposes. It is used a thousand parts to one as medicine, and the rest for cooking. You will now like to know the name in Arabia and in this country.

RUANO

I wish you to tell me all, very clearly.

ORTA

The Latin name is CARIOFILUM, others calling it GARIOFILUM as you did just now. The Arabs, Persians, Turks and most of the people of India call it CALAFUR. In Maluco, where alone it is found, and in all that region, the name is CHANQUE. The names given in Pandetario, such as ARMAFUL, are wrong, and the name written in Arabic CARRUMFEL is an error of the Arabian writer, a corruption caused by the lapse of time. The tree of this *cravo* (clove) grows in Maluco on some islands subject to the king of Portugal, taken in a just war some time ago. The right to these islands was disputed between Spain and Portugal for a long time, and you, being devoted to your king, will acknowledge the justice of our tenure of the islands.

RUANO

I owe so little either to the king of Castille or to the king of Portugal that I am able to say that I have as many mills here as there. Speaking confidentially to you, I owe more to the king of Portugal, for it is him who has granted the great part of what my brother-in-law possesses; and these advantages I owe to the king of Portugal, while the king of Castille has never given me anything, nor have I any expectation from him.

ORTA

You must know that Maluco is within the line of Portuguese conquest, which extends 200 leagues further, as has been proved by observing eclipses. But the devil entered into a

Portuguese,[1] who, because the king would not grant him an unjust favour for which he asked, went over to Castille, fitted out armed ships, and discovered a strait, before unknown, which led by another route to Maluco. He died with the greater part of the people who went with him; so that they were unable to return by the way they came. Another bachelor, Faleiro,[2] who went with him and against his king, went mad and also died in his work of discovery. Then Castillians came to Maluco on other occasions, but were obliged to go back, and most of those who attacked the Portuguese were killed. Those who surrendered were given the means of returning to Castille. Such is the clemency of the King our Lord to vanquished Christians. The king of one of the islands, called Ternate, when it was proposed to him that he should help the Spaniards, said that the clove was given by God to the Portuguese, because each clove contains the five *quinas* of the Kings of Portugal. It may be that he said this by the will of God, though he was an infidel. In the same way Balaam and his ass prophesied, not being a rational animal. I say this subject to correction from the Holy Mother Church. Afterwards this king became a Christian. He received his kingdom from the king of Portugal, and I knew him at Goa. Returning to the cloves, I say that they are only found in these islands of Maluco, which are five in number, and from there they are distributed over all parts of the world. If you ask whether there are trees of this spice in Ceylon, I reply that there are, but they do not give fruit there, nor in any other part except Maluco. The trees are of the height and shape of a laurel. They have many flowers which are made into cloves. The plant grows like a myrtle. The flower is first white, then green, and finally vermilion and hard, which is the clove. I have been told by those who have seen it, and are worthy of credit, that when the flower is green it gives

[1] Magellan.

[2] Ruy Faleiro did not sail with Magellan.

out the most delicious scent in the world. Those who gather or dry this clove find it to take the colour that you see now. They grow from buds, like the myrtles, and some say that the clove comes from within; but that is not so, only the bunches do not come to perfection. They collect them because the branches that form a great cup give out cords with which to collect the cloves, and this is the reason why the trees are beaten and flogged, and they do not give so good a harvest for a year. The cloves are dried for three or four days, and thus they sell them, and keep them to send to Malacca and other parts. The clove which is left on the tree becomes larger, and they like it, in this way, in Java. We, with the others, call it the head. You must know further that nothing whatever grows under or round the clove tree, because the clove draws up all the juice out of the earth.

RUANO

Where is the plant which the Castillians call *fuste* and the Portuguese *bastom*?

ORTA

Without the twigs whence these cloves hang as the flowers hang from the small twigs, and the great clove, which I mentioned to you, is what we call "the mother of the clove," and not because it is so, nor is it male, as Avicenna and Serapiam say, for all is one, but because it is older than the others. For what we call "the mother of the cloves" is not of the same year, but the year before. This was told to me by persons who know. One was a factor from Maluco, who said that from the clove there is much ripe fruit which falls down.

RUANO

Do they make any profit from these trees of the clove, either by planting, or cleaning, or pruning?

TWENTY-FIFTH COLLOQUY

ORTA

No more than cleaning the ground where they collect the cloves; and the trees grow without seed being sown, only from cuttings. They do not grow very near the sea, but a cannon shot distant from it, though on islands surrounded by the sea. These islands where the cloves grow are five in number[1] as I have said, and the principal island is called Geloulo,[2] and for this reason they call the clove in Spain CRAVO GIROFE,[3] because it is from Geloulo. It is also called CRAVO because it resembles a nail in form. Some say that when it is young there are more cloves than leaves, and that the leaves have not so strong a scent as the cloves. These trees grow from cloves which fall on the ground, like chestnuts in our country, though it is not necessary, for the earth always yields these cloves, and rain is never wanting for them to grow and produce fruit. The trees of the cloves come to maturity in eight years, according to information from the natives, and they last for a hundred years. The harvest is from the middle of September to January and February.

RUANO

Do the natives use the cloves in their food or for medicine?

ORTA

According to my information the people of Maluco do not use these trees themselves. The Chinese came in their ships to this land, and took the cloves to their country and to India, Persia, and Arabia. They relate this from memories preserved among themselves. The clove is easily preserved with salt water, and in another way by making it into powder.

[1] Ternate, Tidore, Motil, Makian, Bachem.

[2] Gilolo. Barros calls it Batechina.

[3] The word GIROFE has nothing to do with Gilolo. The *Caryophylion* of Pliny is certainly our "Cloves," the dried flower-buds of the Myrtle bloom *Eugenia Caryophyllata*.

CLOVES

Ruano

You say that the natives of Maluco do not use the cloves; do the other Indian races and the Portuguese use them much?

Orta

When the cloves are green the Portuguese who live in Maluco make a conserve of them with vinegar and salt, which they call *achar*; and they also preserve them in sugar. I have eaten them and they are good. The people of Malacca use the vinegar conserve when they can get it, and the Portuguese women, living in Maluco, distil water with the green cloves, which is very fragrant and a good cordial. It would be a good thing to introduce it into Portugal. Many Indian physicians make a sudorific with cloves, nutmeg, mace, and long black pepper, and they say it draws out the Castillian itch. I have also seen Portuguese physicians use it, but I do not think it is a good medicine. Some people apply pounded cloves to the head, and say that they find it good for headaches. Women are much addicted to chewing cloves to make the mouth smell sweet, and not only Indian women but also Portuguese.

Ruano

Serapiam, quoting Galen, says that it is the leaf of the nutmeg. Is the tree of the clove and nutmeg all one?

Orta

They are from entirely different countries, one from Banda, the other from Maluco. The tree of the nutmeg has round leaves and is like those of a pear tree, and those of the clove tree are like a laurel.

Ruano

Avicenna and some others say that the tree is like *sambacus* but more black.

ORTA

It is not like *sambacus*, a plant which we call jasmine, nor like *sambucus* which we call an elder tree. You will see the difference between the one and the other.

RUANO

It is said that it has been taken to some islands of India, and that the gum or resin from it is like *trementina* in virtue.

ORTA

The statement that it has been taken to some islands of India is true, but there is no such gum in Maluco. I have spoken to several people who have been in Maluco, and they all say there is no such gum. I do not deny that many trees yield gum, especially when they are cut, but up to the present time this has not been tried. Nor with your pardon can I allow the truth of what has been written in New Spain, that the gum of the clove is gum mastic. For trees are of different natures and do not yield gums contrary to their natures.

The leaves of the clove do not come to India, so I do not write about them. The scent of the clove is said to be the most fragrant in the world. I experienced this coming from Cochin to Goa, with the wind from the shore, and at night it was calm when we were a league from the land. The scent was so strong and so delicious that I thought there must be forests of flowers. On enquiry I found that we were near a ship coming from Maluco with cloves, and then I understood the truth. Afterwards, men from Maluco told me that when the clove is dry it gives out a strong scent extending far from where it is.

RUANO

In Serapio and Avicenna I find many names which must be corrupt, such as the names of authors. I should be much pleased if you would tell me what you know about this.

ORTA

I do not know except that some words are wrong. They call Rasis Benzacaria, and Mesue Menseus.

RUANO

Serapio seems not to read without an aspirate; Hachim, it seems to me, should be Aly.

ORTA

No; it is Hachim which means a philosopher, and, as among them, some are called philosophers specially, it may be that they use it as a name.

RUANO

Is the herb which we call *cravos* (pinks) in Maluco or in India?

ORTA

It is not in Maluco. I have seen it in these parts coming from China; but it has not the same scent as that of Portugal. The cause may be that it is very superficial, or that the heat of this country works out the scent. I will not say more, as you know more about these pinks than I do. I may tell you that in a certain part of the island of San Lourenço there is a very round fruit, larger than a filbert with the husk, which smells very like cloves, but is not a clove nor like one.[1]

[1] *Ravensara aromatica.*

PLATE XI.—CLAVOS.

TWENTY-SIXTH COLLOQUY

GINGER

RUANO

LET us talk of ginger, which gives us a flavour on fish days, and excites the appetite in the salads made with it in the preserve, which your slaves call ACHAR. It seems to me that it is to rectify the fish diet, as it is laid down by our doctors.

ORTA

It was certainly unnecessary to speak of this product, because there is no doubtful point of which to write. However, I will treat of what every one knows, on condition that you repeat it to no one in India, but only in Spain. For all do not care for what I tell you in this way. That you may not ask in vain I may tell you that with us it is called GENGIBRE; the word is GIMZIBIH among Arabs, Persians, and Turks. When it is green the Guzaratis, Deccanis, and Bengalis call it ADRAC, and when dry SOCTE. In Malabar it is known as IMGI, and the Malays call it ALLÁ. It is a herb with a root like a flag or like a *lirio espadanal*: but the root is larger, and the leaf a deeper green, and 2 to 3 *palmos* long. When green it is not so sharp to the taste, especially that which is grown in Baçaim where it is milder, or to speak more directly it is not so sharp to the taste, the land being moister. For the great axe of necessity makes things warmer. It reminds me of the common saying that the axe has a warm edge. This is the feeling of authors when speaking of ginger, that it does not become warm as quickly as pepper owing to its humidity. The green ginger is eaten in salad mixed with other herbs, oil, vinegar, and salt, and in pasties of fresh fish. Some eat it in the pasties of meat. They grow in all ports of India, at least those we know. They are sown, and doubtless some grow without being sown, but so few that they hardly count. It is grown in Malabar, and this kind is most liked by the Arabs and Persians. It also grows in Bengal, Dabul, and Baçaim, and along all that coast. From the interior there is little, and none comes to us. There is ginger in the island of San Lorenzo and the Comoro Islands, which are near the coast of Ethiopia, and there are those who say that it is in Trogoldita and Arabia. It is gathered in December and January, dried and covered with clay in holes to prevent it from decaying. It is also enclosed in clay to make it weigh more and to keep it fresh, preserving its natural humidity. Besides, if it is not well covered with clay the worms eat it. It is also more humid and has a better taste.

RUANO

Few questions arise respecting this ginger, Serapio calls it *lingibil*, and some nations use that name.

ORTA

I have enquired, but I do not find where it is so called. Where is this name correct?

RUANO

Galen says it came to us from Barbary.

ORTA

If by Barbary is intended the coast of Barbary it is not true; but if by Barbary is meant the unknown interior it may be, for no country can be more strange as regards India. But this is to speak very vaguely.

RUANO

Dioscorides says that they have it in Trogoldita and in Arabia.

ORTA

It is in Trogoldita, and in the Comoro Islands which are near that coast. It is also in Ethiopia, according to what I am told, but only enough to supply the country. As for what Dioscorides says, that it is in Arabia, asking your pardon, he does not speak the truth, for it is an article that is imported there. He says what is true about its use when green, as we use it on fish days in salads, as I have already said. He also says, for example, that the root is much used, and it may have been more in those days than now, the scent being strong. They also used the root more medicinally, both against the plague and against poison. Some practitioners now prescribe a salad made of the root and other things, against the plague.

RUANO

It is said that the roots are small like that of the *junca avelanada*.

ORTA

It is not so. They are much larger than the root of a reed.

RUANO

You say that they choose the roots that have not been covered with clay. Why then do they say that they are so covered? To prevent them from rotting?

TWENTY-SIXTH COLLOQUY

ORTA

I did not say that to be bored through by a worm was good, but that to be covered only with clay is not so good. I now say that it is better, for it protects them from the attacks of worms, and that is the object of doing it. Those who say that they are taken to Italy in *canteiros* may be right, but it would be better first to cover them with clay. Those that say that ginger causes looseness have reason, for it promotes good digestion; and those who say that it closes the stomach also have reason, because the looseness caused by indigestion is stopped.

RUANO

Tell me whether it spreads out like flax.

ORTA

No such thing. But it is rough like flax. It cannot be called a bush, as Serapio calls it.

RUANO

Well, Musa, a diligent writer, says that it spreads like flax, and that it has a leaf-like cane.

ORTA

It is not a good comparison, for the leaves of ginger are rough, and the leaves of the cane are not rough.

RUANO

He also says that the conserve leaves a sting in the mouth.

ORTA

This statement is not good, or it has been falsified, or he has fixed upon the conserve to indulge his malice. For it is not bad preserved with sugar, and well washed through many waters; first pricking it into holes for the water to enter, and keeping it for many days well sugared. It is then very good, and does not burn nor leave a sting in the mouth. As they produce much sugar in Bengal it is better there. The women also make it well in Chaul, Baçaim, and Dabul; but in Betecala it is not so good because it is not made in the right way, nor with such good sugar.

RUANO

Where was that made which you gave me the other day?

ORTA

In the house. Another kind, quite as good, I will give you in a jar. It came from Bengal. Bring it, girl, and show it.

SERVANT GIRL

Here it is.[1]

RUANO

Both are so good that I do not know which is the best. I kiss the hand of your worship.

[1] Ginger is the rhizome, or underground stem, of *Zingiber officinale*, Roscoe; the word "Zingiber" being derived, through the Greek ζιγγίβερις and Sanskrit *shringabera*—meaning "horned root"—from the Arabic *zanjabil*; referring to Zangibar (?). It is not a native of India: and its proper Sanskrit name is *ardraka*, meaning, literally, "of the wet" season, *i.e.* the "monsoon"; "green ginger" is in Hindustani denominated *adrak*, and "dry ginger" *sont*.

Plate XII.—Gengidre.

TWENTY-SEVENTH COLLOQUY

Two kinds of herbs for dysentery, and of another which cannot be touched without its withering.

RUANO

FROM what I have seen in the hospital there is here, and as regards those patients, that I see you treating, the worst diseases are cholera and dysentery. I therefore desire that we should talk of the herb[1] with which you cure the dysentery. According to what is said in Portugal it is very efficacious.

ORTA

The cholera, as you have seen, is a very strong and dangerous malady, and dysentery, when it has become old (what we call chronic), is very difficult to cure, and when there is a hot humour it is very dangerous. The treatment must be more cautious and more careful than in Portugal, for any mistake made in the course of it is difficult to remedy. The herb, or, more properly speaking, the bush is called CORU in the Canarese language used at Goa, and we call it the herb of Malabar, because the people of Malabar effect good cures of

[1] *Holarrhena antidysenterica*, R. Brown, a Dog-bane, the source of "Conessi-bark," or "Tellicherry-bark"; and of the seeds known by the name of *inderjao*; both being often confounded with the bark and seeds of *Wrightia tinctoria*, Rox., another Dog-bane.

this sickness, and here come some who will show you. Certainly they, as well as we, cure this infirmity well, after the greater part of the material is evacuated. Most of these plants are the size of an arbutus or smaller, the leaves like those of a peach tree, the flowers white with a scent like that of a honeysuckle. We use the dry bark of the root because when fresh it gives out some milk.

RUANO

Should it be classed as hot?

ORTA

So I thought before trying it, and when I tried it I found it to be insipid and cold, and I saw the effect it had, so I classed it as cold and dry as more correct than coldness alone, and so it is classed in this country.

RUANO

With regard to what you say the Licentiate Alvaro Fernandes told me the other day that he had a very dangerous attack of dysentery and took the herb of Malabar by your advice in the way it is administered by the Malayalims, and not in the Portuguese way. It seemed to him to be the most bitter thing in the world, whence I conclude that a thing so bitter cannot be cold.

ORTA

It may well have different effects in different parts like the ZARGATOA[1]; and with regard to the bitter taste it has, it is only slight, just as the opium, though so cold is bitter. It therefore seems to me to be cold; at least we shall be right in saying that it stops. To return to the subject in hand, I say that we make the bark of the root into a powder,[2] and take as much as will go on the cup of a still. It is then mashed well with sour milk. We then take *amcos*,[3] the seed of *aipo* and

[1] *Cineraria campestris*, Flea-wort.
[2] Caçola de alembique.
[3] Wild Cummin. (For continuation see p. 485.)

coentro dried and black cummin seeds pounded—those which come from Ormuz. All this is mixed together in an ounce of crude butter, and so it is put to be distilled. We give four ounces of this distilled water to the patient, mixed with two ounces of rose water, or "root of roses," or plantain. When it is necessary we mix lozenges of Malabar herb and make the liquid in the same way, not using butter, and it is given with the same water as the above. This water is also used in clysters to take at night, and it is a thing with which I have been successful many times. We give these clysters actually cold, as the climate is very hot, and this is also the custom of the native physicians. It does not seem bad to you? If there is urgency we give these waters to drink twice a day, once at six in the morning and again at two in the afternoon. During this treatment we give the patient boiled milk with rice and chicken soaked in this rice water (which they call CANJE), and according as we judge of the weakness of the patient, so we give him food. The Malabars never give wine, nor do we, there being many reasons for this in cases of chronic dysentery. This medicine is very good, and I have often succeeded with it, yet I feel bound to confess that it is not so valuable nor so certain as the herb which the Malayalims give, very roughly compounded, but in the same way as the other, powdered and soaked in sour milk, or in rice water much boiled, and almost dissolved. Others make this liquid of the green herb pounded. This is stronger to take and very bitter. It is given to the patient, seven ounces in the morning and the same in the afternoon, if necessary; as the herb is not pleasant to the taste they give it with some sour milk.

RUANO

What kind of water is found best for the patients?

ORTA

That of the Malabars is found much the best. When we

find that our patients do not appreciate our gentle medicines, we deliver them over to Malabars to be given the stronger medicines. Now the Malayalims give us their water, which is compounded in the royal hospital. If the Malayalims see greater urgency they mix opium with this medicine. Some Arabs cure all dysenteric illnesses with opium rectified with nutmeg. I saw an Arab cured in this way when I was marching with that great Sultan Bahadur, in company with my master, Martin Affonso de Sousa.[1] An honourable and discreet gentleman from Portugal told me that Dom Manoel Telo de Meneses cured several people in this way. For the medicine had run out, and it smelt of opium, the use of which medicine he learnt in Xael, being a captive.[2] But I do not look upon this as certain.

RUANO

And to me it appears doubtful, at least in principle, for there are other things for rectifying opium, better than nutmeg.

ORTA

The Malayalims never care to confess that they give opium. I cured a very honourable gentleman, whose name is well known in Spain, and who was near to death. He had a regard for a Malayalim who had saved his life in a difficult case of dysentery. Finding himself in Goa with a slight attack, he called in this doctor who, to take the shortest way, cured him with a medicine containing opium. I was afterwards sent for and found him at death's door, with all the symptoms of having taken a dose of opium. I cured him and restored him

[1] This was the occasion when Orta accompanied Bahadur Shah in his adventurous march against the Mongol soldiers of Humayun, crossing the whole peninsula of Kattiawar from Diu to Ahmedabad.

[2] Meneses was sent by the Governor Nuno da Cunha to the coast of Arabia to examine into some complaints of the Chief of Xael, of violence from Portuguese. He was himself made prisoner, but eventually released. (Correa, *Lendas*, iii. 844.)

to health. But this Malayalim doctor would never confess that he had administered opium in the medicine he gave, and showed me the medicine he had prescribed, which was so unfit for the purpose that I was of opinion that he had given opium. I was more convinced of this when I had cured the gentleman. I found this medicine to be most efficacious when the greater part of the material has been discharged, and, by other treatment, there is a relapse many times.

Ruano

Is this medicine used for any other diseases?

Orta

For vomiting and weakness of stomach, taken with a mixture of mint water and some powder of gum mastick. In this island there is also a small tree that yields more than the other shrub. It has leaves and flowers like a myrtle, and yields a fruit like myrtle berries, with the same taste but more styptic. It is called Avacari. An old Portuguese, who had been a long time in the country, told me that it grew on a hill in his estate, and that it was very efficacious in cases of chronic dysentery. He said that he had a daughter who had been ill with dysentery for a year, and that other medicines were of no use, while this one restored her to health. I asked him who told him that this plant was good for dysentery. He told me that one of the native doctors gave him the powdered bark in rice water. They say that the root of this shrub smells like clover. I asked the doctors here about it, and they said that it was good for dysentery, and that they mixed it with another herb called Coru, which makes a very good mixture. This is what I know about these medicines, and I will take you to see patients being cured by Malayalims and Canarese, that you may know it all more thoroughly.

TWENTY-SEVENTH COLLOQUY

SERVANT GIRL

A boy has come from the Franciscan Friars with a basket.

ORTA

It will not be things to eat, for it is the Friars who are in need of those.

BOY

Here are the herbs you asked for.

RUANO

Some are roses, and on the other side are they medicinal herbs?

ORTA

No, but they have a strange property, showing that they do not like to be touched. That herb will not consent to have itself touched, for if you do so you will see how it presently shrinks.

RUANO

It is a very notable thing that this plant should be so clean and so modest that it will not consent to be touched. I speculate touching this philosophy. The leaves resemble a polypody frond, the flowers are yellow, and neither Dioscorides nor Pliny mention such a plant.[1] But the author of a book on New Spain says that there is a plant in Peru the leaves of which dry up when they are touched. As I think that you are now weary, we had better go to dinner.

[1] *Biophytum sensitivum*, Dec., an Oxalid, the "Yerba Biva" of Acosta.

TWENTY-EIGHTH COLLOQUY

JACA, JAMBOLÔES, JAMBOS, JANGOMAS

RUANO

WHAT is that fruit which is the size of a large nut?

ORTA

You have been eating the chesnuts that are inside it, and you said that roasted they tasted like chesnuts; and now you may eat the rinds that cover them, which are yellow and have a pleasant taste.

RUANO

They taste like melons, but not so good as the best.

ORTA

Just so, and owing to their viscosity they are bad to digest, often passing without any change. I do not use them much. In Malayalim they are called JACA,[1] in Canarese and Guzerati PANAZ, and they grow near the seaside. In the interior they

[1] *Artocarpus integrifolia*, Willd.; Sanskrit, *panasa*, "laudable," *i.e.* "well"-fruiting, and Hindustani, *chakra*, "circle," *i.e.* "round"-fruited; the "Bread-fruit" of the East Indies.

only dry the chesnuts and eat them roasted or boiled. The tree is large and high, and the fruit grows from the trunk and upwards, and not from the branches like other fruit. I assure you of this; I will here show you a JACA whence these were taken. See here that they are the size of very large melons, and some larger. The rind that covers these chesnuts is very thick and hard, as you see. and of no use.

RUANO

There never was a melon so large or so beautiful as this fruit.

ORTA

It is dark green and all surrounded by spines smaller than those of a hedgehog, but they do not prick as those do. I do not think the JACA should be eaten, except at the end of a meal, and then you might taste the chesnuts of this fruit as you did the other day.

RUANO

I have eaten those olives, for so they seemed to me, and I found them very sharp; for the rest they were like ripe olives of Cordova.

ORTA

They are called JAMBOLÔES and grow in the country on a bush like a myrtle, and leaves resembling those of the arbutus. This fruit, like the JACA, is not considered very wholesome by the people of this country.[1] But this that I show you is much esteemed here.[2] It is not long since it came here from Malacca, where it is abundant. I ask what you think of this fruit. It is the size of a chicken's egg or rather larger. You

[1] *Eugenia Jambolana*, Lam., a Myrtle-bloom. Ibn Batuta also compares the fruit to the olive. It is the "Black-Plum" of Anglo-Indians, and *jambu*, *jamam*, and *jambul*, all meaning "eatable," "toothsome," of the people of India.

[2] *E. malaccensis*, Linn., the "Malay-Apple" of Anglo-Indians, and *Malaka-jamrul* of Southern India, and known also by the name of *karika*. The Anglo-Indian "Rose-Apple" is the *E. Jambos* of Linnaeus.

see that the colour is white and yellow, and the scent like that of rose water, so that both senses are gratified. Now it is necessary, as it seems good to the sight and the smell, that it should be tried as regards the taste, and for this purpose it must be eaten.

RUANO

I have tried it and it tastes very good. It is a taste which does not cause much craving, the fruit being watery, but for me the taste is very pleasant. The appearance and smell remind me of large gall-nuts when they are new (those we call *maçans de cuquo*[1]). Tell me what they call this fruit in the country where it grows.

ORTA

In Malacca the name is JAMBOS, and it is so called also in this country.

RUANO

These are better than the JAMBOLÔES. I had heard much of it, and I say that it justifies its fame, and is agreeable to all the four senses. It is certain that this fruit is fit for a prince in our Spain, and it does not seem to me that it would do any harm to eat it before a meal. I perceive that it is cold and humid. Now tell me about the tree.

ORTA

You can see the trees in my garden from this verandah. The small ones have been planted two years, and in four they will yield plenty of good fruit and several times in the year. The shape of the tree, like that of the fruit, is oval, with fruit the size of a plum. The flower is red and very sweet, with a taste like sorrel. The leaf is like the point of a lance, large and of a very pleasant green colour. The roots of this tree strike far into the ground to uphold the tree when it is loaded with

[1] Oak apples.

fruit, which is the case many times in the year. Conserves are made both of the fruit and the flower.

SERVANT GIRL

The man is here who brings letters from your tenant at Bombay.

ORTA

Send him here.

THE CAPTAIN

The letters are from your tenant, and also this basket of JAMGOMAS.[1]

ORTA

I will read the letters afterwards. We will taste the fruit, and you can open it first between the fingers—so.

RUANO

It seems from a first taste like a plum, and styptic.

ORTA

There are many in Baçaim and Chaul, and I have also seen them in Batecala. The tree is like a plum tree, and so are the leaves. It has white flowers. There are many spines in the trunk. They call it JAMGOMAS, and for the most part it grows in the country. It has also been transplanted, and persons worthy of credit have told me that the best way of planting it is to find the seed in the excretions of a certain bird with which it is mixed. In this way it soonest yields fruit.

[1] *Flacourtia catophracta*, Roxb., of the Savantvari country; the *paniala* of Hindustan, and *talishapatri* of the S. Deccan, the latter being one of the Indian names of the Cinnamon tree.

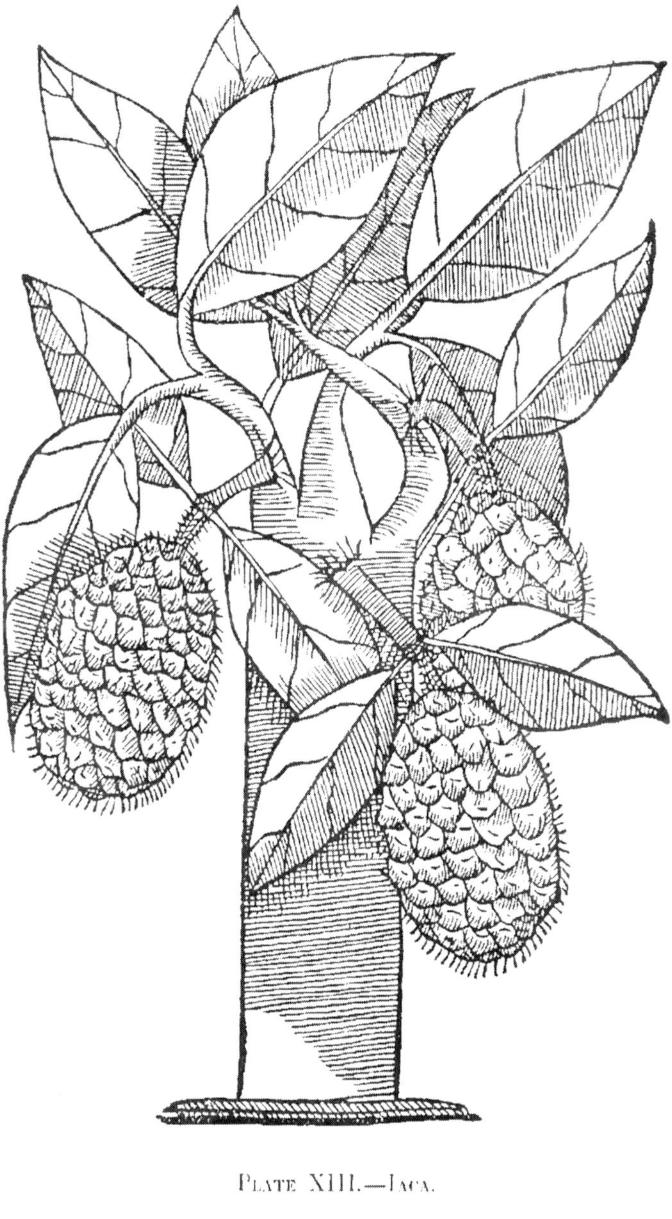

Plate XIII.—Iaca.

TWENTY-NINTH COLLOQUY

LACRE

RUANO

NOW the turn comes for LACRE, which is used so much in this country in closing letters and for seals, instead of wax.

ORTA

Rather the wax is used for want of LACRE, for the LACRE is hard to unfasten, and the seal cannot be opened but must be broken. LACRE is the name for what, in the books of druggists, is called LACA, in Arabic and Persian LOCSUMUTRI or LACRE of Sumatra; not because Sumatra is near Pegu, where it is found, but because it comes from those parts. The Arabs and other people supposed it was from there, and so gave it that name. In Bengal, Balagate, and Malabar where it is, they also give it that name, learning it from the Moors, but the name in its own country is LAC. In Pegu and Martaban, whence the best comes, it is called TREC. There are some who say that it comes from Jamay,[1] whence also comes the ALMISCRE; those of Martaban and Pegu taking it to Sumatra for sale. This is the reason

[1] A province in the kingdom of Laos.

that the Arabs call it Locsumutri. The Pegu people bring back pepper in exchange.

Ruano

Tell me what it is and how it is made, and in what way it is produced, for you will tell the truth as you know it, and I will then discuss what is written in books ancient and modern.

Orta

I was deceived for a long time. For they said that in Pegu the channels of the rivers deposit mud into which small sticks are driven. On them are engendered very large ants with wings, and it is said that they deposit much lacre on the sticks. I asked my informants whether they had seen this with their own eyes. As they gained money by buying rubies and selling the cloths of Paleam and Bengal, they replied that they had not been so idle as that, but that they had heard it, and it was the common fame. Afterwards I conversed with a respectable man with an enquiring mind, who told me that it was a large tree with leaves like those of a plum tree, and that the large ants[1] deposit the lacre on the small branches. The ants are engendered in mud or elsewhere. They deposit the gum on the tree, as a material thing, washing

[1] Lac, from the Sanskrit *laksha*, and Hindustani *lakh*, meaning "100,000," as in the phrase a "lac of rupees," is the resinous exudation produced on various Indian trees—as *Butea frondosa*, *Ficus bengalensis*, *F. religiosa*, *Schleichera trijuga* [*kusumba*], *Shorea robusta* [*sal*], and *Zizyphus Jujuba*, and also on the plant *Cajanus indicus*, the "Pigeon-Pea,"—by the puncture of the females of the lac-insect, *Coccus* or *Tachardia Lacca*, as their common nuptial and accouchement bed, the seraglio of their multi-polygamous bacchabunding lord, the male *C. Lacca*. As freshly gathered from the tree this resin is called "stick-lac"; when the females, known as the colour "lake," are washed out of it, it is called "seed-lac"; and when this is melted and run out in layers, and cooled, it is called "shell-lac," from which "lacquer" and "sealing-wax" are formed. Both the males and their colonies of females live only for the time they are ceaselessly reproducing themselves, and as if only to dower the world with one of its most useful resins, and most glorious dyes, the colour "lake." Lac has been erroneously identified with the *kankham* of the Arabs, the κάγκαμον of the Greeks; which was probably the resin *dikamali*, yielded by *Gardenia lucida*, and *G. gummifera*. See pp. 246 and 247.

the branch as the bee makes honey; and that is the truth. The branches are pulled off the tree and put in the shade to dry. The gum is then taken off and put into bamboo joints, sometimes with the branch. But the best is that which has the least bits of the branch with it. For they presently say this lacre has much branch, or has little. They also told me that some was very dirty and turned into powder, and that this was because it was mixed with earth. I sent to Pegu to enquire about this and found it was quite true. Being in Balagate, where there is some which is brought for sale to the seaport from the territory of Cotamaluco, they told me the same. Here in Goa a boy brought me a branch he had torn from a tree we call MACEIRA and the Deccanis BER,[1] of which I have spoken before, and here it is in my garden. I see it with my eyes. As it has little lacre, the country not being suitable, no account is taken of it. But many tell me that it is the MACEIRA that is suitable for those ants, and this is clearly true, for the wings of the ants are often found mixed with the lacre. This lacre of which we speak, when it is mashed, becomes a fine red, and of this they make those sticks that you see for closing letters. They stain and so increase the necessary colour for the stick with the dye which makes that colour. Of these sticks, or large thin plates, the carpenters or turners dye those they choose on the lathe, drawing off the lacre from the stick to the lathe. With this lacre they fill up the silver and the gold which is empty, to make their best works. Further I know that the tree is not like a myrtle either as regards size or the leaves. It is sometimes as large as a walnut tree, sometimes smaller. It is not called AEC, the name given it by the Pandetario, nor ANCUSAL, these names being corrupt.

RUANO

Avicenna calls it LUC in the amended translation by the

[1] *Zizyphus Jujuba.*

Belunensis, and Paulo states that the tree is like the tree yielding myrrh, that it has a good scent, that it is necessary to administer it with caution, that those are wrong who say that it is like the carrob, and that it is true that it has the virtue of the carrob in some respects. What do you think of that?

Orta

I say that Avicenna did not know the Lac and spoke from conjecture. He may never have seen the stick. I would swear he had never seen the tree if he thought it was like the myrrh tree. Yet I do not know the myrrh tree to confute his statement; but I know that the gum of lacre is formed on overlaid sticks, and the other is distilled from the tree. The myrrh has a scent and the lacre has not, though Avicenna says that it has. It is called Luc by the Belunensis; perhaps he found it so written in ancient originals, for now all the Arabs call it Locsumutri. Those who say, after Avicenna, that the lacre has the properties of the carrob are to be reprehended, for this is false. The carrob is glutinous and styptic. The lacre is aperitive, and, for being very aperient, Avicenna says it must be administered with caution; and, as you know better than I do, at first we used aperients less than we do at present. But what is to be regretted in Avicenna is that he thought the lacre was the *cancamo* of Dioscorides. For Paulo seems to speak by his mouth. And clearly it is false, for our lacre has no smell, while the *cancamo* is fragrant and used as a perfume. Avicenna says another thing calling for reprehension, that, when lacre is wanting, dragon's blood takes its place, which is also a styptic medicine.

Ruano

Why do they call it Locsumutri? Is it, by chance, in Sumatra?

ORTA

No; as I have told you, it is in Jamay, and it is taken from there to Sumatra, pepper being given in exchange. Now this route is not so much used, as the lacre is only sold by the Pegus to the Portuguese, and we sell it to Arabs, Persians, and Turks. We also take it to Portugal, whence it finds its way to Africa and other parts; so that there is no longer any reason for calling it LOCSUMUTRI, as formerly. The Chinese who brought it to Ormuz and other lands did not care whether it was from Sumatra, and as from Sumatra they sold it; but there is none in Sumatra, or if there is any it is a very small quantity, and up to the present we have no notice of there being any in that island.

RUANO

Having now examined Avicenna, we will proceed to examine Serapio. He calls it SAC, meaning LACA, and refers to Dioscorides in the translation of Abtabharic, who says that the gum found in Arabia is like the tree of myrrh. He quotes from Rasis that it falls from heaven on to the branches of GUBERA, and alleges that Isac says that the red stuff which falls over the sticks is used as a dye for cloths. He also says that they bring it from Armenia, a land well known in India. Now tell me the truth about all this.

ORTA

SAC is a corrupt word, and the lacre of Dioscorides is not lacre, of which neither Dioscorides nor Serapiam knew anything. For Serapiam thought it was the *cancamo* of Dioscorides, and said it was like myrrh or storax. You can see that this lacre has no scent whatever. Galen, in the translation of Abathabarich[1]—some suppose it was Paulo—alleges the same. Neither Serapiam nor Paulo say what is true as

[1] There is diversity in the spelling of this name in Serapio (ed. Brussels, 1531), Athabarich, Albatarich, Atabari. He was the Arabian translator of Galen.

regards what Galen wrote. For they say that the gum of a tree which grows in Arabia is like myrrh. But this lacre does not belong to Arabia, for there is a trade with it to Arabia from India. How then can there be any credit for these statements? Rasis also states that the lacre falls from heaven over the branches of the GUBERA. This is false, for GUBERA[1] in Arabic means SORVA, and there is no SORVA in all India. The Nizamoxa has it brought from Persia and Khorasan, and I have seen it in his house.

RUANO

Perhaps NESPRA is intended, for other books say that it is over the branches of NESPEREIRA.

ORTA

It is badly translated, for GUBERA is SORVA and ANZURUT is NESPERA, but there is neither one nor the other in all India. And the assertion that it comes from Armenia is also false, for neither is there any in Armenia.

RUANO

The Italian Friars, who wrote on Mesué, say that no man has ever seen the true lacre in our parts, and that it is not credible that nature can be at fault, for many believe that it is the *cancamo* of Dioscorides, because the descriptions of Dioscorides and Paulo agree. Yet no person has even seen this *cancamo*, and some say that it is what we call *benjoim*, while many good physicians hold it to be dragon's blood.

ORTA

To me it seems that the Friars say well in affirming that nature is not at fault as regards these simples; and they are right in saying this because the lands are better known and the use of medicines is better understood. But in saying that

[1] *Sorbus domestica*, or *Pyrus Sorbus*, The "Service Tree."

it is not here they are wrong, and would be more correct if they said it was here; for it is brought to India every day, and is used by Moors and Gentios. They would be still more correct if they said it was unknown to Avicenna and Serapio, or to those who turned the mention in Dioscorides and Paulo of *cancamo* into it, which we have not got. For what we use is it, and we know nothing of *cancamo* nor of *benjoim*. I note this because it is not in Arabia, as I have told you, speaking of *benjoim*. I have already put dragon's blood in its proper place. Taking you to Spain, the lacre is used there with much boldness. Here the learned Moorish physicians in Balagate use the DIALACA, which they call DALLACA, the AL being the article of genitive; so they call DIATURBIT DALTURBIT, a compound of TURBIT; and so with all the other compound words, where we put DIA they put DAL. You, who are a good Grecian, know that DIA is right, for I have heard you say so, who am not a good Grecian. Of this I will say no more, so as not to get beyond my depth.

RUANO

What you have said appears to me to be right, but I cannot form a judgment because I am not a good Grecian, and Ruelio, a curious and cautious writer, is doubtful on this point.

ORTA

He has reason, but you should speak on it and relieve yourself of doubt.

RUANO

What you say is well said as regards my feelings notwithstanding your argument. I think ours is not the true lacre, for the Greeks knew it as *cancamo*, as did Avicenna and Serapio; and neither the one nor the other have scent, nor are they used to perfume dresses; but, mixed with myrrh and storax, the addition neither causes nor increases scent, but diminishes it. I therefore conclude that we have not got the lacre but the *cancamo*.

ORTA

You say and then unsay, for first you say that nature has not been deficient in these medicines so celebrated by the Greeks and Arabs, and then you say that we have not got them.

RUANO

So I say again, until you give me reason for altering my opinion.

ORTA

It is no less inconvenient that Serapio and Avicenna did not know the lacre than to err in saying that nature has failed. You know that lacre is that which you see being sent from India to Portugal, and over all the regions of Asia and Africa and many parts of Europe, and what it is called here in India. You do not doubt what it is, nor do the Friars nor others. I say, as I have said before, that Serapio was mistaken in thinking that it was the *cancamo* of Paulo and Dioscorides, and Avicenna was much more mistaken, for he gives an account of *cancamo*, and then has a chapter on *cheickem*[1] as if they were two different things. When he makes such a mistake as this, it is not much to be ignorant of lacre.

RUANO

Very well, you have persuaded me of that. Now about *cancamo*, how is it we have not got it?

ORTA

It is a less evil that we should be wanting in one simple than in two. I will tell you what the *cancamo* is, according to my opinion. I have not got certain evidence to enable me

[1] *Keikhem*, or *kamkeham*. Sprengel held the word in Avicenna to be an Arabic transcription of *κάγκαμον*. *Kankham* is the Arabic, and *khuniham* the Persian for *dikamali*, the resin of species of Gardenia. See footnote on "Lac," p. 241.

to come to a conclusion; but when I get better proofs I am ready to change my opinion.

RUANO

I think you are going to say that it is *benjoim*, and to this I cannot agree, for *benjoim* is not in Arabia, as we have already seen.

ORTA

I do not say that, but I think it is *anime*, which has a good scent and is used as a perfume. It comes to Portugal from Ethiopia, a country bordering on Arabia.

RUANO

Certainly that satisfies me, but some say that the *anime* is a kind of *carabe*?

ORTA

This helps me for, according to some, *cancamo* is a species, and Avicenna, reprehending that opinion, says that it is not *carabe* though it has the same virtues, yet it is not that but *anime*.

RUANO

A chronicler of the Indies of Castille says that there is *anime* in Çirvamlha near Maluco, and in Brazil.

ORTA

The Castillians, if you will let me say so, are a people who exaggerate greatly, and in this there is a mistake. It refers to a certain material for caulking ships of which much comes from Sumatra and many other places. But it has not the smell of *cancamo* nor of any other gum. In this way you may speak of *lacre* or *cancamo* until you find another medicine which more truly complies with all its signs.

RUANO

God be praised that I am satisfied as regards *lacre* and

cancamo, and now it all seems clear. But which is the land where this gum is principally called LAC or LOC or LUC?

ORTA

This is the cause of many errors. In the case of the ESPODIO, as we now call it, it had another name, that of TABAXIR, the name where it grows, and the confusion caused much contention among Arabs, Latins, and Greeks, because the medicine is not known under the name of Espodio in the place where the plant grows. This gum, which is wanted and needed for dyeing, was called LOC, and so the gum was known as LUC, and it remained as the name in Arabia, to which land it was brought from where it had been bought by the Chinese. After a time it got the name of LAC among the natives of India, and those names which are so much alike no doubt originated in that way.

RUANO

So much is established in the affirmative that I can no longer raise a negative, and especially as regards two reasons so like each other. Now I want to ask you a question relating to trade. It is to obtain license to send 100 quintals of lacre to Portugal or Ormuz. It seems to me best to take it to Portugal, because they tell me that it is now worth four times less than what it used to be worth here.

ORTA

I will tell you. The Captains of Ormuz have an establishment and factory in Baçora, a city of Mesapotamia, whither merchants of Aleppo come to make purchases. Lacre may be sold to them at very good prices. The factor of the Captain of Ormuz who had it for sale, without much hope of getting much for it, was offered a good price by a very great merchant of Aleppo who saw it. He gave, as a reason, that there was a very rich dyer in Aleppo. He was sent for by the Governor (called by them Pasha) who told him that the great quantity of money he had belonged to the king. His house was searched and it was found that he had 100,000 *venezeanos*. The Pasha said to him, "You are a dyer, and a rich dyer may not have more than 1000 *venezeanos*, how then is it that you have 100,000?" So they took all from him. As no justice can be had against Moorish Kings, this man made his complaint to the Grand Turk. He, by agreement, discovered in his land some dye in earth or on trees or bushes, which was better than lacre, consequently lacre is no longer needed in those lands, which yield to the Grand Turk more than 100,000 *cruzados* a year. For the service thus done to the Grand Turk all his money was restored to him, and many privileges besides. For this reason very little lacre is wanted in Ormuz for Persia, and none at all for Turkey or Arabia.

RUANO

What is that dye? It cannot be *kermes*, for that is in Spain and other parts.

ORTA

I am unable to say, but what I have told you happened as I related it. One thing I pray of you, that, when you return to Spain you will not allow them to pass dragon's blood for lacre, nor let them believe that lacre is what the Arabs call QUERMES, for one is a gum and the other a seed, of which there is plenty in Spain.

THIRTIETH COLLOQUY

LINALOES

RUANO

I INTENDED to ask you about the LINALOES,[1] but I have not done so because it seemed to me that all that can be known about it has already been written.

ORTA

On the contrary, my dear Sir, it would be more true to say that little is known about it; yet not so little but that the tree has been seen by some Portuguese, or I should say a branch of it, and the whole tree by many of our slaves, as I will explain presently.

RUANO

Will you refer to what is said on the subject by Greek, Arabian, and Latin writers, and tell me where they are right

[1] *Aquilaria Agallocha*, Roxb., growing in Tenasserim, and on islands near Mergui, and in the valleys of Assam and Silhet, and the chief source of "Aloes-wood," "Lignaloes," "Eagle-wood," or "Calambac," the *Agallochi* of the Arabs. A similar product is yielded by *Aloexylon Agallochum*, of the Leguminosae, and by *Excoecaria Agallochum*, a Spurgewort. See p. 262.

and where wrong, and above all your own opinion, and what has come to your knowledge?

ORTA

I say that you shall be served by me as regards what I know, although there can be no doubt about these simples or what relates to them.

RUANO

Galen, according to Serapiam, quoting the translation by Albatari,[1] states for what it is useful.

ORTA

The best modern authors all say that Galen does not mention this drug, and where he is otherwise quoted it is in books which falsely attribute statements to him, therefore this Grecian author may be left apart.

RUANO

Dioscorides calls it AGALOC, and says that it comes from India and Arabia and that it is sometimes used for incense.

ORTA

The true LINALOES is only found in these parts of India, and the land referred to by Dioscorides may be traced to Arabia for, as I have often mentioned, all the merchandize of that land goes by Gida, or Torso, or Ormuz to Cairo, Alexandria and Aleppo, and so to many other lands. If some is heard of in Arabia it is not this special LINALOES, or if so it was adulterated. As to what Dioscorides says about its being used as incense, this cannot have been the statement of so grave an author. It would be less wrong to say that for want of LINALOES they used incense. A quintal of incense is worth

[1] The name of the Arabian translator of Galen referred to in the Latin edition of Serapio (Brussels, 1531) occurs in several forms: Athabarich, Albatarich, Atabari, Albatari.

1 *cruzado* and a half of the best, and it would be much cheaper in the land of Dioscorides. An *arratel* of the best *aguila* is worth 3 *cruzados*. Now judge whether this permutation of Dioscorides holds good.

RUANO

It seems to me that in this you have much reason, for those which remain in Arabia ought not to be so dear as these.

ORTA

Do you know how far it is to come from India to Dofar, Caxem, Xael, or Barem?[1]

RUANO

Let us examine Avicenna as an authority.

ORTA

He gives two chapters—one, 742, to XILALOE; another, 14, to AGALUGEM.

RUANO

How is that? Are there two medicines?

ORTA

No, there is only one; and he made two because he was doubtful, so that he explains better the second time than the first, and says that there was not more than one medicine. The names he gives in Chapter 742, when stating what lands they come from, are to be made out. ALMUDILUM may be presumed to mean Melinda, because in that land there is a black stick which sinks to the bottom in the water. Another, Avicenna says, is brought from Sofala. Now it is found in Encuama, and it is not inconvenient to call Sofala a region of India, for so we called it formerly. Another name is ALCAMERI which is Cape Comorin, a well-known promontory

[1] Ports in Arabia.

called CORI by the ancients. Other names of LINALOES are ALSEUFI[1] and ALBERI, another in the region of Catai and Seni, and others the derivations of which I confess I do not know.

RUANO

And what do you say of these parts? Do the true LINALOES come, or are they native here?

ORTA

The true kind is not native. It is true that in the parts of Cape Comorin and Ceylon there is a plant that smells like it, which we called *aguila brava*; but the scent is that of other plants we have here. This plant was taken as an article of trade to Bengal under the name of *aguila brava*, but afterwards the Bengalis were undeceived and would no longer buy it. This plant is what they call ALCAMERI and ALSIFICASI at Cape Comorin and Ceylon. If these derivations do not satisfy you now, at another time they will seem good to you.

RUANO

Please me by saying whether it is from Catay or Seni?

ORTA

From Cantam (most famous of all China) and Seni is Chincheo.

RUANO

Is the LINALOES in that land?

ORTA

No. How many times have I told you that the Chinese who navigate those seas bring the good LINALOES from Sumatra and Malacca, and the bad from Ceylon, buying it well and selling it. What they buy is very good. They

[1] A misprint. In Avicenna it is ALSAMFI.

take it to their country, for the LINALOES in China is an article of great price.

RUANO

Avicenna also says that in boiling it, in the land where it grows, the plant loses its virtue. Tell me what you think of this, for many say it besides Avicenna.

ORTA

Nothing of the sort. It is not the custom to boil this wood, of this you may rest assured. Presently I will explain the way of selecting it and how it is treated, and here you can answer that the plant is not boiled.

RUANO

Serapio states this from Dioscorides, and I can well believe that it is true, for he says that it has a rind rather than bark, and that it has bitterness with styptic properties, and that it is brought from India and Arabia.

ORTA

This is not wholly true. They speak the truth when they say that it comes from India and Arabia, but it is not true that it grows in both those countries. It grows in India[1] and is brought to Arabia as merchandize. It is wrong to say that it has skin rather than bark, for it has bark like any other plant. But the scent is not good except that of the pith (what the Portuguese call *cerne*[2]), but as the bark and wood become dry the scent gets stronger.

RUANO

There are many species of LINALOES INDO, and it is said that some grow in an island called Fuma,[3] which is better, being black and variegated. Then it is said that another kind comes from parts of India called Model, Sief, and Alcomori,[4]

[1] India beyond the Ganges.
[2] The core or heart of a tree.
[3] Fimua in Serapio.
[4] Sief and Alkumeri in Serapio.

from which places the journey by land takes three days, that the best is the kind that sinks in the water, and which does not burn when put in the fire.

Orta

Those names are very corrupt, and I confess to you that I do not know the interior very well. I cannot, therefore, criticise Serapio on this point, nor tell you whether he says well or ill. If by Fuma he means Sumatra he says well, but the derivation is very doubtful, and as to Model I cannot tell what place he alludes to. Alcomari and Sief are Cape Comorin and Ceylon, the voyage to them by sea taking three days, as he says. That this is right may be proved because those countries have a kind of Linaloes which we call Aguila Brava. And with this *aguila brava* the Baneans are burnt when they die, these Baneans never eating anything that can have died. The Chinese, who bring Linaloes from Sumatra, touch at Ceylon and Cape Comorin, and when they sell it they would say whence it comes. Serapio also says that they bury it, and the sign that it is good is that it comes full of earth. This test may be applied to any fragrant wood. The other statement is that besides this there is the *abel* brought from Sofala. It is not to be wondered at that the sticks from Sofala and Mozambique should be black, or that they make scented things from them, for it is said that combs and boards are so made. As for the statements that they cut the branches and bury them for a year, that the earth does not injure them in the slightest degree, and that they come by the rivers, there is a certain amount of truth in them, as I will explain presently.

Ruano

They also say that it has a red fruit like pepper.

LINALOES

Orta

Nothing is said on[1] that point where I have been able to verify. The Muhammadans do not speak of it; for neither Rasis, nor Avenrrois, nor Isac mention this plant except as to its uses, not as to whence it comes.

Ruano

Pliny says, according to Ruelio, that there are two places whence the *casia* comes; from the countries of the Nabateos and Trogoloditas; and that the merchants sell the *aspaltum* for it.

Orta

Pliny does not make a good road in this. It may be that *aspalto* may have been adulterated and sold as *linaloes*. You do not suppose, as some do mistakenly, that *aspalto* is what we call *betume judaico*, for that is a cheap drug brought from the strait of Mecca, and sold here as pitch for ships.

Ruano

I also want to know this. Sepulveda, and the Friar who wrote *Modus faciendi* and the author of *De proprietatibus rerum* says that it comes from the terrestrial paradise.

Orta

I have never deserved to go to the terrestrial paradise. But I may say that the rivers which are said to come from the terrestrial paradise are far enough from where it grows, and that these authors have no excuse for repeating such a fable.

Ruano

A modern author is surprised that it should go to the bottom, when it is so light.

ORTA

He is wrong, for the centre or core is quite light, and some of it floats on the water, being very good, and for this I will cite other experiences and better proofs before we have done.

RUANO

The Italian Friars say that Avicenna is wrong in making two chapters for this drug when there ought to be only one, and they reprehend Savanerola for making a difference between *xilaloes* and *linaloes*, there being really no difference, but one word being Greek and the other Latin. Savanerola does that in his fifth treatise, *de lignis*.

ORTA

I do not want Friars as reprehenders except in the pulpit. I say that Avicenna makes two chapters, as of a thing that is not well known, and the second is more copious than the first. To the first he gives the letter *alif* of the Arabic alphabet, and to the last he gives another letter H(*ayn*), whence it may be deduced that when he got more information he wrote a fuller chapter. With regard to the fault found with Savanerola it should not be supposed that, though he was doubtful, he did not know that *xilaloes* and *linaloes* were the same. What he wrote was not from bad writing or forgetfulness, as often happens, for such a man would not err in a matter of importance.

RUANO

Sinforino says that he had never seen the true *linaloes* anywhere but at Liam in France.

ORTA

Come here, and pay well, and take it. To tell the truth the very good kind (which is called *calambac*) is worth more in

Malacca than here. Much comes from Malacca to this country, and is taken to Arabia, Turkey, Persia, and is used by the Moors and Gentios here. I will choose a sample for you to take to Castille, from the house of a merchant who has 10 quintals here, he to whom I gave the *benjui de bonina* the other day. When you want to go you will find the ships of Malacca at Cochin, and you can choose and buy them as you please.

RUANO

I kiss the hand of your worship for what you intend to give me so graciously, and for the advice you have offered me. The Pandetario says that the best LINALOES is on the higher slopes of the mountains, either owing to the force of the winds, or the age to which the trees attain, or the long time since they have come from the river below. It is also said that he falsified the plant, putting *camalea* in its place. So says Acacio Felici, that one of his species is from an island called in India Camar, and another Caemer, and a third Sares. I do not wish to repeat what he says about boiling it, because you have said that it is false.

ORTA

They do not grow on mountains, but in valleys; and they do not come by the rivers, or very few, for the best grow where they are and are brought by the river for sale, and sometimes by land, but the route by the river is most used, being the safest. For the tigers are very numerous in the woods, and are called *reimões*. The land where they grow may be Sumatra or Ceylon as I have said; and the *camalea* is a false attribution, saving your reverence, for it is not in that land.

RUANO

Valerio Probo,[1] a diligent writer, who has written some

[1] Or Cordo. "The errors of the press in the first edition are very numerous in this Colloquy" (Portuguese Editor).

additions to Dioscorides, says that the worst and vilest kind is of Rhodes. Have you seen it?

ORTA

No, as you may see, nor are included in this genus the *agaloco* or *linaloes*. It is well to know what is correct.

RUANO

Ruelio and Antonio Musa are writers, who being more modern and being persons having some knowledge of Portuguese navigation, should be more accurate.

ORTA

I well believe that they are more truthful, but I shall say what I say, and I say what I know.

RUANO

Well Ruelio says that the *agaloco* is in our land, and that the first Grecian who called it *linaloes* was Aecio, I do not know why, and the whole company of physicians and druggists followed him. He says that Galen does not mention it, and that Paulo calls it *lenho Indico.* Aecio says that in the east it is principally found in the parts nearest the sun, also that it never has the scent until it has first been powdered with certain insects, then buried in the earth and pounded, and with rottenness the scent is attained. He adds that the natives break it in pieces and bury it, and when the scent comes they sell it to the merchants. There are four kinds, namely (*Indico* the best known and most celebrated of all), the *Safico* (after a town called Safo), *Hispcon*, and *Higron.* The Arabs say that its fruit is a red berry, smaller than pepper, which is called *piperela*, and the *linaloes* is not subject to rottenness.

ORTA

In this he speaks correctly, and in saying that it is not known up to the present time he is right. But I say that it is now known in this land, and from this time it will be better known whence it comes,[1] and I will tell you how the scent is produced. Of the four kinds I only know one, namely that of India, and it may be that the other kinds are not really *linaloes*, but only scented wood. I do not know and have not heard what fruit it bears, but the statement that it is not subject to becoming rotten cannot be entirely true. This plant powders very slowly in the pith (or core as we call it), so that in those parts it is less subject to decay. It is wrong to say that to make it into powder it is eaten by insects and that the scent then comes out. Afterwards he says that it is not subject to decay. From this it is to be understood that the core does not decay, and the second and other part next the bark does decay.

RUANO

Antonio Musa says many things in agreement with your views. He further says that your Portuguese find it in the forests, cut down the trees, and leave there until it decays, and that they are worth much in Sumatra whence they come, and that in spite of this much good *linaloes* does not come here, and when it is on the open hand the scent is strong, and that this is the kind called *calambac* taken to China and Canton, a very celebrated port, and that it is long before it burns when put in the fire.

ORTA

For the most part what he says is very true, but there are not a great many trees in the forests, for they are not numerous and it is dangerous to go there on account of the tigers. I

[1] Count Ficalho thinks this statement of Orta rather audacious, for he did not himself know whence it came any more than the others.

must tell you that a great deal of the good kind comes to our hands through the captains of Malacca who navigate in those parts, if they wish to buy well. Here I will show you some of it.

RUANO

Now I would know the names in the different languages, and the appearance of the tree.[1]

ORTA

They call it AGALUGEM and HAUD in Arabic and the Guzeratis, and Deccanis UD which is nearly the Arabic. In Malay GARRO, and the very fine kind CALAMBAC. The tree is like the olive, sometimes much larger. I do not know the fruit, as I have told you, nor the flower, for I have not seen them, owing to the difficulty there is in procuring them. It would suffice if I should see the leaves and branches brought from Malacca coming from higher up the river. They say that when it is cut there is no scent. But it is not necessary that it should decay before the scent comes out, as some say. You must know that the bark is thick, and that the scent is not there but in the core or pith, as it is called in the cork tree. When it decays outside there is a concentration on the pith and the grease of the plant and the scent becomes very strong, all the virtue centering on the inside. Owing to this, I say that it is necessary to decay first to produce the scent. When the plant is very dry the central part is better and the scent stronger because there is nothing to work against it. Persons who smell the plant presently say that it has a strong scent. For throughout the plant there is a difference in the wood. The Malays clean the plant, which is a drawback when they bring it for sale.

[1] The names by which Eagle-wood, or "Aloes-wood," is known in India are:—Sanskrit, *agaru*, meaning "without weight"; Arabic, *agalugi*; Persian, *agalurchi*; Hindustani, *agar*; Telugu, *krishna-agaru*; and Malay, *kalamba*. See page 251.

RUANO

Is the CALAMBAC of the same species?

ORTA

All is one. The finest is called CALAMBAC and the others as I have said.

RUANO

How is the best kind known?

ORTA

That which is very black with grey veins, and heavy with much oil in it, is the best. As a test put it in the fire, and that which sweats most and has most grease is the best. Besides these points the Guzeratis and Deccanis require that it should be large, like a gem or a pearl, for they say that the largest have most virtue. That you may try this work I have taken this piece of CALAMBAC which they do not value much.

RUANO

Certainly I have not heard so much in all my life, and it surely seems to me to be good, and I well know that I owe much to you for it, and many thanks.

ORTA

What I wish you to know is that it has given me much trouble and some cost to acquire the little I know of this drug, and it gives me much satisfaction to know that you appreciate it.

THIRTY-FIRST COLLOQUY

CATE[1]

RUANO

COMPLAINING of relaxation and softening of the gums, your cook told me that I should eat (chew?) BETRE, ARECA, and CATE well mixed together; she added that CATE was the best of all. She provided it, and I found it to be styptic and rather bitter. She also said that she knew very well that drinking water should be taken after it. This she brought, and I know not anything as good as that, but with a bitter taste. Then she told me that, with a mixture of BETRE and ARECA, she knew it would be very good. Certainly it seemed to me to be a very good medicine to dry and open. We know where it is, what it is called, and how it works. But I want to be informed respecting the appearance of the tree, for what the medicines are used in this land, and some of the writers who refer to it.

ORTA

This plant is most abundant in the territories of Baçaim, Manora, and Damam, cities of our King. It is also in the

[1] *Areca Catechu.*

mainland of Goa and in many other parts, but not in such quantity as in the places I mentioned first, whence it is taken to Malacca and China in large quantities, as well as to Arabia, Persia and Khorasan. As a medicine the demand is not large, but a great quantity is used in China and Malacca to chew with betel. They all call it CATE, in Malacca CATO with some variations. The Arabs, Persians, and all the people of India call it CATE. In its own proper country they call it UPLOT, also PUCHO, owing to the great trade in it to Malacca, where it is so called. The tree which yields this CATE is the size of an ash tree, and the leaves are small like those of the gorse (or *jounas*, as they call it in Portugal). We may also compare it to the tamarisk, for it has many spines and leaves all the year round, and the branches are very strong and heavy. It never decays, according to the natives, neither from the sun nor from damp, and among themselves they call it "the plant that is always living." It endures blows, and for this and because of its weight, they make certain sticks of it, with which they beat the husks from the rice. They call them *piloes*. They tread it into a very large stick made like a pestel, and this stick, which they put inside to stamp, is made like the handle of a pestel, and 6 *palmos* in size. They call this tree in its native habitat HACCHIC. As I do not know the language of this country so well as I know Portuguese, I do not know the reason that they call it CATE. Well, this is enough on the subject.

RUANO

All you have said explains things very well, but I want to know about the flower and fruit.

ORTA

It has flowers, but they tell me there is no fruit.

THIRTY-FIRST COLLOQUY

RUANO

Tell me how these lozenges are manufactured and the shape they are given.

ORTA

They cut very small twigs from the tree, boil them and pound them into shapes like lozenges or wafers, or like the flour of NACHANI,[1] which is a small black seed with which they make bread like rye. With that flour and the ashes of a black stick found in that land, or without them, they make slices and dry them in the shade, that the sun may not draw out their virtue. These people all use it, and also the Chinese, being so discreet and well informed, affirm that it is a very good medicine. I have tried it for dysentery and for pain in the eyes and found it very good. As to what is known of mention made of some writers, I will tell you a thing which I believe to be perfectly true, if you will give me leave.

RUANO

I shall be much obliged to you.

ORTA

I say that CATE[2] is what Galen and Pliny and Dioscorides, Avicenna and Rasis call LICIUM, because it was first found in Lycia, a province of Turkey, or because the best kind was found there in former days. The Arabians, Avicenna and Serapiam called it HACDADH.

RUANO

How is it that you say it is the LICIUM? Is it not called

[1] *Eleusine Coracana*, Gært., a dark-coloured millet, the *korakan* of Ceylon, the *ragi* and *nachni* of the Bombay Presidency; these names being corruptions respectively of the Sanskrit *rajika*, and *natchini*.

[2] *Berberis Lycium*, Boyle, and other species of *Berberis*, the λύκιον Ἰνδικόν of the Greeks; the watery extract from the wood of which is famous all over India under the name of *rusut*. The Hindustani name of the wood of these trees is *chitra*, the Sanskrit *daruharidra*, *i.e.* "wood turmeric," the Persian *pilzahra*, the Arabic *ambarbaris*, and the Chinese *kanki*.

CATE by the Arabian writers, and this being so why do you call CATE the LICIUM?

ORTA

I call it so because all the modern and ancient writers, Greeks and Latins and Arabs and Indians, all prefer the LICIUM of India to all others; still more because it is this with which is taught all that here is known; still more because all the things for which LICIUM served are used in this country for CATE; still more because it has the conditions for making good LICIUM. It serves for weakness of the eyes, strengthens the gums and teeth, kills parasitical creatures, and is good for the throat, for rheumatism and for dysentery. You ask why the Arabs do not call it CATE. To this I reply that many things lose their correct names when transferred to an alien language. Now it may be that, in speaking with Arabian physicians, they will give me another name that is used in their language. You say that all speak of this simple. Well, Galen says that it is a thorny tree and that the best comes from India; also that there is plenty in Lycia and Cappadocia, and that their virtue is astringent and dry. Others contradict this. Pliny gives the advantage to India and says that it comes in leathern bottles of rhinoceros and camel hide, and tells how they are made, to which all agree. Why, then, do you ask how it is made? I will not talk of this any more.

RUANO

So as not to hinder so good a discourse I will not ask about the leathern bottles, but will remember what you have said.

ORTA

I will answer everything. I say that Dioscorides praised the Indian kind most, and described the tree, which did not differ from the CATE, at least only a little, for he says it is a small tree when it is a large one. He says it is like box and that most of it grows in Lycia and Cappadocia, and when he speaks of its uses he says the same as the others—that it has the virtue of opening and soothing. Referring to the chapter in Dioscorides, it will be seen that the Indians use it in the manner that he describes. Avicenna calls it HACDADH. He says that the kind from India is stronger and better than that which comes from Mecca, but I believe that the kind from Mecca is the same that comes from India. He adds that when it cannot be had, they put in its place ARECA and SANDAL.

RUANO

As to this your cook says one better, that she puts in BETELE and ARECA and the same CATE.

ORTA

These they had for use from the beginning of the peopling of this land. Rasis says that it is made of the juice of *berberis* made very thick for boiling, and Serapiam says the same, calling it HACDADH.

RUANO

Do the modern writers agree with this?

ORTA

Sepulveda says that they make it of the juice of honeysuckle and Valerio Probo says the same. Laguna says that we have not got the true LICIUM. Antonio Musa also says we have not got it, but, from the indications of Dioscorides, it seems to be the box tree. The Friars desired much to find the true LICIUM, for FELUZALANGE, which, by the advice of Avicenna, may be put in its place and is the tree of LICIUM according to the translation of Belunense, is also wanting. To put in its place the FAUFEL, which is *areca* and *sandal*, according to Avicenna, is more difficult, for Avicenna says it is not well known what it is.

RUANO

There is much red *sandal* in Portugal, and cannot they bring plenty of *areca*?

ORTA

Yes. But the Portuguese druggists show little diligence in obtaining medicines and much in making money. If they would apply to the India House they would get it in abundance.

RUANO

So that if we cannot find our LICIUM we must profit by the other Indian one, and be careful to call it LICIUM, since the best is in Lycia and Cappadocia, which ought, it would seem, to be preferred to all others.

ORTA

I do not say that Indian LICIUM should take the place of Lycian LICIUM, but I maintain that when the Indian kind fails, that of Lycia should be given, for this is the view of all the writers. When that fails the medicine made of the juice of berberis or honeysuckle may be used, or of styptic wild plums. As to what you say that the kind of Lycia should *par excellence* be called LICIUM, this is not so, if you will pardon me for contradicting you. The use of this medicine was first adopted in this country and was the Indian kind, and for want or defect of that of India it had to be rejected; and this is the truth and the other view is not; for in no other region is the CATE so much used as in India.

RUANO

I will take it from this country and will use it, for it has the effects which you describe. It will be well that you should tell me if there are not many leathern bottles of camel and rhinoceros hide in this land, as Pliny says that they carry it, that we may see the quantity of it by this leather.

THIRTY-FIRST COLLOQUY

ORTA

I have not seen leathern bottles of camel skin in this country. There are some camels in the Deccan and Guzerat which take the kings and captains to war, but not so many camels die as horses, and of skins of the latter leather bottles are made. As for rhinoceroses, which the Indians call GANDAS, they are not domesticated in our land. It may be that there are wild ones in Bengal or in Patane, and in the lands of the Patane some are made tame. I have never seen a rhinoceros, but I know that in Bengal they use the horn against poison, taking care that it is the unicorn. But it has not got one, according to the testimony of those who well know. For the Nizamoxa would weigh two hundred times of gold against a little proved unicorn, and much more for a rhinoceros.[1] I know that in 1512 one was presented by the King of Cambay to the King Dom Manoel, now in glory, who sent it to the Pope. This animal is mentioned by Pliny (lib. 8, cap. 20), and by Strabo.

RUANO

From what you say it seems to me that there is no unicorn in India, for you do not speak of any, and the King, your friend, has not got one, a reason for thinking there are none here, and we do not know where this animal lives.

ORTA

There are so many uncertain things said about this animal, from its not being well known, that it is not worth while to recount them, for the persons who relate them are not eye-witnesses. But I will tell you what I heard from a person of authority. He told me that between the Cape of Good Hope and Cape Corrientes he saw some animals which, though they were disporting themselves near the sea, were terrestrial. The

[1] This is not clear. Count Ficalho thinks the meaning may be that he would buy the horn of a rhinoceros if he judged it to be a unicorn.

shape of the head was like that of a horse, yet it was not a sea-horse. It had a horn which it used, raising and lowering, and turning to right and left, in a way that made it like a finger, and this animal fought bravely like an elephant, but with its horn, which measures 2 *palmos*, and it is said to be an antidote against poison, this being the common fame.

Ruano

They say that these animals do not want to drink if they can put their horn into the water.

Orta

Not only do they say that it is a good drink against poison, according to general belief, but persons worthy of credit say that arsenic was given to two dogs, one receiving a double dose. To the one with the double dose, the scraping of the horn was given him to drink. This one lived; the other that had taken the smaller dose of arsenic died. I know nothing more of this animal, but I saw some of its horns, and how they are fastened to the forehead. I pray to God that we may come to know more, and that what we discover may be for His service; and that what I write may be short rather than long, for I would only record what would be most useful.

THIRTY-SECOND COLLOQUY

MACE AND NUTMEG[1]

RUANO

WE know the names of mace and nutmeg, the country whence they come, namely Banda, and also the names in Arabic and Latin. According to the order, as we cannot first speak of the nut, which is fruit, we will talk of the appearance of the tree, of the leaves and flowers, and of any medicinal qualities connected with them.

ORTA

These names I can tell you. I asked for them in Malay, Malayalim, Persian, Arabic, Turkish; but for you is only necessary the Arabic and Latin, and that of the land where it grows, which is Malay. Well, I say that the tree of the nutmeg is the size of a pear tree, and the leaves are round like some pear leaves. Banda, the native place of the nutmeg, belongs to the King our Lord. Some writers on Peru say that it is in Maluco, but this is not so; for the tree there does not bear fruit, and it is the same in Ceylon. It is like a pear

[1] Mace, *jatri*, is the "aril," and nutmeg, *jaiphal*, the "albumen" or kernel of the fruit of the Myrtle-bloom, *Myristica officinalis*, Linn., the *M. fragrans* of Houtt.

tree, or, to be more exact, like a small peach tree. The rind is hard, the outer skin being harder than green pears. Removing the thick rind, there is a very fine rind like that which encircles our chesnuts. This goes round the nut. The nut is like a small gall nut. The delicate skin which encircles it is the mace. We need not refer further to the thick outside rind except that it is very good made into a conserve with sugar, and it has a pleasant scent. This conserve is very good for the brain and for nervous complaints. It comes from Banda in jars of vinegar, and some people eat it as a salad; but all that comes to this land is in the form of conserve with sugar, a very beautiful fruit, leaving an agreeable scent in the mouth. You must know that when the nutmeg begins to swell, it breaks the first rind, as our chesnuts burst their prickly covering, and the mace becomes very red, appearing like fine *gram*. It is the most beautiful sight in the world when the trees are loaded.[1] Sometimes the mace splits, and that is why the nutmeg sometimes comes without the mace. When the nutmeg is dried it does without the mace, which changes in colour from red to a pale orange. The mace is worth three times as much as the nutmeg, and this is the truth, well known to those who come from Banda. This Banda is very unhealthy. Many go there and few come back; yet people are always eager to go there because there is much profit.

RUANO

Galen knew this mace and nutmeg, and Dioscorides, and some other Greeks, and Pliny.

ORTA

Galen makes a chapter in Book VII. of the simples, and says it comes from India. As for the others, it appears to me that they did not know the mace for several reasons,

[1] "Few cultivated plants are more beautiful than nutmeg trees."—WALLACE.

though they have the word MAÇIR.[1] One is because the temperature is said to be between warm and cold, being warm and dry at the end of the second or between it and the third. And it is said that it is good for dysenteries and for those who have issues of blood, which is not a thing that Galen would have said or have known. Avenrrois says that it is one of the medicines that Galen did not know. Many modern writers believe that the MAÇIR of the Greeks and the MAÇA of the Arabs are different medicines. This is why Avicenna makes two chapters, 456 for MAÇA[2] and 694 for TALICIFAR, and he did this imitating the Greeks whom he always held in much respect and veneration, thinking that they could not be wrong. Yet Dioscorides says it is the bark of a root, not the rind of fruit. Pliny says he does not know the MAÇIR. Much less could these Greeks have known the MAÇA when they are silent about the nut, because the one could not be without the other. They knew so little of MAÇIR that Galen would trace it to India, and Dioscorides to Barbary, where it seems no one knew the MAÇA. About this there ought not to be any doubt.

RUANO

Well, there are not wanting modern writers who say that the CHRISOBALANUS[3] of which Galen writes is the nutmeg of the Arabs.

ORTA

They have not reason, and there are many things against them in the description both as regards colour and taste.

[1] *μάκερ.*

[2] Avicenna knew of the Mace and Nutmeg. He called the former *besbasah*, بسباسه. Its common Arabic name is *shajrati-jauzut-tib*, *i.e.* "tree of nut-fragrant." Avicenna's *besbasah* looks like some corruption of the Persian *burjaah*, "scented."

[3] The Chrysobalanos of Galen was the "Black" or "Chebulic" [Cabul] Myrobalan of India, the fruit of *Terminalia Chebulica*, Retz.

RUANO

Did not the Arabs know something of the mace and nutmeg?

ORTA

Yes, sir! Avicenna especially spoke very distinctly.

RUANO

Well, Serapio says that the Greeks knew these medicines.

ORTA

That was because he was afraid to say anything against the Greeks. Do not be surprised at that because even I, when in Spain, did not dare to say anything against Galen or against the Greeks. Yet when seen in the proper light, it is not strange that medicines should be known in one age and not in another, new things being constantly found. Certainly if the Greeks knew the *Pao de China* they deserve much praise, and it would be much celebrated by them. Avenrrois boldly says that there were certain medicines which the Greeks did not know.

RUANO

Now will you finish with these ancients, and give the names of nutmeg and mace?

ORTA

I am content to do so. In the country where it grows, the nutmeg is called PALA and the mace BUNAPALA. In the Deccan the nutmeg is JAPATRI and the mace JAIFOL. In Arabia and by Avicenna the nutmeg is JAUZIBAM, which means the nut of Banda, and Avicenna calls the mace BEFBASE, the derivation of which name no one can tell. These are the Arabian names most used by all, but many Moors, Arabs, Turks and Khorasanis use corrupt names which have been altered in the course of time. Then the corrupt words appeared in books, and Avenrrois, being a very good

Moor, called it GEOZA. Serapio also has many corrupt names. If these, being Moors, are wrong, it is not to be wondered at that Matheus Silvatico should err. Maça is so called from the similarity of the word to MAÇIR, because the Greeks paint it vermilion.

RUANO

If the nut in Arabic is called GEAUZI, ought not the Arabs to mention the nut of India, which is called COQUO?

ORTA

The Arabs make mention of all the nuts. The COQUO you allude to is called GEAUZIALINDI, and the nutmeg GEAUZI, no more. This is the reason that the city which we call Badajoz should be called GUADAL-GEAUZI, or the river of nuts, from which Badajoz is corrupted.[1]

RUANO

Who told you that, and how do you know it?

ORTA

A Jew who was in Portugal, born in Cairo. He brought to Portugal the news of Sultan Bahadur's death.[2] His name was Isaac of Cairo, a discreet man and learned in many languages. I asked him if Guadalupe meant "the river of wolves," as Lucio Siculo Marineo would have it. He answered me that it did not, but that the meaning was "the river of love." I said to him that a very learned man had written that

[1] Fray João de Sousa (*Vestigios da lengua arabica en Portugal*, 1789) has another derivation for Badajoz بال ل العيش, *belad el aisch*, "Land of Provisions." Count Ficalho thinks that both are wrong. Edrisi writes the name بطاليوس (*Batalios*).

[2] Isaac came from India, by way of Suez and Cairo in 1537, with news of the violent death of Sultan Bahadur (Correa, *Lendas*, 792-846).

this was the meaning. He replied that if it could all be in one language it would be better, rather than make one part in one language and the other in another. As Guadalupe is all in Arabic, the meaning is the "river of love."[1]

Ruano

Though this is not physic, I also rejoice to know more about it than about other things, to relieve the boredom.

Orta

Why do you not remain without it? I knew that an oil is made from the mace, which is a very good medicine for the nerves, and is much used for it in Banda.

[1] Sousa gives وا ل الغبي, *uad el ub*, "River of the Bosom."

Plate XIV.—Nuez Moscada.

Plate XV.—Macer.

THIRTY-THIRD COLLOQUY

Purgative Manna

Ruano

THINGS which are very necessary and much used ought to be well known. For this reason I want very much to know about the *manna* you use so much. For this it is my prayer that you will be very careful of what you tell me, that I may not have to reply. I do not want to know whether the Greeks may have written about it, nor what they call it, for these things have been sufficiently discussed by modern writers.

Orta

Certainly I found this medicine to be very good with a pleasant taste and smell, and that the effects we expected from it were also very good; but I have not obtained much information about it. I only know that it is brought in three ways to Ormuz from the province of the Uzbegs. The chief kind is the first, which is what you see in the druggists' shops in flasks, like sweetmeats, and with the taste of honeycomb. They call it Xirquest or Xircast, which means the milk of a tree called QUEST, for XIR is milk in the Persian language, so that it is the dew that falls from these trees. or the gum that exudes from them. We have corrupted the word to SIRACOST.[1] Avicenna belonged to that country of the Uzbegs, native of a city called Bocora,[2] as I have told you, so that it is natural he should know the name very well. This may be well seen in Avicenna, where he treats of it. The other kind is called TIRIAMJABIM or TRUMGIBIM, as the Belunense has it. They say that it is found among the thistles and in small pieces, somewhat of a red colour. It is said that they are obtained by shaking the thistles with a stick, and that they are larger than a coriander seed when dried, the colour, as I said, between red and vermilion. The vulgar hold that it is a fruit, but I believe that it is a gum or resin. They think this is more wholesome than the kind we have, and it is much used in Persia and Ormuz. For here we do not use it for children until they are past fourteen years, but I tell you that for all it causes purging without doing any harm whatever. Another kind comes in large pieces mixed with leaves. This is like that of Calabria, and is worth more money, coming by way of Baçora, a city of renown in Persia. Another kind is sometimes seen in Goa, liquid in leather bottles, which is like coagulated white honey. They sent this to me from Ormuz. for it corrupts quickly in our land, but the glass flasks preserve it. I do not know anything more about this medicine.

[1] Bokhara.

[2] The chief sources of the Manna used in Europe are the Ash-trees, *Fraxinus Ornus*, Linn., and *F. rotundifolia*, DC., of Calabria and Sicily and other countries of Southern Europe, and it is still known throughout India, as throughout Europe, by its Egyptian [*mēna* of the Malabar Coast] and Hebrew and Arabic [*mānā* of Hindustan and the Deccan] names, meaning "a gift given us." It is also known in India by the name of *shirkhist*, meaning "dried juice." The *turanjabin*, *i.e.* "Persian Manna," of the Indian bazaars is the product of the leguminous "Camel Thorn," *Alhagi maurorum*, Tourn., of the continuous desert region from Marocco to Rajputana; and *gazargabin*, of the Tamarisk, *Tamarix gallica*, Linn., of the same region, most of this variety of manna being imported into India from Arabia. Mannas are also produced by a great many other trees in hot and dry countries.

THIRTY-THIRD COLLOQUY

Ruano

Certainly I will listen to you for a long time if you will say more. Though you may not want to say more, at least tell me where this Uzbeg province is.

Orta

The province of Uzbeg is in Tartary, so called from a city called Tartar which is in it. The men who are natives of it are called Tartars and the rest are Uzbegs, as you may say Toledano or Spaniard, Lisbonez or Portuguese. These Uzbegs formed part of the Mogores, it being a short time since they were detached from them. These Uzbegs are very valiant men, and great archers both on foot and on horseback. They take pay from foreign kings. I knew one serving under Idalcam who was called Meliquetartar, and another with Sultan Bahadur, the great King of Cambay. These Uzbegs border on China according to what I am told, and possibly they are the Parthians so detested by the Romans. But I confess that I do not know the cosmography of the land very well.

Ruano

Tell me about that manna that comes by way of Baçora. I want to know whether this is Babilonia, originally so called, or Bagada, which is further off.

Orta

Both Baçora and Bagada[1] are in Mesopotamia, but it is not Babilonia, though it is vulgarly supposed that Bagada is Babilonia. It is, however, very certain that the true Babilonia is distant ten or twelve leagues from Bagada. It is much defaced and neglected by men. Jorge Gonçalves told me this, a discreet merchant, a great searcher after truth, and very well

[1] Bagdad.

instructed. He was informed by one who was a native of Babel itself; and he said that Babel was reached by the Euphrates, and that Bagada is on another river called Digilá and not Tigris, but that this name Tigris is used now.

RUANO

This Baçora and Bagada, whose are they now, the Turk's or the Xatamaz?[1]

ORTA

At first they belonged to other kings, but now to the Turk.

RUANO

What titles have the Xatamaz and the Turk?

ORTA

The Xatamaz is called Xa,[2] which means king *par excellence*, and all the other kings are called Paxa,[3] which means "foot of the king." Such are the King of Ormuz and the King of Lar and others. The Turk is called Honencar, which means "maker of Lords," for "*hon*" in Persian is a Lord, and "*ecar*" signifies to make, but his title is not conformable to this meaning, for it is very humble.

RUANO

How is he humble?

ORTA

Faquir Suleiman,[4] which means the poor Suleiman.

RUANO

Well, with all these humilities, he may go to hell.

[1] Shah Tahmasp of Persia, son of Ismail, founder of the Suffavean dynasty.
[2] Shah. [3] Pasha. [4] Soleiman II.

THIRTY-FOURTH COLLOQUY

MANGOES [1]

RUANO

THEY praise the mangoes here so much, when they are in season, that they say that they can compete with peaches. For myself it is necessary that I should know all the fruits of this land, and this not that I want it only for myself, but that you may be induced to write on the subject.

ORTA

The more that is said about the taste of this fruit, the more is asserted, but I do not wish to praise it except in order to describe it to you. There are some Portuguese who are so pertinacious that they would rather die than confess that there is any fruit equal to those of Portugal. Yet there are here many fruits which have the advantage, as are all the fruits of thorny trees; the lemons here are so large that they are like citrons, and very tender and well tasted, especially those of Baçaim, and the citrons are much better, the limes also much better, while the oranges exceed ours in every respect; still more those of Pegu, Martaban, Brinjam,[2] and

[1] *Mangifera indica*, Linn.

[2] A small port of Malabar near Cape Comorin.

Ceilam. I do not speak of China and many other places because they are outside India.

Ruano

I confess to you that all the fruits of thorns are better, though the degree of superiority is not very great, except the oranges which you praise so highly, when like those they brought you from Cochin. Those, I confess to you, have a great advantage over ours in softness and juiciness. Moreover, their inside rind which contains the bitterness does not cause bitterness to the fruit in the slightest degree, and on cutting an orange here one eats it with the greatest delight in the world. This is not so with our oranges because there is always the bitter rind. Not only are those here not bitter, but I also tried the outer part, and then it gave me no bitter taste, or very little. Of the other fruit of India which you praise, you have not good grapes, figs, pears, or pippins, or many other fruits we have in Spain, so that your fruits may be pronounced to be good for want of better. For myself I prefer, before the fruits of Portugal, your mangoes and your *durioes* which you praise so highly.

Orta

I do not praise the *durioes* except from hearsay, for I have not seen them. As for the mangoes I may tell you this. In Ormuz, at the time when they were selling grapes, figs, very good pomegranates, peaches, and apricots, the mangoes were not bought by reason of the very high price asked for them.

Ruano

This will be as you say. I shall rejoice to be here until January, to hear and believe.

THIRTY-FOURTH COLLOQUY

ORTA

The season[1] for them in temperate lands is April, and in other lands later, in May and June, and sometimes they come with a turn, as they call it in our land, in October and November. Boy, go and see what those two vessels are that are coming in. I see them from the verandah.

BOY

I will go at once with the message.

ORTA

The mangoes grow in the season I mentioned, and those of Ormuz are the most praised. Those of the kingdom of Guzerat are also very good, especially some they call GUZARATAS, which are not so very large, but with a splendid scent and taste. The stone is very small. Those of Balagate are large and very toothsome. I saw two that weighed two *arrates*[2] and a half. The parts of Balagate where the best grow are Chacana and Quindur, also Amadanager and Dultabado,[3] principal cities of Nizamoxa. They are also very good in Bengal, Pegu, and Malacca.

BOY

Sir, it is Simam Toscano, your tenant in Bombaim. He has brought this basket of mangoes to present to the Governor, and he says that when he has made the boat fast he will come here to rest.

ORTA

He comes most opportunely. I have a mango-tree in that island of mine which has two gatherings, one at this season, and another in the end of May. As other fruit may exceed

[1] Monçam, "Monsoon" of Anglo-Indians, from the Arabic *mausim*, literally "season," and also a mark, a station.

[2] *Arratel*, 1 lb.

[3] Ahmednagar and Dauletabad.

this in scent and taste, so much this exceeds others in coming out of season.[1] We will first prove this fruit. Boy, get out six mangoes.

BOY

There are twenty mangoes here, and six are damaged. You may take the six of little worth, and I will take the rest to his Excellency, who ought to be given the best.

ORTA

Give them here. They ought to be cut with a very sharp knife, that the slice may not be injured, and I want to taste them first, for allowance must be made on account of the season.

RUANO

If now allowance must be made, I say to you that at another time they will surpass all the fruits of Spain.

ORTA

Well, I want to show you how to eat it in another way. Boy, cut these mangoes, and prepare them in slices, because in that way they have a better taste, and the chief thing is to soak them in sweet-smelling wine, like nectarines.

RUANO

Truly these two, which have scales, are like skinned peaches, for their colour is between vermilion and clear green, and the smell is proper to them.

[1] Count Ficalho here footnotes:—"Dr. Birdwood, in the Bombay *Saturday Review* of 28th July 1866, mentions a famous mango-tree at Colaba which gave fruit twice in the year, at Christmas and in May." This was the Mango-tree in the garden of Mrs. Hough, who consulted every botanist who came to Bombay as to the cause of its fruiting at Christmas; and on her consulting me in 1857,—it was the 8th of December,—I told her it was obviously "the Benediction of Mary," but I ultimately found that the tree had some thirty or forty years before been blown during Christmastide, when it at once burst out into flower and fruit; which led to the habit in which it had ever since indulged.—GEO. B.

THIRTY-FOURTH COLLOQUY

BOY

Here they are.

ORTA

Try them both ways, with wine and without wine.

RUANO

With or without wine it is good in both ways, so good that I think it will be necessary for me to remain here all this year to try the others, and in the year that is coming. It does not seem to me that I can now leave with my brother.

ORTA

This is only talking. I pray that you will not write down what is passing here now, for I would not be thought guilty of such levity as to indulge in trivialities.

RUANO

I am not so strict a physician as you would make me, for I also pride myself on being a man of the court. Now tell me further in what ways it is the custom to eat the mango.

ORTA

In conserve of sugar, in conserve of vinegar, in oil and salt, stuffed inside with green ginger and garlic, salted, boiled—in all these ways I have seen them served, and you may try them in this house.

RUANO

Of what complexion [1] are they?

ORTA

Cold and damp. This is clear in conformity with the canons of the second of Avicenna and the sayings of Aristotle in the fourth of the METAUROS and in other places. As I keep remote from these scholastic matters, I will not give you more about it, except that they place them with peaches.

[1] "Humour."

They are at first pricking and styptic, afterwards acid, at last sweet, and the nearer to the stone the more acid, whence it appears that they are cold and damp.

RUANO

All these reasons appear to me to be good. But it is commonly said that they are warm, and some physicians in Portugal said so to me, and they say that those who eat them get boils. There may be some reason in this.

ORTA

I have had a discussion with some physicians, and I am not satisfied with this reason nor with others they gave. For the boils of this season come with the excessive heat there is then. So that the mangoes are not the cause of the boils, though they occur when they are in season. And it is subject to decomposition, to have these fruits heating the stomach, when they will cause fevers, colics, bleeding, or erysipelas, which are infirmities of heat. The same might happen with peaches, plums, cherries, melons, being also fruit in the class of cold and damp.

RUANO

Are the stones of any use in physic?

ORTA

No, except that, as I heard it said, when roasted they are good for the flux. I have tried them and they seemed to me to be efficacious, for they are like the acorns of cork trees, which in our country are called *landes*. They say that the stone, or rather the kernel of it, kills the worms when green, and they have reason, for it is bitter.

RUANO

If the fruits are good for all these things, it is not much

that, as you say, the Banians will not eat meat. And now there is this proposition. I would ask who are those Banians or Bramenes who, you say, do not eat meat. Are they what they call Genosophists, for these use the same clothes that the authors describe? And according to them there are, in many parts, those who come from Guzerat and the Deccan, who have not much to learn in Egypt and those parts, where they say that their doctrines prevail, for they say that they came to Arabia, Persia, and Egypt.

Orta

These are so, but now they care more for trade than for learning. There are many classes of them, but all agree that they must not kill, nor eat what has died. They maintain this precept to such an extent that they buy birds in order to let them fly away. They will not eat radishes, nor onions, nor garlic, nor spinach that are vermilion, on account of the colour. They give water with sugar to the ants, saying that they offer alms to the poor. They give water to the birds, and they come to drink every day. Many of those that die leave a certain portion to those persons who travel in desert places, and they give water to travellers. In Cambay I saw a hospital for birds, where they cured them, and removed their ailments; and there I saw parrots and many other birds cured. When they were cured they did not keep them longer in the house, but let them fly away. They do not drink wine, nor vinegar, nor *nimpa*,[1] nor orraca,[2] nor wine of raisins.

[1] Fermented liquor from the creeping palm, *Nipa fruticans*, Roxb., one of the sources of *tari*, "toddy," or palm wine, and of *arak*, "arrack" and "rack," distilled from palm wine. Our "Fool-rack," that is *phul-arak*, literally "flower of arak," means the "strongest *arak*." Our phrase "Gooseberry fool" refers to the similar Indian "dish" made with the "flowers," *phul*, of the Mango. Our word "punch," the strong drink, is simply the Hindustani *panch*, "five"; that is, the five ingredients of which it is composed, *arak*, lime-juice, spice, sugar, and water.

[2] Arrak.

RUANO

Do other Gentios of this India hold the doctrine of transmigration of souls?

ORTA

Yes, the Bramenes of Balagate and Cambay, and those of Malabar, and others of whom I have no certain notice. All these wash their bodies before they eat, and are more venerated than the Baneans. These serve the kings as treasurers, writers, collectors of rents and ambassadors.

RUANO

And have those who are called Bramenes here got the same customs?

ORTA

These, and those inhabiting the sea-coast, called the Concan, eat all kinds of meat except that of cows, and pork raised at home. All believe in the transmigration of souls, and, besides that, they have faith in many things worthy of much laughter, which I will not repeat, so as not to waste time. The Baneans fast much, and eat little at night, only sugar and water or milk. Some are very religious, and fast for twenty days without food, as I was told by a person well worthy of belief.

RUANO

Avicenna says that the Indian experimenters[1] state that they do not eat milk or fish because it causes leprosy. Tell me whether this is said by the physicians of this land, or any others that you know of.

ORTA

The Gentios, for the most part, eat milk with some fish, but I do not know whether they say that it does so much harm. The Indian physicians with whom I have conversed

[1] Esprementadores.

do not condemn that diet nor abuse it to such an extent. However, most of the Gentios eat fish with grease, whence I deduce that this saying of Avicenna was only what he had found written as common fame. It may have been said by some ancient physician who repeated these things as the sayings of learned men in India, to sell his work better, for any distant and unknown country was called India by the ancients.

Ruano

For you, as you have noted, this land where you live is India, called so both by the natives and by yourself, but how do you know it is right? For it does not seem to me to agree with what a writer on New Spain says, that the West Indies and Brazil are like the East Indies. Also Ethiopia was called India by the ancients. There must be another reason why this is called India, and also the other western one.

Orta

Your author knows that the Castilians like their affairs to be grand, and for this they fill the mouth with "Western Indies"; but not only are your lands not Indies, they were never known to the ancients, nor was Brazil. They may have liked to call them Indies as being distant and unknown. But this, our India, has been called so from the time of Alexander down to this day. Of this Alexander they have many histories, more than we have, and among them he is the most celebrated who is called Ezcader. The river Indo, from which is the name India, is not more distant from Goa than 200 leagues, and it is called by the natives Diul. If the men of that land, and those of Persia or Arabia, want to ask whether one is a Moor or Gentio, he uses these words, "Are you Moçalmam or Hindu?" If he is a Gentio he says he is "Hindu," if a Moor he says "Alhandulila,"[1] which means "thanks to God." For *moçal-*

[1] *Alhamdu lillah*, الحمد لله.

mam means safe. By that you will see what proud names the Muhammadans put to their things. It is common fame in Persia, Khorasan, Arabia, and Turkey, that this land is called Industam. Arabia Arabistam, Christendom Franguistam, for *istam* means a region, and *Indu* India.

RUANO

All this seems to me very good, except the word *Franguistam*. For I believed, like many others who come from there, that they called the Portuguese *Franges*, because *Franges* means bubos, and the name was given in contempt, as much as to say bubo-people, or leprous.

ORTA

Bubos are not called *Frangue* but *Fringui*, and they are not looked upon as infamous by the natives. For the disease originally came from Brazil and your so-called Indies. There are not wanting those among your historians who say that the Castilians brought it from the West Indies in 1493, a year before they went to Naples, to help in the war of Dom Fernando of Naples,[1] and that they gave it to many cortesan women, who gave it to the Italians, and for this reason it was called the Neapolitan disease, and seeing themselves made infamous by this name, they called it the French disease, and because there were many Castilians and Spaniards with it our Portuguese called it the Castilian itch. There is nothing more to be said about it.

RUANO

Why do they call the Portuguese in this country *Frangues*?

ORTA

I will tell you. They do not give that name only to the

[1] Fernando, King of Naples, son of Alfonso V., King of Aragon, but illegitimate. He was besieged in Naples by the French, and a Spanish army came to the rescue under Gonsalvo de Cordova.

Portuguese, but to all Christians from the West. The reason was that the first Christians known in Asia were French, so they called Christianity *Franguia.* They call it so in Ormuz and in all these lands, and all who dwell in their countries. I, when I came from Portugal, asked a Christian, who had been a Jew, being a Spaniard and resident in Cairo, how many Christians there were in Cairo in the time of the Soldan, and how many Jews. He told me so many thousand Christians, so many Arabs, so many Franks and Jews. Asking him what he meant by Franks, he said they were the Christians of Europe, and that *Franguia* was Christianity. And so I make an end of your questions.

Plate XVI.—Mangas.

THIRTY-FIFTH COLLOQUY

PEARLS AND CHANKA SHELLS

RUANO

ONE of the medicinal stones is the pearl. May it be called a stone or not? It is usual to call it so in medicine.

ORTA

It is called *perla* in Spanish, *perola* in Portuguese, *unio* in Latin. The larger pearl is in Latin *margarita*, in Arabic and in Persian *lulu*, in Malabar *mutu*, in other parts of India *moti*, in Portuguese and Spanish *aljofar*.

RUANO

What are these names derived from?

ORTA

Respecting the Latin, Spanish, and Portuguese words, I will presently give you answers, but as to the others you must pardon me, for I do not know. *Perla* and *perola* are from *prefero, preferes*, because of their pre-eminence they are preferred to all others of their kind. *Unio*, so called because it would be a marvel to find two equal in size and shape and being. *Aljofar*, because it means in Arabic "of Julfar," the principal place where they are found. For the best are from Julfar, a port in Arabia near the strait of Ormuz. The best are from the fishery of Barein, Catifa, Julfar, Camarao, and other ports on that coast. Because the place best known to us was Julfar it was given that name, taken from the port of Julfar.

RUANO

I am glad to know that derivation; but why do they call the good pearls *oriental*, perhaps because their colour is golden?

ORTA

No. It is because they come from the East, the strait of Ormuz being to the east with respect to our Europe.

RUANO

Are there any other kinds besides the *aljofar*?

ORTA

That is the best and largest, and it is also found near Cape Comorin and Ceylon. That fishery belongs to the King our Lord. It should yield much, for there is so much zeal for the faith among more than fifty thousand Christians who do the work there. This Christianity was the work of one man, no less virtuous than learned, named Miguel Vaz, who was Vicar-General of India. This conversion to Christianity was afterwards increased by Master Francisco,[1] theologist, who was a principal of this holy company jointly with Father Ignacio, whose virtues and sanctities, if they were written down, would make a large book. This Christianity is now fostered and

[1] Miguel Vaz was an intolerant and cruel fanatic. He initiated the missions near Cape Comorin in 1541, sending there Father Diogo de Borba. Francisco Xavier arrived in India with Martin Affonso de Sousa, and went, five months after his arrival, to the south missions in Tinnevelly and the island of Manaar. He baptized more than 40,000 pearl fishers wholesale. Orta calls him "Master Francisco."

encouraged by fathers and brothers of the company of Jesus, and is honoured by the martyrdom of some of the religious of this holy company. The ALJOFAR obtained from this fishery is more unequal, but amongst them there are some very good and also large. But, on the whole, they are not so large as those of Barein[1] and Julfar, nor do they fetch so high a price. There are also pearls in Borneo which, though very large, are not so good. They also come from China, but neither are they so good. Those which come from the lands and islands of your king, or from Europe, you know more about than I do. And that I may not be contradictory without clearly seeing good reason for it, I will not say that the writers of Peru are wrong when they say that there are green pearls, and many other things of the same kind.

RUANO

So many and such good ALJOFAR comes from these lands that my brother, the factor, brought some to sell here, and says that he doubled the outlay twice over. But I do not know, as you say, that there is trade in pearls with Portugal.

ORTA

All may be true. For the ALJOFAR that comes from here, and the PEROLAS are large and round, and in all respects perfect, and those that come from the Indies are scored, badly shaped and not round, and with dead water.

RUANO

And are they worth more or less than the good ones?

ORTA

Those badly shaped fetch a lower price here in India than in Spain, for in Spain, whether they are round or not round,

[1] Bahrein.

alive or dead, a good or a bad shape, makes a great difference. The PEROLA that has these good qualities is worth there ten and one or two here. Here it is not so near the Canaras as are the inhabitants of Bisnagar and their territories, where the perfect ones are worth ten and the imperfect of the same weight have a value of five or four. So that your brother can double the profit here, and taking ALJOFAR from India make his money there.

RUANO

This is good; but I am a doctor, and I want to know how they fish, and how the physicians use them here as medicine, whether they are pierced or not, naturally and without art, as some of our doctors write, saying, "Take *margaritas* pierced or not pierced." Please tell me, if in this it does not give you trouble, which are the best PEROLAS you have seen, and whence come the ALJOFARS used in medicine, and the price of an ounce of them.

ORTA

The shells are found when they fish for them in the proper season. Those which are highest have the largest pearls, and those in the lowest depths have smaller ones. They are put to dry and then opened. The pearl is found in the meat after it has been dried to some extent. They find in one shell sometimes many, at others few, according to what the shell has, and not only one, as some say, they sometimes find more than 200 seed pearls. Some say that the pearl is pierced naturally, but they would speak of grace and find fables that suit their palates. The Gentios use the pearl as a medicine to some extent, and the Moors very much, putting it into all their cordials just as we use it. The best of the shells that the pearl-fishers bring up are white and smooth, and are called by the natives CHERIPO. They make spoons of them and drinking-cups. There are also pearls in the oysters we eat but not so good.

The largest pearl found near Cape Comorin weighed 100 grains of wheat. I saw many much larger from Borneo, but not of such a good shape. Another weighed 160 grains of wheat, or 40 *quilates*, which is the same. The weight of 100 grains of wheat is 5 *quilates*, which they call a *calanja*, worth 1500 *cruzados*. As to higher prices I will not quote them, for it is better to be a philosopher than a merchant. The pearl-fishers sift the pearls in sieves of tin, and those which only pass through larger holes are worth much more. The merchants of this country have these sieves and fix the prices by them. This is a very subtle way of counting, and your brother will be glad to know, for the rules are very artificial. Pearls which are too small to bore are sold to the druggists or to be taken to Spain. They are worth less than a *ventem* an ounce.

RUANO

Does a pearl fall off in weight in the lapse of time? For they tell me so, and for that reason it is not good as treasure.

ORTA

Yes, it diminishes. I have not tried the experiment, but what is said, and what may be held as most certain, is that the ALJOFAR shell-fish in the waning of the moon is that which dies and the other not, and this may be taken as very certain.

RUANO

If this pearl had not been made so clean and polished as we make it, would it yet have cleanness and life and polish? Tell me this if you know, for you are not such a philosopher as you make out, and you also want the pearls and stones, like the rest of us.

ORTA

Yes I know, and will tell you. Take rice slightly mashed with salt and rub the pearl well with it, and you will find it as clean as the best in the world.

RUANO

And the other kind of which they make things which we call MADRE-PEROLA and CHERIPO?

ORTA

There is another they call CHANQUO,[1] of which they make boxes, tables, and counters, for though it is rough outside, inside it is very smooth and beautiful. This CHANK is an article of trade to Bengal, and used to be worth more than it is now. The large ones, which we call BUZIOS, go to Bengal and are worked up very beautifully, remaining very smooth and white. For this only a small quantity is used, the rest being wanted for bracelets and other ornaments. It was the custom in Bengal that no person of distinction who was a virgin could be corrupted if she had bracelets of the CHANK shell on her arms. After the arrival of the Patans this custom was neglected, and the CHANK became cheaper in consequence. You see here a chess table and board at your service, where you may see the Chank when you like.

RUANO

Very many thanks, but tell me what the black squares are.

ORTA

Tortoise shell. They make many other fine things of tortoise shell, but I do not talk about them because they are not medicine. For we have now been talking about many things which have nothing to do with our subject.

[1] *Turbinella rapa*, the "Chank" or "Chunk" shell of commerce, the Sanskrit *shankha*, "soothing," and so called from the muffled murmurings it makes when held to the ear, as of tranced seas at blissful rest along far distant shores. It is a most sacred object; the emblem of Vishnu and other gods of the Hindus, the trumpet of their epic heroes, and the ensign of the State of Travancore. It is also cut into prophylactic jewelry. A *sankha* with its spiral opening to the right is among Hindus priceless.

THIRTY-SIXTH COLLOQUY

MUNGO

Melon of India and Water Melon

RUANO

ALL things pall upon one, how luscious soever they may be, when one eats too much, and so it happens with me, in talking of medicinal simples, when you talk much of them to me, although there may be things to note. For this reason it is well that there should be something to excite the appetite like capers or olives, and I found the mangoes so delicious that I should be pleased to discuss some other fruit of India.

ORTA

I can give you melons[1] of India or water melons to eat.

RUANO

Not one of those melons that I saw in the house for they

[1] *Citrullus vulgaris*, Schrad., the *tarbuza* of the Persians, the *kulungari* of Southern India, the *batich* of the Arabs and *abatich* of the Israelites [Numbers xi. 5], and *pateca* of the Portuguese.

deceived me. I smelt the finest melon in the world, but when I came to try it the taste was like mud, and it was one of your buyers who took me in. I asked her if it was good and she said yes, and as I had seen cucumbers here like ours in Portugal I thought the melons were also like ours.

ORTA

She told you according to her taste, or as a person who had not eaten melons in Europe. But I must inform you that in Diu there are melons very good to eat, guaranteed to have the taste and smell of those in Portugal. So there are in many parts of Balagate, and those in Ormuz are as good as those in Spain. But they are not the melons that I want to talk to you about. The others are what the Portuguese call *patecas*, which are large and round, or rather it is better to say oval. It is not eaten, cut in slices, as we eat a melon, but cut lengthways. When ripe the seeds are black, when green they are white. Though they are not sweet like our melons, they are suave, very cooling, humid, and very good for fevers and affections of the liver and kidneys, as we know from experience. They promote the urine very much, and the people in good health are accustomed to eat them four hours after dinner, which is the most heating time. To me it seems better to begin with them at dinner time. The seeds of these melons are provocative of sleep, and are the best cooling seeds[1] we have here, so that we do not need others. With regard to this kind of melon it cannot be doubted that its complexion is cold and humid, while as regards our melons being abstergent, cleaning, and sweet, it is doubtful whether they are to be classed as cold. As, however, you have them in sight, sit down to eat and try these melons which we call PATECA. Bring that melon or *pateca*.

[1] The four great cooling seeds of ancient Pharmacy were those of the melon, cucumber, water melon, and pumpkin.—FICALHO.

THIRTY-SIXTH COLLOQUY

SERVANT GIRL

That melon, sir, is not fit to eat. But here are patecas which come from Chaul, and other better ones from Dabul. Here they are.

ORTA

Throw the seeds aside and try what a good thing a *pateca* is.

RUANO

It is one of the best fruits I have seen in all my life; and at certain seasons I should prefer it to our melons; for it is not more than an alterative, in great part it seems to aid the urine, and some part for the dysentery. It does not seem a thing liable to turn bad, as happens to our melons, cucumbers, and *cogombros*,[1] and I will take these seeds with me to sow in Spain. Tell me its name in all the languages, and why it is called *pateca*.

ORTA

According to the Arabs and Persians this fruit was brought to their countries from India, and for that reason they call it BATIEC INDI, which means "melon of India," and Avicenna so calls it in many places. BATIEC only means a melon, and the name of the Indian land is CALANGARI.

RUANO

Who told you it was called BATIEC INDI? Was it by chance mentioned by some Arabic writer?

ORTA

The name is common, and is used by the physicians who know the Arabic language. Serapio wrote another name, it being so called in his country, and must have been a corrupt form. Avicenna clearly has BATIEC INDI in the fourth book, and the chapter on pure tertian fever. He has great praise

[1] A smaller Cucurbitaceous plant.

for it, as you know better than I do, though I have had more practice than you have. If it pleases God that you should return to Spain and should there sow it, you must remember what an excellent thing it is for coleric fevers, and for many other infirmities.

RUANO

I heard it said that there are in many parts of Spain very fine melons which they call BUDIECAS, which may be these PATECAS with the name corrupted, calling them BUDIECAS for PATECAS.

ORTA

I saw those melons in some parts of Castille which they call BUDIECAS, and others they call SANDIAS, but it is a fruit very different from this one, and cannot be considered as the same species as that called BATIEC INDI. The PATECA has not the same leaves as melons, and very different from the BUDIECA, which grows higher and not extended along the ground like the PATECA. They tell me that in Africa they have the same kind as in India, which may well be, but I only put faith in what I see.

RUANO

You, when you tell me that this is not medicinal, nevertheless find curative properties in it, and you tell me things that I esteem as effecting cures. Do the physicians of this land know about this melon of India?

ORTA

No one knows it except those to whom I have told it, not because there are not many learned men here, but because they do not appreciate a thing so lowly. Yet I have asked some great physicians, both Arabs and Gentios.

RUANO

And how did you know the way to ask this of the Arabs?

THIRTY-SIXTH COLLOQUY

ORTA

Because I know all the infirmities of the third and fourth of Avicenna, and all the simples of the second in Arabia. This was very useful for me, in curing that King who was my friend, and his sons, work which at first was troublesome for me. I profited by this in the liking the King had for me. He taught me the names of illnesses and medicines in Arabic, and I taught him the same in Latin, which pleased him very much. For this reason I was also taught by the Arabian and Khorasani physicians in his employment.

RUANO

And do the Gentios understand also?

ORTA

Very well, for they are men who cure according to experience and custom. It is so good to them to deceive the Portuguese that they are easily deceived by them, and the worst of it is that some Portuguese, either to content the people, or to clear themselves from the cure of their illness, do not care about working to try cures, but go according to their idea of them. In order to make things pleasant for the people they practise in the way to make most money, and on that principle use their medicines.

RUANO

Do they use ours?

ORTA

Very often. But most of them not correctly. For they say there is bleeding, and they never bled before we were in the land; but they used cupping-glasses, sawing,[1] and leeches. According to what I knew from the physicians of Sultan Bahadur and Nizamoxa, they were never accustomed to look at

[1] Çarrafar.

waters, except by watching what we do, and copying like apes. Whence it happens that when they see the urine white they take it for a good sign, and when it is red and thick, they look upon it as bad. These and many other things I had from their own lips; and why not from those who know so little should there not be something good? I can tell you that they cure the dysentery well, can tell whether there is fever or not from the pulse, and whether it is weak or strong, and what is the humour that offends, whether it is blood or heat or phlegm, or melancholy; and they give a good remedy for obstruction.

Ruano

Do they give syrups or distilled waters, and is the custom ancient among them?

Orta

They do not use them in Balagate, but I know those here who practise, and they say that they give violet syrup, *lambedor*[1] water, against flux, plantain or barley, cordial slices, sugared rose and almond water, but none of these things was it the custom to use before we came. I only know that in Balagate the Moors and Gentios used the seeds of endive pounded, and drunk with fresh water, for all kinds of fevers. They did not distil water before we came, but they had the custom of giving boiled water of legumes and seeds to drink, and juice of herbs roughly prepared. They walk along one street curing all out of one flask that they carry.

Ruano

Galen does not come here, who cured at Thessaly; but according to what your words show, you are on bad terms with these people, and I fear they may give you poison.

[1] "Lambedor," another kind of syrup.

THIRTY-SIXTH COLLOQUY

ORTA

Before all things I am well with myself for they say that I am not very covetous, or to speak more correctly and without prejudice, I try to cure them and to effect as many cures as I undertake; but I first ask what the patient would like to do, and if it is the medicine that I know is good and will do no harm, I say that he should use it if he wishes to be cured by it, and if it is bad to refuse it. If I do not know whether the medicine is good or bad (as often happens) I also refuse it. These doctors are also wrong in the classification of the medicines, for they call pepper and cardamon cold and opium hot. As for anatomy they do not know where the liver is, nor the spleen, nor anything else.

RUANO

Do you confess to me that you take any of their things?

ORTA

Yes, many, but first I try the medicines of my doctors, and when they do not approve themselves to me I take them to the Brahmines of this land.

RUANO

That girl who you brought from the Deccan asked me for MUNGO, and said that in her country it was given to eat, with the rind torn off and boiled. Is it given in that way?

ORTA

They give it to eat in that way if it is wanted, but it is better boiled with bread and chicken, for she is from a country where they eat bread and not rice; that is Balagate where they have little and in small *measure*.

RUANO

Have they wheat in Balagate or Cambaya?

ORTA

Much; but they do not till the land with manure and labour as we do. They sow on the face of the earth after very little tillage in November, and in the middle of January they gather much and very good, sometimes without there having been any rain at all, merely the dew and richness of the soil, which is very good for this crop.

RUANO

And what medicine is it that this girl speaks of?

ORTA

It is a green seed which becomes black when very ripe, of the size of a dry coriander seed. Horses eat it and sometimes people. The Guzeratis and Deccanis use it in fevers. All men who have fevers do not eat for ten and sometimes fifteen days. At the end of that time they are given boiled water of mango, in which there is some little sustenance. Afterwards they give it to eat, with the rind torn off, and boiled with rice. They do not give wheaten bread until after many days. I will relate to you what once happened to me. Travelling with the Sultan Bahadur, in company with Dom Martim Affonso de Sousa, the latter was taken ill with a fever. The King sent for me and asked me how I would cure Dom Martim. I told him that I should have to bleed the patient, to give him syrup made of lime juice, pomegranates, and sugar, and purge him with a little rhubarb and manna I had brought with me, for other medicines they had none in the camp. He replied that the Portuguese did not know so well as the Guzeratis how to cure fever; for that the Guzeratis did not cure in any other way than to give nothing to eat. In order to avoid contradicting him I said that he was quite right, and that my patient had had nothing to eat for three days, but that now I was going to give him the syrup, and put him on some special diet. He replied that four days were very few, and that he ought to be without food for at least twenty days. He confessed to me that the Portuguese were very good physicians for other infirmities, but that they did not know so much about fevers as the Guzeratis. I did not want to contradict him as he was the greatest king in Mourama. Further, as their physicians are not learned, there are none that cure according to our rules. Some years afterwards I found myself in Cambaya, the principal city of Guzerat, where a very rich Moor resided, a native of Tripoli in Barbary, who could talk Portuguese. He called me in to cure his son of fever, after he had been ill for four days. I cured him, first giving him chicken to eat, for he had eaten nothing whatever for four days. Afterwards I bled him; and without purging, he was free of fever. The father told me that the treatment of the Guzeratis for fever was what the King stated. I answered that a shoemaker did not fit every one with the same pair of shoes: that the cure was for the Gentios who eat nothing with blood. But for his son and for rich merchants who were accustomed to eat much meat and drink wine, a different treatment was necessary when they had fever. What I said seemed to him good, and that my treatment was better. From that time, during the time I was there, all the Moors wanted to be cured by me.

Ruano

I pray you of your kindness to tell me how you would cure the Nizamoxa your friend. Would the cure vary much from our way, and contain anything that you chanced upon with him and seemed appropriate? for these things dispel in great measure the errors which occur in the work of curing.

Orta

He had the wish to be cured in our way, but the custom of the country is much against it and it had to be withdrawn, especially because the educated physicians at his court liked to gratify the natives and to oppose me. So that when I was present they cured in one way, and when I was absent in another. If it does not bore you I will relate what happened in curing the hereditary prince, then a man of the age of thirty, strong and with a good constitution and appetite. He had learnt the Portuguese language from me. He asked me in Portuguese what would be the best cure. I said that he should be bled and then take some rhubarb with whey. He said he was content to be bled, because in the last few days he had been bleeding through the nose. Wishing to do it, I was hindered by a physician of his who was the lord of many lands. Though both the father and the son were my friends, they decided to do what the other physician ordered. For they said that he was much exhausted by women, the patient having told me the opposite, and this was done owing to the envy of the doctors. At the end of fifteen days the same man fell ill of a fever, and my vote was that he should be bled. But the physicians and the father would not consent for the reason already given. They further said that what he had was small-pox, which is a very dangerous illness in that land. I said that he had not got the symptoms of small-pox, but that if he should have them it was better to bleed him during the three first days in conformity with the rule of Avicenna, and to give him some drink made of tamarinds. They replied that it was true that Avicenna said what I had repeated, but that the custom of their country was different, also that the Deccanis held tamarinds to be bad for small-pox, so that neither quotation nor reason would prevail with them, at which the father and still more the son were grieved. Presently they began to give infusions of figs, fennel, and ferns and saffron to throw back the small-pox which had never been there. By this treatment he became very hot and a very slight rash came out on his sides, which were not small-pox. But they told me he had small-pox, and that the fever would now end. Fourteen days had now passed and the fever had not ceased, nor were there further signs of small-pox, but they would not bleed nor purge, nor give him anything to eat except *mungo* and rice water. He was dying of hunger, and complained to me of this every day in Portuguese. I secretly advised his father that he should be bled, as it was better late than never, and be given a fat fowl to eat, for he was a great eater and drinker of our wine when he was well. This appeared good to the father, and still better to the son. So I bled him copiously twice, and gave him plenty to eat, without the knowledge of the physicians, but in the counsels of the King, until success was in sight. At the end of twenty days he was well, without fever or any rash, the physicians being content with this success, boasting of their method of treatment and asking for rewards. The father replied that for their treatment they deserved a roasting, and that if I had not been there his son and heir would be dead. After fourteen days I explained the treatment by which he had been cured, and when they heard it they put their fingers in their mouths saying *Ala quibir*,[1] which means "great God"; but not for this were they ashamed or confused.

RUANO

The King and his son must have been grateful to you.

ORTA

Yes, they were.

RUANO

It seems to me that neither the Arabs nor the Greeks wrote on the subject of what you call *mungo*,[2] yet it is much used here.

[1] *Al-Kabīr*, "the Great," *al-Khabir*, "the Aware," and *al-Kadir*, "the Almighty," are all among the hundred names of God, *Allah* being the first.

[2] *Phaseolus Mungo*, Linn., var. *radiatus*, *mung*.

ORTA

I know that they have it in Palestine, according to what a Moor told me who came from there. Avicenna also describes it in book ii. chapter 489. This entailed upon me a good deal of trouble for he calls it MESCE, and the Belunense, amending, has MES, but I found from doctors and other learned persons that it should be MEX, and the letter at the end should be pronounced very lightly, for that is the way they pronounce it. I know well that this does not explain much, nor to relate to you the stories I have told you, but Galen often tells them, whose shoe-latchets I am not worthy to unloose. You must, therefore, pardon my prolixity, for when once a man begins to chatter he is apt to do so a good deal. But you need not write down more of this than is necessary.

RUANO

Of the superfluity may God take care of it, of the rest I may write. But tell me whether Avicenna mentions this MEX in any other place.

ORTA

In the first book, chapter 7, he says that birds should not be eaten with MEX,[1] and he says well that MEX should be digested first, or else the MEX makes food indigestible.

[1] Orta mistook the word. It is MEST or acid milk. Avicenna gave the advice not to eat the meat of birds with sour milk, as it would be indigestible.

THIRTY-SEVENTH COLLOQUY

THE MIRABOLANOS

RUANO

I HAVE detected myself in great carelessness, for one of the principal things about which I wanted to ask you was the MIRABOLANOS so much praised by all the Greeks and Arabs, ancient and modern, and yet I have never asked you until now.

ORTA

Also, if you please, it is praised by the Spaniards, for they say *that man who saw so much, has eaten MIRABOLANOS.*

RUANO

Leave these adages and proverbs, and tell me the names of it in the native languages, and why we call it MIRABOLANOS amongst ourselves, and what the Arabs and Indians call it, also whether the physicians of this country use it, and for what illness it is serviceable.

ORTA

It is clear that the MIRABOLANOS of the ancient Greeks was not ours; neither Dioscorides nor Galen nor Pliny knew our MIRABOLANOS. But they gave that name to another medicine from which they made oil. MIRABOLANOS was in Greek as much as to say a fragrant and oily nut or acorn. As these nuts appeared to be nuts or acorns, for that reason those we used were so called also, although to me they are more like plums; but that has nothing to do with it. Serapio, translating Avicenna from Arabic into Latin, put this name so as to look better, and where Avicenna had given the name of DELEGI he put MIRABOLANOS.

RUANO

Well, Serapio calls it ALILIGI, being as much an Arab as Avicenna.

ORTA

This must be an error of the scribe, or time has corrupted these names. For the Arabian physicians of whom I enquired, told me that these names were used, namely, DELEGI by all, and the citron-coloured ones AZFAR, the Indian AÇUAT, the Quebulos QUEBULGI, the Belericos BELEREGI, and the Emblicos EMBELG, whence we derive the names.

RUANO

Why do you not mention the Negros, those of Seni mentioned by Avicenna?

ORTA

The Negros are no other than the Indians, so called because they are darker than all the others. There was no need for me to ask about them. Many modern doctors have written about them, especially your Italian Friars, who call them Negros because they are darker than any of the others when they are ripe. This is proved in the texts of Serapiam and Mesue, for they show it very clearly as you can see. I will not say more about it.

RUANO

Well, so far it is well. But the Friars, to whom you refer, say that there are no more than four species, leaving out those of Seni given by Avicenna. I asked you to give me the reason of this, and you, evading the question, gave me no answer on that point.

ORTA

Those of Seni are the *Emblicos* which are counted by Mesue and Avicenna who call it SENI, and in this way make five species.[1] Avicenna does not mention the *Belericos*, and that is the truth. For Serapiam says that they have a thin bark, so that they are *Emblicos*, the bark of which is the thinnest of all.

RUANO

Serapiam has another species on the authority of Mesarunge, which he calls of *Damasquo*, and says that it is good against melancholy. How do you answer that?

ORTA

I say that there are no MIRABOLANOS there, except that some MIRABOLANOS INDOS may arrive in conserves, and, because they arrive there, are called of DAMASQUO, and as to their being good for melancholy I concur, but they are no other than what we call INDOS.

RUANO

Serapiam himself says, quoting Mesue and Albasor, that those of Seni are a species of olive.

[1] Acosta names five species, *Chebulos, Indos, Emblicos, Citrinos, Belericos.* See p. 320.

MIRABOLANOS

Orta

There are no olives in all this land. It is only because the *Emblicos* are eaten with salt in this country, or in another way with vinegar, when they are called Achar, that they are called olives. But they are more like round plums, and in this it does not take much to be deceived.

Ruano

It seems to me that all are from one tree, some ripe and some unripe, so that when some are dry others are green. Some are of opinion that the *Quebulos* and *Citrinos*, at least, are from the same tree. It is true what Mateo Senense says, that this is neither proved nor disproved. You who have seen them are able to undeceive them all.

Orta

They are all wrong who say that the different species come from the same tree. For there are five trees of five species of Mirabolanos. You will further be surprised to hear that one kind is in one land, and the others sixty or one hundred leagues from it. For in Goa and in Batecala there is one, and in Malabar and Dabul, and in all Cambaya there are four kinds of Mirabolanos. Then there are the *Quebulos* in Bisnaguer and the Deccan, and Guzerat and Bengal, and may be in other parts. All these plants are wild, and not cultivated. Those which they take dried to Portugal are, for the most part, from Dabul or Cambaia. It is found by experience that fruits from lands furthest north are least liable to corruption, as I gather from the Gentio physicians in this island of Goa. I found that there are three kinds of Mirabolanos here which are used for purgatives, when it is desired to cause little strain and to administer small doses. They call these three kinds, in the language of the country, Tinepala, which means three kinds, namely Arare, the name used by the people but not by the druggists, who call it ARITIQUI. These are round, and purge the bile. We call them *Citrones*. Others are called by the natives ANVALE, and by us EMBLICOS, others REZANVALES, being those called by us INDOS, others GOTIM, which are round, and known to us as BELERICOS. The QUEBULOS which clear the phlegm, found in Bisnaguer, Cambaia, and Bengal, they call ARETCA. You here have four kinds, namely, three used in Goa, and one in Cambaia, Bengal, and Bisnaguer. To make up the five I must tell you that the MIRABOLANOS called ANVALE is found in Goa, but as it is not used as a medicine it is not counted. The kind which the natives call ANVALE, and we call EMBLICOS, is used in tanning skins, like the sumach for ink, and when green to give an appetite. So here you have the five kinds, and now what do you want more of me to serve you? I want from you to tell me whether that which the writers of Peru say, namely, that there are MIRABOLANOS in many lands, is true or not.

RUANO

I never saw those in Spain who came from Peru, and I would rather that you would tell me the appearance of each species, and about the trees and leaves.

ORTA

I say that the ARARE, which we call CITRONE, is round, and it has a leaf like a cork tree. The ANVALE or EMBLICOS has a leaf like a fern. The REZANVALE or INDOS has eight points and a leaf like a willow. The BELERICOS, called GUVI, has a leaf like a laurel but more grey. The QUEBULOS, called ARETCA by the natives, are large and round, and when they are ripe somewhat long. They have angles, and the leaf is like that of a peach tree. All the trees are the size of a plum tree. This is what you may know and see respecting the appearance of the trees of all the species.

RUANO

Your explanation is so clear that it will not be necessary to say more about it. Now I want you to satisfy my doubts as to what properties the Indians give these trees, for all confess they are cold and dry. Serapiam, quoting Xarach, says they are all warm. What do you say to that?

ORTA

I say that they are cold and dry, though this is not what the Indians, Arabs, and Latins say. Because their taste is piquant and oily, like green *sorvas*[1] but more acid. They are all heavy, and all these things are arguments for declaring their complexion to be cold and dry.

RUANO

As to the mode of preparation, tell me how this is done by the Indians, for Serapiam says they should be prepared with plums to remove the prickly taste.

ORTA

They do not make the preparation in that way because they only want to compress or repress. They do not use them to purge, but in cooking, and they take a much greater quantity than we do in Portugal. They also use them in conserves, namely, the QUEBULOS which they prize highly. They make them in Bisnaguer, Bengal, and Cambaia; and also use the CITRINOS and INDOS for conserves in Batecala and Bengal. Without any doubt this is a medicine which they hold in great esteem, and no physician is dishonoured in using it. I made these three species into conserves because they would be very good merchandize in Spain; and I will make two jars of them for you, which I had sent to me from Bengal,

[1] Berry of the service trees.

coming at a good hour. Know that I also ordered the water of green MIRABOLANOS to be distilled to use for drink over some piquant conserves. I also ordered syrup to be made when necessary. The water works very well over these green MIRABOLANOS. I use the CITRINOS and BELERICOS at the beginning of a meal, for it causes looseness in the stomach. Being acid it is also a styptic. They also use the juice of the MIRABOLANOS a good deal for dysentery, when they are green. Now you may try many of them in my house.[1]

[1] The following trees yield the Myrobalans of commerce :—(1) *Terminalia Bellerica*, Roxb., the "Belleric Myrobalan," the Sanskrit *tushu* and *baheruka*, and Arabic *heleyluj*, found throughout India and Burmah ; (2) *T. Chebula*, Retz., the "Chebulic [*i.e.* Cabul] Myrobalan," the Sanskrit *huritaka*, the Hindustani *hurda*, the Mahratti *hulda*, the Persian *helilah*, and Arabic *helilah-cabuli*, found all along the slopes of the Himalayas, and in Central India, of the fruit of which seven varieties are recognised, including *helilah-asfar*, or "yellow-*helilah*," *helilah-jawi*, or *helilah* dried when no larger than a "*barley*-corn," *helilah-zingi* or "nigger"-black *helilah*, etc. ; (3 and 4) *T. citrina*, yielding "Citrine Myrobalans," and "Indian Myrobalans"—its unripe fruit ; and (5) *Phyllanthus Emblica*, Linn., the "Emblic Myrobalan," a Spurgewort, the Sanskrit *amalaka*, the Hindustani *amlika*, and Mahratti *aonla*. The Myrobalanus, or "Glans Ægyptia" of the older Pharmacists, was the seed of the leguminoid *Moringa pterygosperma*, Gaert., the *Ben*-nut of the Arabs.

THIRTY-EIGHTH COLLOQUY

MANGOSTAES

RUANO

THAT gentleman who talked with you on the affairs of Malacca complained to me, saying, that you seem to have a dislike of the fruits of that land, because you have barely mentioned the *doriam*,[1] being the most highly praised fruit there is in India. Of the things of China you say nothing, though from there come also much esteemed fruits, such as the LIXIAS[2] and other very good fruits that grow there.

ORTA

I do not speak of the things of China because there is so much to tell that there would never be an end. I spoke of some of the medicine of that country, such as *galanga* and *pau de China*,[3] because they were medicinal. I did not fail to refer to the other fruits, and of the DURIÕES of Malacca I said more than usual, because it has a large tree of the size of a

[1] Colloquy XX.

[2] The *litchi*, *Nephelium Litchi*, Sapindaceae, Camb. Orta was the first European writer to mention it.

[3] Colloquy XXIV.

walnut; and in general I have always said that the most delicious fruits in the world are those of Malacca.

RUANO

They praise a fruit called mangostaes. Tell me what you know about it.

ORTA

What I know about the mangosteen is that it is one of the most delicious fruits in this land.[1] It is the size of a small orange and the rind is separated from the inside. The inside is murrey colour and clear. Drawing the rind away, in the inside there are seeds like those of small oranges. The tree is the size of an apple tree, not being very large. The leaf is like a laurel, the flowers yellow. They say that the scent of this fruit is not sweet and causes loathing. I do not know with what it may be compared, as I have not tried it.

[1] This passage is quoted in Yule's *Glossary*, p. 557. The Mangosteen, the *mangusta* of the Malays, is the *Garcinia Mangostana*, Linn., Guttiferae. The pulp of the fruit is of an exquisite flavour, rivalling that of the grape, but it is hard to get at through its thick and tough, almost woody, rind.

THIRTY-NINTH COLLOQUY

NEGUNDO OR SAMBALI

RUANO

YOUR servant girls praise very highly a tree which is here. They say that we wash our feet with the knowledge of it, and that it is astonishing how useful it is for so many things.

ORTA

I think it is in this garden. Let the girl who praised it come here.

SERVANT GIRL

What does your worship want?

ORTA

Which is the tree you have been praising so much?

SERVANT GIRL

The NEGUNDO.

ORTA

It is a tree that has many excellent properties, and the more the branches are pulled off, the more it increases. It is a very efficacious medicine, mitigating pain to a great extent, where there is no wound, either by boiling it, or the same herb warmed, or fried in oil and put on the top. Truly it seems to need no physicians, for upon entering a house to cure some pain, there is sure to be some person saying that the NEGUNDO should be used boiled or fried in oil. Many people also say that it should be put on a wound, and that in one night it will work in such a way that the pain will be gone and the wound closed up. This is stated by many persons and not one only. Women consider it to be very good for preparing a mother to conceive, and they say that drinking it has the same effect. I consider it to be a better medicine and stronger than MACELA. It has a very pleasant smell; in chewing it there is a little burning like mustard, by which it is clear that the complexion of the plant is warm. The tree is usually called NEGUNDO, and in Balagate some call it SAMBALI. The people of Malabar call it NOCHE and use it with the juice of tamarinds. The leaf is like that of the willow, barbed in the same way, and like it rather downy at the back. The tree is the size of a peach tree, the flowers a greyish white, the seed black, and the size a little larger than a peppercorn. There was a druggist in India, much relied upon by a chaste and virtuous governor, who was anxious to repress lustful desires. He asked the druggist if he had anything to effect this. The druggist replied in the affirmative, and that it was a tree called AGNOCASTO. The governor used NEGUNDO for many days, for there was not wanting a physician who said that in truth it was the same as the tree called the AGNOCASTO. When I was told this I read the chapter on AGNOCASTO and compared it with the tree called NEGUNDO; I found them to be as different as possible. I then said that AGNOCASTO was not the same as NEGUNDO, and any one who affirmed the contrary had not seen the book, for there was no druggist who had seen the AGNOCASTO in this country. Afterwards I met a druggist in this country who was learned and accurate in what he said.[1] He told me that there were many AGNOCASTOS in Portugal, and that they differed from NEGUNDO[2] in their leaves and everything else.

[1] Dimas Bosque.

[2] *Vitex Negundo*, Linn., Verbenaceae, the Sanskrit *nirgandi*, and Persian *panjangusht*, *i.e.* "five-fingered," and so named from its frequent "five-foliate" leaf. But its leaves are more frequently "three-foliate," and it is undoubtedly identical with the *V. trifolia* of Linn., the true Ἄγνος, and Λύγος of the Greeks [Pliny, xxiv. 38 (9)], and "Agnus Castus," "Chaste Tree," "Salex amerina," "Abraham's Balm," and "Hempe Tree" of our older Pharmacists; see *The Nomenclator* of Adrianus Junius, 1585, p. 148*b*. It may be that its Greek and Latin names were derived from its Persian name *panjangusht*! Dr. Waring says that there are few plants in India that have such varied medicinal uses. It is especially used in the treatment of Beri-beri.

Plate XVII.—Negundo (male).

Plate XVIII.—Negundo (female).

FORTIETH COLLOQUY

NIMBO

Ruano

I WANT to remind you of the tree with which you cured your valuable horse, for you told me to recall it to your memory.

Orta

You have good reason, for it is certainly a tree of great medicinal virtue among the people who know it, and they call it Nimbo. Its efficacy came to be known in Balagate, for I saw ulcers of horses which were very difficult to clean or to wash quite cured by it. The bad places healed quickly only by placing the bruised leaves on them, mixed with lemon juice. The same cures are effected on the sores of men, and they say that the cures by means of the juice of this herb are marvellous. Many persons praise it here, and they tell me that in Malabar it is much used in the way I have described. The juice of this herb is also used for rheumatism, and it seems that this is right, for it is very bitter.

NIMBO

Ruano

I remember that when you told me about the cure of the horse, you said that, in this city, you did not know of more than one tree of that kind, and that you would show it to me when we went to hear Mass at San Domingo. I saw it, and it was about the size of an ash tree with leaves like an olive, but serrated round the edges and green on both sides, not grey nor hairy, but more pointed than an olive leaf, and the tree with dense foliage. I say this to you that you may see that I can describe the tree well, but I do not know about the fruit.

Orta

You certainly describe the tree very well. You must know that it has a very useful fruit like small olives, and yielding an oil which is excellent for the nerves, as many people have found, anointing themselves with it when warm. It is much used in Bisnager and Malabar, and they have brought it here to Goa for sale, making good profit. I know nothing more of this tree, and if I should obtain further details I will write.[1]

[1] The Nimbo of da Orta is the *Melia Azadirachta* of Linnaeus, the *Azadirachta indica* of Jussieu, and *nimba* and *limba* of the people of India, and "Neem" of Anglo-Indians. Orta was the first European writer who treated of it, and of its medicinal qualities. Acosta gives an account of it, and a good drawing of a branch. See also Hanbury and Flückiger, *Pharmacographia*. *M. Azedarach*, Linn., is the *maha limbo* of the people of India, and "Persian Lilac" of Anglo-Indians; and *M. composita*, Willd., the *nimbara* and *limbara* of the people of India. *Azadirachta* is the Persian name of the tree, meaning "Free-growing tree." *Nimba*, the Sanskrit name, means "Health-giving."

FORTY-FIRST COLLOQUY

OPIUM

Ruano

I SHOULD like to have accurate information about the Amfiam which is what the people of this land use, and we call Opium. Whence comes such a quantity as is used here, and how much is taken each day?

Orta

The Amfiam is the opium, and as for its being much used to eat among many people, it is really eaten in small quantity, though much is required in trade to supply all the things it is in demand for. If it is not used there is danger of death ensuing, so that in the land where it is wanting its price is very dear, and there is a very strong desire for it among those who use it (as he who guards the wheat for May). The men who eat it go about sleepily, and they say that they take it so as not to feel any trouble.

Ruano

And they do not take it as a luxury as they told me, for

this is against all medicine and reason if it is efficacious for the work of Venus.

ORTA

There is much reason in what you say, for it is not efficacious in that way but rather harmful; so that those who take it for that are not kings, nor powerful persons, nor rich merchants, for they all know the truth very well. These only take it in small quantities, and for other effects. All learned physicians tell me that it makes a man impotent, and soon makes Venus take leave of him. I knew a Portuguese in Balagate who came there elevated and became impotent from its use; and the Portuguese there certified me of this.

RUANO

But so many people use it for fleshly lusts, they cannot all be deceived.

ORTA

I will tell you for what it serves, if you will give me leave, for this subject is not very decent, especially when discussed in Portuguese.

RUANO

I spoke because things are not dirty until they are said to be dirty, when not being clean they intensify.

ORTA

The imagination respecting the effect of anything helps much in carnal lusts, and as it is superior it gives expulsive virtue; and this imagined power, as it is stronger so it more quickly effects the act of Venus. For the imaginative dominates the expulsive virtue which forces the genital seed, and how much the imagination prevails, so much the quicker does the member expel the seed. As those who take opium are beside themselves, the act of Venus comes more slowly.

Many females do not give the seed quickly, and when the man is slow, the female also reaches the act of Venus more slowly, so that they both complete the act at one time. The taking of opium assists in the completion of the act of Venus more slowly. The opium also opens the channels by which the genital seed comes from the brain, by reason of its coldness, so that they complete the act simultaneously. I know that this is very well known to you, but if it is written down in common language, it does not appear a very honest practice.

RUANO

Those who practise this have some reason, though not a very decent reason. Now tell me the names of the *amfiam*, and by whom they are given.

ORTA

All call it AFIOM, namely all the Moors, who took the name from the Gentios, and it got corrupted into AMFIAM, so that the Moors call it AFIUM or OFIOM; and as the Arabs took many words from the Greek language, in which it was called JHUNANI (almost the YAVANA or IONICA), the Greeks called it OPIUM. For in Arabic the letter F includes the letter P, one taking the place of the other. So they called it OFIUM or AFIUM.

RUANO

How many kinds are there?

ORTA

There are many forms of it, different in the various lands. In Cairo (where they call it MECERI) it is white and fetches a high price, and may be what we call TEBAICO. At Aden and various neighbouring places in the Red Sea it is black and very hard, the price varying. In Cambaya, Mandou, and

Chitor it is softer and more of a yellowish colour, and is worth more in many lands, because it is customary to eat it there, so that it is worth more in the countries where it is more used. I say this of Cambaya opium, as I called it, most of which comes from a country called Malvi.[1]

RUANO

How is it made and brought, for it smells like spurge laurel ?

ORTA

It is no more than the gum of the poppy which I knew in Cambaya, seeing them sell in the market the heads of poppies so large that one of them weighed a *canada*, and I also saw some small ones like ours. Asking for the name they told me it was CAXCAX, and in truth that is what they call it in Arabic. They told me that these poppies make the AMFIAM, cuts being made in them and the AMFIAM runs out. As for spurge laurel there is none in all Cambaya, nor have I heard of any in India, so you may rest assured that it is not raised here.

RUANO

They will be black poppies, for Avicenna says that when it is found necessary to cause a stupor or to paralyse a member they do not use white poppies, for those that cause stupor are cultivated. He also says that opium is made of black poppies.

ORTA

I did not see a black poppy in Cambaya, nor did I hear it said that there were any, so that Avicenna must have been deceived or the black ones are made in other countries.

[1] Malwa.

FORTY-FIRST COLLOQUY

RUANO

I am much surprised at that, it being so narcotic and stupefying. Now I want to know the quantity a person takes in a day.

ORTA

My information is a weight of 20 to 50 grains of wheat. I knew of a Secretary of Nizamoxa, Khorasani by nation, who took 3 *trollas*[1] every day weighing 10½ *cruzados*.[2] But this Khorasani, though he was well instructed and a great writer and annotator, always was drowsy and sleepy. Yet, when put to work, he spoke like a discreet and educated man. From this you can see how strong the custom is.[3]

[1] 96 *rattis* equal to 180 grains troy. Three *trollas* would weigh 540 grains; but the weights varied at different times and in different districts.

[2] A cruzado weighed about 52 grains.

[3] Indian opium is obtained from the capsules of *Papaver somniferum*, Linn., the "White Poppy" of Greece, Egypt, Syria, Persia, and India; but it is also obtained in the Levant and in Persia from *P. Rhoeas*, Linn., the "Red Poppy" of temperate Europe and Asia. The plant yielding it is the *μήκων*, and its juice the *μηκώνιον*, of Theophrastus, and the "opium" and "meconium" respectively of Pliny. The Sanskrit name of the juice is *ahiphena* contracted to *aphena*, meaning "serpents' froth," but it is found only in modern Sanskrit writings; and of the plant *madrika*, meaning "stamped," the reference being to the seal-like form of the dried stigmas crowning the capsule. The Persian name of the plant is *koknar*, *i.e.* plant of the "cough-capsule," the Arabic, Hindustani, and Mahratti *khashkhashk*, the Hindustani also *afiyun*, and the Malay *apyun*; while old English names of it are "Balewort," "Cough-capsule," "Joan-silver-pin"; and an old pharmaceutical name of opium itself, "Manus Dei."

FORTY-SECOND COLLOQUY

PAO DE COBRA

RUANO

HERE, in this house, I have seen given to the children a root or stick called PAO DE COBRA which is said to be good for rheumatism. I beg that in a few words you will tell me what it is, from what country it comes, whether it is considered detrimental by the people, or whether it is useful in any way.[1]

ORTA

It is not only efficacious in the treatment of poison from serpents or cobras. It has been tried for rheumatism, small-pox, measles, and cholera, which is called MORDEXI in these parts, and it has the common fame among the people where this stick is grown. They also say it is useful in fevers which are difficult to treat, at least so I was told by a Franciscan Friar worthy of credit, who had twice given it to a man suffering from a returned fever. On these two occasions he had administered it, one ounce in water, and it caused a vomit of much bile, and from this he concluded that it was profitable in cases of old fevers.

[1] Acosta gives drawings of two plants which he says are both called *Palo de Culebra.*

FORTY-SECOND COLLOQUY

RUANO

How is it known that it is good against the bites of serpents?

ORTA

In the beautiful island of Ceylon, where there are many good fruits, forests and beasts for the chase, there are yet many of those serpents vulgarly called COBRAS DE CAPELO, but called by us in Latin REGULUS SERPENS. Against these God has given this PAO DE COBRA. It is found to be good against snake bites because in that island there are small beasts like ferrets[1] which they call QUIL. Others call them QUIRPELE. They often fight with these serpents. When one of them knows that it must fight with them, or fears that it may be so, it bites off a piece of this root and rubs it, or rather anoints, with its paws which are wet with the juice. It puts this on the head and body, and the parts where he knows that the cobra will bite when it springs. It then fights with the cobra, biting and scratching until it is dead. If it does not succeed in killing the cobra, the snake having more force than its antagonist, the QUIL or QUIRPELE rubs itself again with the root and returns to the combat, and at last conquers and kills its enemy. From this the Chingallas took an example, and saw that this root would be good against the bites of cobras. The Portuguese believed the good things that the natives said about the stick, and in time they gained some experience founded on reason, after having profited from this antidote, also having seen the effect with their own eyes, and that the fights between the cobra and mungoose were true. To give you more faith in its credibility I will relate to you what this Franciscan Friar, who is worthy of belief, saw in Negapatam, which is on the mainland near Ceylon.

RUANO

I shall be much obliged if you will tell it to me.

[1] Mungoose, *Herpestes* sp.

PAO DE COBRA

ORTA

Many Portuguese have these mungooses in their houses, tamed and domesticated, to kill the rats, and to fight the cobras de capello which the Yogues[1] bring who seek for charity. These Yogues are Gentios who go about begging all over the country, sprinkled over with ashes, and are venerated by all the common people and by some Moors. Travelling over so many countries they know numerous medicines and modes of treatment, some true, others false. Some of the Yogues bring these cobras with their teeth drawn, for in that condition they can do no harm. By this means, and for the gain they make, they hold the cobras in their hands, put them round their necks and on their heads, and hold them to be enchanted, but I consider this to be a lie. It happened that a Portuguese in Negapatam sent to a Yogue to bring a cobra to have a fight with his mungoos. But the Yogue did not wish to do this because he had drawn several teeth of the cobra, in which its power lies. The mungoos, seeing that a fight was intended, first went under the seats to see if it could scent some stick or root of the PAO DE COBRA. Not finding any it made itself wet with its own saliva, and came to fight with the cobra. The cobra darted at its head and wounded it badly two or three times, and the mungoos returned the blows with interest. They were parted, both being badly wounded, but the cobra worst. The Yogue, finding how the battle went, and that the cobra was alive, for it was healed afterwards, brought out another cobra whose teeth had not been drawn, and delivered it to the Portuguese who wanted to make the animals fight again. The Yogue had to stake most, as his cobra seemed to be dying, and he had to produce another. So the Portuguese gave him more than he had given him before, and the Yogue was satisfied. For his new cobra came better armed and the Portuguese, with his mungoos, got ready for the war. He brought the root and the mungoos

[1] Yogis.

bit it for some time, and then anointed itself with its paw, wetting it where it had been bitten. This it did on its head, loins, and stomach. The mungoos being prepared for the fray, the Yogue came with the serpent, which raised itself up nearly half its length and made a spring on the mungoos, wounding it on the body, then springing back it struck again, and so they went on, sometimes the cobra striking the mungoos, at others the mungoos biting the cobra. Finally the mungoos jumped at the cobra's head biting a little further back, and tore the cobra so that, being tired, it was killed, and the Yogue made desperate.

Ruano

Certainly this was a good fight, and it must be true, the monk being a man worthy of credit. Tell me whether this stick is found anywhere except in Ceylon, and describe its appearance to me.

Orta

Of this stick there are three kinds in Ceylon. The most esteemed kind, of which I spoke, is called in Ceylon (the land of the Chingalas[1]) Rannetul, a bush rising to two or three *palmos*. It gives out few branches, only four or five, and those very slender. The root is the profitable part, and that also is very slender, finer than the finest of our vine roots, and some root of this stick is always above ground. If it is broken or bitten off at any part, presently other roots spring out where the place is that was broken. The fruit of this stick is like the elder, only vermilion and harder. It grows in round clusters, like honeysuckle, the vermilion grains being smaller and more open as I said. The flower is a very deep vermilion and apart from the leaf, which is like that of a peach tree, the green colour being deeper. The colour of the root is between white and grey, and is very smooth to the touch, and very

[1] Cingalese.

bitter. This stick is found in many parts, both in Goa and on the mainland. It is drunk in water, being first well mashed. We also give it in wine or in some cordial water. It is also ground like sandal. It is called BOQUETI AVALE in Cingalese, so the ambassador told me. In Ceylon there is another stick or root used as an antidote. The tree is like a pomegranate and not larger, and the leaves yellow and very beautiful. The whole stem is thorny. The bark is white and thick, cracked, very smooth, and bitter, but not so much so as the first. The stick and the root and the bark are what is given, all mixed, but they say the root is the best. This tree grows like a pomegranate, and is with the other trees of the forest that are neighbours to it, but leans to them in the same way as a gourd, and so the highest branches embrace the rest. I ordered cups to be made from this tree for the sick, when they had been touched with poison, and I believe that it did them good. These cups also serve to make a posset of treacle as some of our doctors prescribe, to prevent a poison from doing harm. They say that some of this stick is in the island of Goa, but I have not tried experiments with it. When the Viceroy Constantino[1] was at Jafanapatam, which is an island off Ceylon, they presented a fagot of sticks, with their roots, as a thing much esteemed for an antidote. The root has a pleasant smell; it is slender, hard, and black. They say that there are many of these sticks and roots on the mainland of Goa. The leaf is like that of a mastic tree, as fine and long, with white and grey spots, not green. The branches are slender and extend far over the ground, more than four or five *covados*. The leaves are very few, and the branches few and slender, unable to bear much weight. The licentiate Dimas Bosque told me about this in his last letter. He is a person of learning, very truthful, and of very good judgment as regards the cures he has effected.

[1] Constantino de Braganza, brother of the Duke.

FORTY-SECOND COLLOQUY

Ruano

Tell me whether, in the parts of Malacca, they shoot with poisoned arrows, for they say that there is a root which is a very good antidote. I want to know what it is.

Orta

The forests there are full of tigers and the people have little curiosity. I have never been able to obtain a description of the tree, so that I cannot tell you anything about it; except that a few persons who have seen it say that it is the same as the *Pao de Cobra*[1] of this country. So it seems, as the roots are said to be alike. It may be, but I cannot affirm it, as I have not been well informed.

[1] The tubers of *Aristolochia Serpentaria* are the "Ceylon Snake-root" of modern botanists. The *Ophiorhiza Mungos* of Linnaeus, the *mendi* of Ceylon, a Chinchonad, is used for snake-bites in Ceylon and India. Among many other "Snake-roots" and "Snake-woods" may also be named the Dog-banes *Ophioxylon serpentinum*, Linn., of Ceylon; and *Rauwolfia* sps. of the West Indies and tropical America; the Solanaceous tree *Strychnos Nux-Vomica*, which is useful in staying the victim of a cobra's bite against collapse; and the Asclepiad *Hemidesmus indicus*, R. Brown.

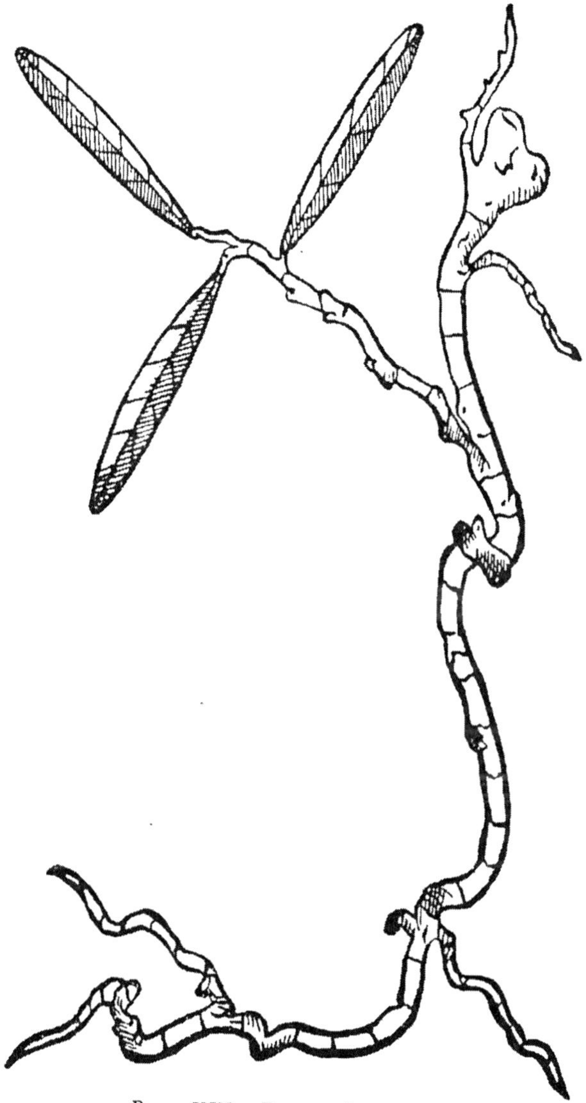

PLATE XIX.—PALO DE CULEBRA.

FORTY-THIRD COLLOQUY

DIAMONDS

RUANO

I SAY of the diamond that in Latin and Greek it is ADAMANS; in our Castilian we call it DIAMANTE; in Portuguese, DIAMAM. It will be well, because it is the king of stones, that we should discourse over it, for it has eminence over pearls and emeralds and rubies, if we believe Pliny.

ORTA

In this country and in all the world among lapidaries they think more of an emerald or of a ruby (which have more value if they are perfect and size for size) than of a diamond. But as they do not find other stones when perfect and of good water so large as diamonds, it happens that they often fetch a higher price. The value of stones is no more than the will of buyers and the need for them. For of greater virtues and more tried is the *pedra de cevar*[1] which staunches the blood.[2] They sell them by maunds[3] (which in Cambaya are 26 *arratis*),

[1] Or *pedra ïmam*, the magnet.

[2] The carnelian.

[3] That is the diamonds. The maund was about 26 *arratis*. The Bombay maund is equal to 28 lbs.

and emeralds are sold by *rattis*,[1] weighing three grains of wheat each. Other stones are sold in Europe by the *quilat*,[2] which is 4 grains, and in India by the *mangeli*, equal to 5 grains. Pliny did not speak of their value as a jeweller or lapidary.

RUANO

Let us give the names, and say how they are used in medicine.

ORTA

In Arabia, to which country all the Moors send them, they are called ALMAZ, though Serapio gives them another name (chap. 391). All the Gentios, in the land where they find the mines of these stones, call them IRAA. The Malay name, where they are also found, is ILAM. As for their employment in physic, it is not the custom to use these diamonds; yet I have seen Gentio doctors who give them to break the stone, administering with a syringe. But it is not given inwardly, as there is an error among the people that it is poison because of its great penetration, and boring the entrails.

RUANO

Is not that so? Laguna, with many others, looks upon it as poison. Do they have it in common use?

ORTA

Speaking the truth with you, I may say it is not so. I have heard, in these lands of the negroes, of lapidaries who swallowed diamonds and confessed to their lords that they had swallowed them, and waited and expelled the diamonds from below without any harm done. Of this I cannot myself testify.

[1] A *ratti* = 175 grains troy.

[2] A carat.

FORTY-THIRD COLLOQUY

Ruano

They told me that it is poison when made into powder, which seems reasonable, for it reaches the parts of the stomach and bowels, and bores through them.

Orta

It will not be a poisonous powder, because the virtue attracts the parts of the stomach and will not draw it to itself, and it will drop lower, being heavy, for it is a stone. I knew a woman who had a husband ill from chronic dysentery, and, abhorring the long continued illness, she gave orders to buy ground diamonds, and gave them to him for so many days without his dying that she got tired of it. Afterwards she left off administering it, because it was certified to her that he could not recover. So, without taking any more, he died a long time afterwards. I knew this from the person who went to buy the diamonds. So that those who say that diamonds are poison deceive, for it is not a thing written by authentic doctors.

Ruano

Well, now, I want to ask you about some errors, and this will be by mentioning what the ancients have said, respecting which I have felt some doubts. They say that originating in the minerals of crystal near the mineral of iron, the crystal does not take the colour of iron but is clearer than crystal. They also say that ADAMANS means a force not easily subdued, and put on an anvil they cannot be broken by the force of a hammer, but resist it and throw out the parings of the iron. But if the diamond is first dipped in the blood of a he-goat, it softens, especially, as some say, if the he-goat has first eaten celery and other opening things, and has drunk some wine. They further say that in this way it can be worked, and in any other it cannot. They also say that it is never found larger than a hazel nut. "Why do you not assert that it is false?" says Pliny, when he talks of the he-goat's blood poisoning a man. But some say that the text of Pliny is corrupt, and that it should be, "It does not poison a man," and that this about the he-goat's blood happened by revelation, for in any other way he could not be known.

Orta

Do not go on any further, for my memory is not good enough to answer so many objections, not to call them fables. You must now know that diamonds are found in three or four places. In Bisnaguer they are in two or three rocks which yield much to the King of Bisnaguer. As in Spain the tunny fishery provides large rents for the King, so in these mines the diamonds yield a great income to the King of the country. The stone, which has a weight of 30 carats, belongs to the King. For this a guard is placed over the diggers, and if any person is found with any, he is taken with all he has. The other rock is in the Deccan, near the territory of Imadixa (whom we call Madremaluco),[1] a land of a Gentio lord, with better diamonds, but not so large. These are called, "of the old rock," and they come to sell them at a well-known fair in the Deccan at a city called Lispor, within the territory of Madremaluco. There the Guzeratis buy them, and take them for sale to Bisnaguer, where these diamonds of the old rock fetch a very high price, especially those they call Naifes, being those which nature has worked; while the Portuguese value those most which have been polished. The Canarese say that as a virgin is more valuable than a woman who is not one, so the Naife diamond is worth more than the cut one. There is another rock in the strait of Tanjampur, in Malacca, also an old rock. The diamonds are small but very good, except that they have a blemish, and they weigh much. It is a

[1] Imad-ul-Mulk.

thing that is not good for those that buy but good for those that sell. In none of these parts is there crystal, nor in all India; for crystal prefers very cold countries like Germany. But in India there is the beryl, which is like crystal, and is obtained in large pieces for making jars and porringers. I gave 200 *crusados* for one, and they did not like giving it for that. But this beryl is not in Bisnaguer but in a few places between rocks. There is much of it in Cambaia and Martavam and Pegu, where there is a good market for diamonds, as there are none there. The beryl is also in Ceylon, where there are no diamonds. As to what you say about diamonds being so strong as to resist a hammer and anvil, and to break the hammer, I tell you if you try it with a diamond of price you will not have that experience, for when you try you will smash the diamond all to pieces with a hammer. They are easily broken in a metal mortar, and in that way they are made into powder to work up others. I have seen this with small diamonds. The lapidaries here had one large diamond which, owing to its bad shape, they wished to make into three or four, so they broke it in that way. It is true the diamonds are only to be worked with other diamonds put on a wheel, and they cannot be bored, though a modern doctor says that they can. The way to know a diamond, whether it is one or not, is to touch it with the point of another diamond or with a fragment of one. If it is not a diamond it makes a scratch, for there are other diamonds so strong that they would wound the other diamond. Diamonds will not be bored nor clipped, nor does the fire injure them.

With all this I would have you know that the diamond is well known to lapidaries, and they say that it has live water, while the topaz, sapphire, and crystal have dead water. But to soften it with the blood of a he-goat is a fable taken from the fact that the he-goat's blood breaks the stone in the kidneys and veins. I have myself tried it. As to what you say that no diamond is larger than a filbert, neither Pliny nor any other writer is in fault. They only speak of what they have seen, but there are diamonds four times the size of a filbert. The largest I have seen in this land was 140 carats, another 120, and I have heard that a native of this land had one of 250 carats. I know he had it, and made a large profit though he denied it. Many years ago I heard from a person worthy of credit that he saw one in Bisnaguer the size of a small hen's egg, and this might be. What astonishes me most is that a substance should be set far within a mine to grow during many years; and yet I see that they do so in two or three years. For in one year they dig in the mine to a height of a cubit, and at the end of two years they return to dig in the same place and find diamonds as before. Many people say this, but others tell me that the large diamonds do not occur in the face of the rock but well within it. But it seems doubtful whether they can be produced in so short a time, even small ones. As to what you say about their being overcome by lead on account of the quicksilver, it is not reasonable, seeing that the diamond overcomes iron and all other metals and stones. For if so it would cut a diamond as a knife cuts a turnip, and he who wrote it was dreaming or, to speak more plainly, he did not say what was the fact. I told you so in reply to your questions, and I asked further on for the doubts you entertain. If my answers are not agreeable to you, you should not be surprised, for he who speaks against common beliefs is detested.

RUANO

A chronicler named Francisco de Tamara says that there are diamonds in Peru. What do you think of that?

ORTA

I never heard it said by persons who have seen them in Peru, but you should know best being in Spain. I saw, in this author you quote, many fables concerning the working of diamond mines in these our lands, saying that they are guarded by serpents that may not be extracted, and that the people who own the mines throw poisoned meat in a certain place for the serpents to eat, while they in another place extract the diamonds at their will. It would be well, if Francisco de Tamara wants to invent fables, that he should tell them of his Indies and not of ours.

RUANO

A Dominican friar named Domingo de Baltonas also says that there is a diamond mine in Spain.

ORTA

I knew that friar in Salamanca, according to my recollection, and held him to be good and religious. He must have quoted what he found written by others, but I never heard of it.

RUANO

They say that there is another kind of diamond in Arabia which is not so strong because it is broken by the blood of a he-goat, but it has not the same splendour and is therefore cheaper.

ORTA

I never saw nor heard of such stones being found in Arabia. There are stones which we call topazes, the colour of gum mastic, and some sapphires of a water like diamonds, but I never heard that they came from Arabia. They are not diamonds but are called so because of their resemblance. Stones which have not the same indomitable strength are not diamonds. I have not seen those that are said to be in Cyprus or in Macedonia, and it seems to me that the Turks do not make much of them if they are in their land. I say this because the greater part of these diamonds are taken by the Turks.

RUANO

They say further that they all draw the iron, and what astonishes me most is that the *pedra de cevar*,[1] a diamond being present, does not attract iron.

ORTA

As to a diamond attracting iron, I will presently show you the contrary when I try the experiment. That the *pedra de cevar* does not attract iron, a diamond being present, is an absurd fiction. I proved the contrary before a number of people, both with diamonds of the old and the new mine, and I will do it again if you like. Of the other virtues of which they write, if they are not contrary to reason, they may be true. A stone which God created with such strength may well have these other gifts. But to say that when one is put behind a woman's head unknown to her, and when she is asleep, she will embrace her husband if he has been faithful, and if the contrary will avoid him, is a thing I am unable to believe, though they tell me that it is stated by writers of authority; for they say the same of certain herbs and we know that to be nonsense. But I will tell you one thing about diamonds of the old mine, which is that they have points that fit one with the other, fasten themselves, are fastened, and cannot be unfastened. Thus I saw a diamond warmed, attracting a straw like amber. As I have seen these things, and you can see them if you like, I have faith in them. I do not want to discuss writers who falsify stones, for it is not the province of a philosopher much less of a theologian. For this writer to make a diamond of a sapphire of water, he showed that a sapphire is not a diamond whatever it may look like.

RUANO

That we may not dwell too long on things unconnected

[1] A magnet.

with medicine, I will ask you whether you have seen the *pedra armenia* in *these* lands, for we are without it in Europe, though we have *lapis lazuli* in some places.

ORTA

I will let you see the *pedra armenia* at once. Girl, bring me that key.

SERVANT GIRL

Here it is.

ORTA

Draw out the cloth on which large stones are fastened.

SERVANT GIRL

It is here.

ORTA

Now see the *pedra armenia.*

RUANO

Many signs point to what you say, being to some extent blue and a clear green. But how do you know it is *pedra armenia*?

ORTA

The Moors, great physicians who attend the Nizamoxa, gave me these stones, with which they cure melancholy. They call it so in Arabic—HAGER ARMINI, which means the same as *pedra armenia.* I asked them where they had these stones and they replied in ULTABADO, a well-known city in Balagate. Enquiring whether the stone was found in Turkey and Persia, they replied that it was, but in small quantity. They did not know whether it came from Armenia or not. As a medicine it purges very little, according to my experience. I asked many Armenians here in this country, but they did not know whether it was in their land or not.

RUANO

The PEDRA DE ÇEVAR is very common, but I want to ask you what you know about it, for Laguna and others say it is poison, and that it turns a man into a lunatic.

ORTA

The PEDRA DE ÇEVAR does not turn a man into a lunatic, nor is it poison. For they have it here among the Gentios, who eat it in small quantity to prevent them from growing old and to preserve their youth. For this the King of Ceylon, an old man, ordered cakes to be made of this stone for him to eat.

RUANO

How do you know this?

ORTA

Besides its being common fame, Isaac of Cairo says that he ordered them to make it. This Isaac of Cairo is a Jew who came to Portugal with news of the death of Sultan Bahadur.

RUANO

Antonio Musa says that the Portuguese who navigate to Calicut find that the ships have wooden nails by reason of the mountains of PEDRA DE ÇEVAR, which would draw out the nails if they were iron.

ORTA

These are fables, for the Portuguese never saw anything of the kind, and at Calicut, as well as all along that coast, the vessels have iron rather than wood. It is true that in the Maldive Islands they use wooden nails, but the only reason is that wood is cheaper than iron.

FORTY-THIRD COLLOQUY

Ruano

They also say that the mine of Pedra de çevar is near an iron mine, and that the former mineral draws the iron to itself.

Orta

No such thing. They are found in different places, the Pedra de çevar where there is no iron.

Ruano

A Parisian philosopher says that the Pedra de çevar moves the iron of itself, a virtue which is not imprinted in it, but which moves to it. For this reason the Pedra de çevar does not weigh more with much iron than with little.

Orta

On the contrary, we tried the experiment here, I and some other persons, and at this you need not be surprised, because men make all sorts of statements.[1]

[1] The Pedra de çevar or Pedra imam is the loadstone or magnet.

FORTY-FOURTH COLLOQUY

PRECIOUS STONES

RUANO

NOW it will be well if we discuss the precious stones that enter into the composition of cordial lectuaries, which are so necessary in physic.

ORTA

I will only speak to you of medicinal stones, and of those which are in India, for if we discussed them all we should never come to an end. I must tell you that the medicines which enter into the composition of the lectuary *de gemis* are usually called *fragmenta preciosa*.

RUANO

I will ask you afterwards to tell me about some of the others.

ORTA

I will tell you in few words. But as to those others it is necessary to advise you to take back 10 *crusados* of them to give to the Spanish druggists, that henceforward they may buy the true stones, for they are not very dear. First the sapphire, which deserves to be worth a great deal, but which may be bought with little money. Its blue is very delightful to the eye. There are two sorts, one very obscure, the other very clear, which is called a "water sapphire." This has not so high a price, and sometimes they set them with colour to look like diamonds, by which some people have been taken in. One and another is found in Calicut and Cananore, and in many parts of the kingdom of Bisnaguer, also in Ceylon, and much the best in Pegu. These stones being so delightful to the eye, yet a large and clear one of the first water scarcely reaches to a price of 1000 *crusados.* I say that on the authority of what I have heard from those lands. When you are on your way to Cochin you can buy, in Calicut and Cananore, lots which, when washed, leave a good quantity. I also bought them in a lot, for they cost little money.

RUANO

Tell me about jacinths and garnets.

ORTA

There are some in such quantities that it is only necessary to spend a little money to buy a sackful. You will find many in Calicut and Cananore. A score of those that have been washed go for a *vintem.* Those uncleaned are much cheaper. Garnets are not only in the places I mentioned but also in Cambaya and Balagate, where they are sold at very low prices.

RUANO

Next comes the sardonix.

ORTA

That stone is not in our country. Some of them come from abroad, but there is much divergence as to what stone it is. My advice would be to leave it in its place, when you do not find jacinth or garnet. The jacinth is also brought to Lisbon from a place called Belas, and they may be found in many corners of Spain if they are sought for. Some consider these two stones, the jacinth and garnet, to be species of the ruby.

RUANO

What can you tell me about the ruby and carbuncle?

ORTA

I say that under this name of ruby there are many kinds. The principal is called in Greek *antrax*, and in Latin *carbunculus,* which is as much as to say "a live coal."

RUANO

I want to get one for myself, not to use as medicine, but I have heard that they light up at night.

ORTA

Do not you believe it. These are the sayings of dotage.

RUANO

Have you not seen it yourself, or heard that it is so?

ORTA

I have never seen it. But it is true that a lapidary told me that he counted on a table a few very fine rubies from Ceylon, very small, such as we call score rubies, because they are sold at twenty the *vintem.* One got between the folds of a table, and at night, in the dark, the table seemed to have a spark of fire, so that it was like a candle. A very small ruby was found, and when it was taken up the spark no longer appeared on the table. I do not know whether this is the truth or a lie. But I know that the lapidary who told me this professionally, told lies sometimes, as he found them profitable in his trade, and he got so used to it that he occasionally related marvels of his own accord.

RUANO

Now when a ruby is very fine as regards colour and water—I mean when it is 24 carats and quite perfect as to colour—should we call it a carbuncle?

ORTA

I think so. I have seen some called TOQUES, which have a price according to their size, shape, and water, and the clearest I saw were those the price of which was 20,000 *crusados.*[1] It was owned by a great Lord of the Deccan, who showed it to me, as I was a great favourite of his, and he made me give my word that I would not mention it to any one in that land nor to the King of it. He told me it cost him 6 *maunds* of gold, equal to 5 Portuguese *arrobas.*

RUANO

This would be a good stone to use in pharmacy, if the druggists are liberal.

ORTA

No. For the druggists you can find many as cheap as the jacinth, and indeed a certain quantity has been sent to Castile. There is another kind which we call BALAX, which is somewhat red and with a lower price, and another called ESPINHELA. This has a colour more inclined to flame, but it is of less price, because it has not the water of the true ruby. There are other white rubies in many shades. Others are slightly encarnadine, or of a whitish cherry colour. Others half-white half-vermilion, others half-sapphire half-ruby. All these I have seen, and if

[1] 43,000,000 reis=£8600. Four times that in present money.

you like I can show them to you before you go. There are many other kinds of these rubies about which I do not care to talk to you, nor about their prices.

RUANO

Can you excuse me for asking the cause of this variation in the colour of the rubies?

ORTA

That which I have heard said, that is most conformable to reason, is that the ruby in the rock, when it is near its birth, is white, and that as it matures it reaches to the perfection of vermilion. As this perfection cannot be acquired at once, sometimes stones are found becoming red, as I said, and sometimes with bands of yellow and white. As they say that the ruby and sapphire come from one rock, there are other stones that become a mixed blue and red, like a true composition of dark blue and vermilion, and almost red. In some languages of India they call such a stone NILACANDI, which is as much as to say "ruby and sapphire."

RUANO

Tell me the names of these stones in Arabic and in the language of this land.

ORTA

The Arabs and Persians call the ruby JACUT, and the people of this land MANICA. The jacinth and garnet have special names, as when one says "yellow ruby" or "black ruby" for garnet. The sapphire is NILA.

RUANO

It is very desirable that you should tell me about the emerald, the best stone of all, which is an ingredient in the lectuary *de gemis*, and they call it *ferruzegi*.

FORTY-FOURTH COLLOQUY

ORTA

Emeralds are not so cheap, and for emerald the *ferruzegi* should not be understood. The true emerald is rare and fetches a very high price. The proper rock where it occurs is not known, so that those who come to where it is washed do not find sufficient. He who says that *ferruzegi* means an emerald does not understand Arabic. The intention of Mesué was to enter emerald in this composition, though Cristofero de Honestis, the commentator of Mesué, felt differently. The reason is that the name of an emerald in this country and in Persia is PACHEC and in Arabic ZAMARRAT. Serapio, in the chapter on emeralds, has ZABARGET instead of ZAMARRAT, but it should not be ZABARGET, as the Pandetarian[1] shows.

RUANO

What does FERRUZEGI mean?

ORTA

You must know that in Arabic *p* and *f* as letters are like brothers, as I have said before, whence Mesué in Arabic has this FERRUZEGI, which means a turquoise, or of a turquoise, for PURUZA is a turquoise in Arabic, of which there is a great quantity in Persia.

RUANO

Truly for this alone one would wish to come to India, but if I did not find you perhaps I should not say that. From this time forward when I find FERRUZEGI in Avicenna or in any book of the Arabs, I shall understand it to be a turquoise, and I will not allow the druggist to fling green enamel into the lectuary *de gemis*, nor any other green stones. For I remember, the other day, one came here to sell a jewel with many small emeralds, and you said that they were all false

[1] Matheus Silvaticus.

and that they were made of bits of glass in Bisnaguer and Balagat, at least the largest of them, and that it was a thing so common among them that no one noticed it. Henceforth when I come to an emerald I shall say, beforehand, that it is not to be put into the lectuary without being quite certain what it is, "*A viride vitro libera nos Domine!*" A modern doctor says that our emeralds of Peru are of much more use for medicinal purposes.

ORTA

I say that those stones of Peru, called emeralds, were brought to this land. At first they fetched a high price, but afterwards they were found to be false[1] and cannot be sold. You will have to be on your guard against them.

RUANO

Tell me whether the turquoise is used in medicine?

ORTA

Some people have told me that it is, others that it is not, among the Gentios. Among the Moors all say that it is used in medicine.

RUANO

Tell me about the chrysolite, amethyst, beryl (you say there is no crystal in this country), and the jasper.

ORTA

Of the jasper I need not speak to you, as there is plenty in your own country, and you know more than I do about it. There are small porcelains[2] of jasper or green stone, which is like emerald, and it may be that the stone found in Genoa, and called emerald, may be this stone. But it has been sought,

[1] This is not true. Those from South America are true emeralds of excellent colour and water.

[2] Word used for a cup, irrespective of the material it is made of.

a few times, to obtain better authority for this, and the Genoese may speak in due time. In Balagate they gave me a porcelain for 200 *pardoas*, and if it was emerald, a thousandth times that amount would not be taken for it, according to the value they put on emeralds. I spoke to you of the beryl when we discussed the diamonds. There is a good deal of it in Cambaia, Bisnaguer, and Ceylon, and in other places. Chrysolite and amethyst come from Ceylon, and they have both these stones and many others in Balagate of Nizamoxa. All that Balagate is surrounded by many kinds of stones. That called ALAQUECA[1] by us, but QUEQUI in Arabic, is worth a Castilian real for one *arratel* of the stone worked in small pieces. This stone has a more certain virtue than all the rest, for it staunches the blood very suddenly.

RUANO

Cat's eyes are, I believe, very good. Where do they come from?

ORTA

The best are in Ceylon, and are worth more there than in Portugal, for I saw one taken to Portugal which was worth 600 *cruzados* here, but in Portugal it would not fetch more than 90. It was brought back here to be sold for its value. For this reason the cat's eye is not taken to Portugal for sale.

RUANO

What properties has it?

ORTA

The people of this country say that it has the property of keeping a man's wealth in his possession without its diminishing, and rather it will be increased.

[1] Carnelian.

RUANO

Whence are those rubies of the localities of which you have not told me?

ORTA

Some few are from Ceylon, but not very good, others from Pegu, and they say that they come from the lands of Burma, which is much farther off. This is the best information I have. If I am wrong in anything you must pardon me, for I do not know everything thoroughly.

FORTY-FIFTH COLLOQUY

BEZOAR STONE[1]

ORTA

I AM much surprised that you have never asked me about the bezoar stone, for it is so praised by all the Arabian writers, and with much reason.

RUANO

I did not ask you about it because you spoke highly of it at the time when we had practice with the cholera patient, and I waited for you to speak of it, for to me it seems to be a false thing to a great extent. Not that it is so much praised as the medicines which are antidotes to poison, which we call BEZEDARICAS *par excellence*. At the same time I shall be much obliged if you will tell me, in a few words, of its origin, falsification, the use it is put to by the people of this country, and whether it is much esteemed.

ORTA

They call the sheep (or more correctly the he-goat) PAZAM in the Persian language, and it is met with in Khorasan and

[1] The name of the Bezoar, or Bezoard, called also the "Snake-stone," is derived, through the Portuguese, from the Persian *padzahr*, meaning "expelling poison." The Persian for wine is *zahr-i-Kush*, "the poison of delight."

in Persia itself. I saw one here in Goa, large and of a reddish-yellow colour. They told me there were others smaller, of the same colour and of different colours. This stone is engendered in the paunch of these he-goats on a very fine straw which is in the middle, and so it goes on twisting and forming a rind like that of an onion. Thus it is formed into a round column, but not always the same shape. That straw is often found in the stone, as I have seen it, and at other times it is not found there. For the most part it is loose and the colour of a brinjal. There are large and small ones. They esteem them according to their size, for it is thought that the greater the size the greater the virtue. I have had one which weighed nearly 5 *oitavas*, though in Portugal it was not valued highly, where they sold it for 2000 reis; and but for the diligence in pushing the sale, they would not have been sold at all, for they worked hard to sell them.

Ruano

Are you certain how they are generated?

Orta

Yes, because in taking out the stone, some fine pieces of straw are found on it, and persons worthy of belief tell me that this is so with regard to all those at Ormuz. I afterwards found them when serving in the fleet off the *Ilha das Vacas*[1] (near Cape Comorin), when many he-goats were killed for the fleet. For the most part they contained this stone in their paunches, and the people who sought for them found a great many. Afterwards those who occupied the island had the custom of killing a great many. Those of Benfiala touched

[1] In 1543, serving under Martim Affonso de Sousa. The Dutch called the "Ilha das Vacas" the Isle of Delft, in Palk Strait, an island six or eight miles long, with water in a small lake and abundant pasture.

at that island to discharge part of their cargoes, on account of the reef of Chilam, so that they always got many bezoar stones.

RUANO

Then it is not only found in Persia?

ORTA

You are quite right. It is also found in the island I have just mentioned, and in parts of Malacca. But those in Persia and Khorasan are much the best, and the Moors know the difference between some stones and others. They open them on their hands to see if they are false, and blow on them; those which are blown away by the wind are held to be counterfeit. This stone is called PAZAR, from PAZAM, a goat so called. So that when they ask you for some medicine against poison, they call it PAZAM, and so some call the unicorn[1] and treacle.[2] This name of PAZAR is used by all Khorasani, Persians, and Arabs, and we, in Europe, corruptly call it BEZAR, while the Indians have a more corrupt form BAZAR, which means "stone of the market or fair." For BAZAR means a place where they sell anything.

RUANO

And for what is it used, and who use it most?

ORTA

The people of this country use it against poison, and the Moors of Ormuz and Khorasan take as much as 30 grains for all illness caused by poison or melancholy. All wealthy persons purge themselves twice every year, in March and September, and after purging they take 10 grains every morning for five days, in rose water, and they say that with that it preserves their youth. I may say to you that in many

[1] LOCORNIO, a misprint for UNICORNEO.

[2] Triaga.

old melanchoitic infirmities, such as prickly heat, leprosy, itch, ringworm, I was told by a governor that it was found to be good. I have used it for other diseases and found it to be very good. I therefore believe that it would be good for quartan ague.

RUANO

If a man took a great quantity of it would it do harm?

ORTA

Though this may not be a poisonous medicine, nor composed of poison like *triaga*, yet I think it more prudent to take it in small quantity. In Ormuz very little is given. A factor of Ormuz, a friend of mine, said that the Moorish physicians told him that a patient would die if he was given much bezoar stone; and this is reasonable, for things they use properly have their effects in smaller quantities. Certainly a man of Ormuz, who is worthy of credit, who was the agent of the captain, had much conversation with respectable Moors. They told him there were Moors who were much debilitated, and who wanted to take bezoar stone for such weakness, and that they were so feeble that it seemed as if they could not survive, and he said to the physician that they could not survive, and the physician said that after taking the bezoar stone he would not know them. He heard afterwards that they became very strong, for which they thanked God who had created such a stone.

RUANO

Mateolo Senense says that this stone touches the flesh, preserves those who have taken poison, taken outwardly that it is good for ulcers, that it draws out poison, and is good for the bite of any mad dog or other animal.

FORTY-FIFTH COLLOQUY

ORTA

As to its preservative effect in application outwardly it is not so used in practice. For some poisonous wounds it is true that it is efficacious, given in powder. I know that it is useful for poisonous bites, and for postumes of the plague where they are open. They say that it is preservative from the plague, and that it cures it. As in this country bladder complaints and measles are very virulent and often fatal, many hold that this bezoar stone should be given every day, in quantity of a grain or two, with rose water, and thus the poison is extracted.

FORTY-SIXTH COLLOQUY

PEPPER[1]

Ruano

IS it not unreasonable that, after such labour as the Portuguese have gone through to have all the pepper in their own hands, they should consume the smallest part while Germany and Flanders use the rest? I want you to tell me where the greatest quantity is grown, and the names of the countries where it is found, also what it is called in Arabia, how it is gathered, what the tree is like, and whether it is used in medicine.

Orta

The greatest quantity of this pepper is in Malabar, or along the coast from Cape Comorin to Cananore. There is also a certain amount in the parts of Malacca, which, however, is not so good, being less fruitful. There is some in Sunda and Java, and in Queda and other places, and it is consumed in China, in its own country, and is also taken to Martaban and Pegu. In Malabar also it is largely consumed in the country itself, where much more is used than in any other land. Some is used by the people along the sea-coast, some taken to

[1] See footnote at end of this Colloquy.

Balaguate on oxen; and a large quantity goes to the Red Sea, against the orders of the king, but nothing can be so well guarded but that much will be taken to the westward by the Moors. These are the places where the pepper is raised, and though there are some trees to the north of Cananor, they are so few that we do not take account of them, as the people of the country consume all that is gathered. For these trees do not grow in the interior nor in other parts, so that the trade is good, and I have seen much taken inland by oxen.

Ruano

Tell me the names in the countries where it grows.

Orta

It is called in Malayalim, where there is the greatest quantity, Molanga, and in the parts of Malacca where, as I have said, it is also grown, Lada. In Arabic it is Filfil, a word used both by the Arabian physicians and the people. Avicenna, according to the translation of the Belunense, calls it Fulful, and the long pepper Darfulful and Falfel, and Serapio, who was also Arabian.[1] It seems to me that Filfil, not Fulful or Felfel, is the more correct form, it being so easy to corrupt a written word when taken from the mouth of the people. There is nothing more to detain us on this point, except that in Guzerat and the Deccan pepper is called Meriche and in Bengal Morois and the long pepper Pepilini.

Ruano

Tell me the appearance of the tree, how it grows and how all is grown on one tree, for in this Greeks, Latins, and Arabs all agree, as well as the writers who have treated of the subject recently.

[1] This sentence is very confused, but Ficalho says that the meaning is clear.

PEPPER

ORTA

All agree, with one accord, not to tell the truth, although Dioscorides may be pardoned because he wrote with false information and at a great distance, with intervening seas not navigated as they now are. He was copied by Pliny, Galen, Isidore, Avicenna, and all the Arabs. But those who write now, such as Antonio Musa and the Friars, have the greater fault, because they merely repeat in the same way without taking the trouble to ascertain things so well known as the appearance of the tree, pepper, the fruit, how it ripens and how it is gathered.

RUANO

Are all those you have mentioned in error ?

ORTA

Yes, if you call saying what is not true an error.

RUANO

This being so, tell me what you have seen or heard from persons worthy of belief, and afterwards I will come with my doubts.

ORTA

The tree of the pepper is planted at the foot of another tree, generally at the foot of a palm or cachou tree. It has a small root, and grows as its supporting tree grows, climbing round and embracing it. The leaves are not numerous, nor large, smaller than an orange leaf, green, and sharp pointed, burning a little almost like betel. It grows in bunches like grapes, and only differs in the pepper being smaller in the grains, and the bunches being smaller, and always green at the time that the pepper dries. The crop is in its perfection in the middle of January. In Malabar the plant is of two kinds, one being the black pepper and the other white; and besides these there is another in Bengal called the long pepper.

FORTY-SIXTH COLLOQUY

RUANO

It seems to me that you abolish all the writers, ancient and modern, by this that I have heard you say. For Dioscorides says that the tree of the pepper is low, and produces a long fruit like a sheath, which they call long pepper, and inside this sheath there are small grains like gram, and that this is the perfect pepper, for at the proper time these sheaths open and discover some close clusters and the grains which we know, and that they are gathered before they are quite ripe. He says that they are sour and these are the white pepper. They are ingredients of the medicines they make for sore eyes and against poison that has been drunk, and against the bites of venomous beasts. The long pepper is strongly biting and rather bitter, owing to having been gathered before it is ripe, and is therefore efficacious for the things I mentioned. The black pepper is more suave and sharper, and more agreeable to the taste, from having been gathered at the right time, and also more aromatic than the white kind, and so it is more profitable for tempering the food. The weakest of all is the white pepper, owing to having been gathered before it is ripe. The black pepper is heavier and better. The people of the country call it BARCAMANSI because some empty grains are found amongst it. This is what Dioscorides says on the subject, at present it being unnecessary to enter upon medicinal qualities. At the end of chapter x. he says that the root is like that of *costo*.[1] Pliny says that the trees are like junipers, and that they grow only on Mount Caucasus according to what some say, also that the seeds are like those of the juniper, and that one seed divides or goes apart from another in a small part of the pod, like figs. The prices of them was 25 *livras* for long pepper,

[1] Costus, the Sanskrit *kushta*, and the *kut*, pronounced *koot*, and *kustum* pronounced *koostum*, of the Indian bazaars, formerly attributed to the Ginger-wort *Costus speciosus*, is now known to be the root of *Saussurea Lappa*, C. B. Clarke [the *Aplotaxis auriculata* of De Candolle, and *Aucklandia Costus* of Falconer], Compositae, of Cashmere.

of black 16 to 18, white 17, a *livra* equal to 3 *cruzados*. He says that pepper in its own country is wild and not planted, and that in Italy he heard of a tree which was like a myrtle, also that there is pepper in the part of Arabia called Trogoldita, which is called in the language of that country BARCAMANSI. Everything else about its use is copied from Dioscorides, so it need not be referred to here. Avicenna has two chapters, one on FULFUL, the other on DARFULFUL, which is the long pepper, and both Avicenna and Galen do no more than copy from Dioscorides, and so with Serapiam, who only has what he found in Dioscorides and Galen. Something that Paulo Egineta wrote is not relevant. These are the remarks on the subject made by the ancients. Turning to St. Isidore, he must, as a saint, be considered a high authority. He says that when the people of the country find that the pepper is ripe for gathering, they set fire to the wood for fear of serpents and burn the serpents. The pepper turns black owing to the fire applied to the wood. But I, to tell you the real truth, look upon this as a fable; so I wrote it first and then spoke. St. Isidore cannot have said this because he believed it, but to relate what others said. So I do not care to make excuses about these things, for I do not believe them. But I must tell you that I do not know for what reason you discredit such ancient doctors, and of such high authority, whose statements are confirmed by modern writers such as Mateas Silvatico, Sepulveda, Antonio Musa, the Spanish Friar, the Italian Friar, and so many others who have written on pharmacy. On this account I require you, in the name of God, to tell me only what you have seen and heard from persons well worthy of credence, confirming what you say by reasons which you know so well how to give, and finally we will consider how it is used in medicine by the physicians of this land, then I will put any necessary questions to you, and I regret if I have spoken too freely.

FORTY-SIXTH COLLOQUY

ORTA

In the first place, your worship must understand that pepper does not grow either on the skirt or on the slope of Mount Caucasus, as Pliny says. For there the price of pepper is higher than in any other country. This you must know, for you know how far Mount Caucasus is from Malabar or Sumatra, places where there is the greatest quantity of pepper. Nor is it like the juniper, for it is a climbing plant, while the juniper stands by itself, nor are its leaves like those of the juniper. Their shape is as I have already described to you, and the bunches grow like those of grapes. When they are green, with the berries apart and unripe, they put them into vinegar and salt. This I know very well from the testimony of my eyes. In the same way I know that the tree of long pepper grows in a land very distant from Malabar, the nearest point being 500 leagues off, for it is in Bengal and in Java. This long pepper is worth at Cochin, where there is the greatest quantity of black pepper, 5 *cruzados* the *quintal*, and four years ago at that place, when there was a greater demand for long pepper at other places, the *quintal* was worth 15 to 20 *cruzados*. The usual price of black pepper at Cochin is 2½ *cruzados*, but in Bengal 12 *cruzados*; while the long pepper sells in Bengal at 1½ *cruzados*. This is enough to show that the long pepper does not come from the same tree as the ordinary kind, much less is it needful for a man who has seen a thing with his eyes to give further proof of it. The white pepper comes from a tree of its own, and, to tell you the truth, there are not many but very few in Malabar or in Malacca. They put this pepper on the tables of the lords as we put salt. It is esteemed in both parts of Malabar as good against poison and for the eyes. It would be well if all that Dioscorides said was as true as that this pepper is good against poison. You will now see that these three trees are different, namely those of long, black, and white pepper. The long pepper is called PIMPILIM in Bengal. The tree of the long is no more like that of the black pepper than a bean is like an egg. The black and white pepper trees are very like each other, and only the people of the country can tell them apart, just as we cannot tell the black from the white vines unless they are bearing grapes. If you do not want to believe me, believe in these three seeds, that one is of long, the other of black, the other of white pepper. As for pepper being called BARCAMANSI no such name has ever been heard of in any of these countries, nor anything like it.

RUANO

Truly I find myself corrected, as I do not see it as the others do, it being made so clear.

ORTA

You see here the green pepper grown in clusters on this branch of a tree, and you see there another done with vinegar and salt, which you should taste before all.

RUANO

I see it all well, and now that I am corrected I see that the new writers never investigate satisfactorily. Laguna complains of the Portuguese because they do not describe these things and only care about skinning and robbing the Indians.

ORTA

It is true that the Portuguese are not very curious, nor are they good writers. They are greater friends of doing than of talking. They labour to acquire by their lawful trading, but they do not ill-treat the Indians; on the contrary peaceful Indians are much favoured by the Governors. The root of pepper is not like *costo,* nor is *costo* a root, but a branch. That you may not be surprised when the common people do not know these things well, I will tell you a story about what happened to a druggist in the time of a Governor who was very curious about medicines, to whom I spoke of the three kinds of pepper, and told him their names. As to the long pepper being another tree, I confessed that to be the truth; and when I said that the black and white pepper were separate trees, he told me I was mistaken. On this the Governor related to me that when he was wintering in Mozambique, he found that his ship was making much water and was unfit to go to sea. So he unloaded the ship and, to pass the time, he examined the pepper and found amongst it some white from having cast off the outer rind, a thing which often happens with old pepper when much rubbed. I replied that there might have been, in such a quantity of pepper, some that was white, and it might be found in Mozambique, but it would be found much better in the India House in Portugal, where the pepper is older and more rubbed. The Governor saw that the druggist did not believe me, so he wrote to the King of Cochin to send and tell him the truth about it. The King sent him a sack of white pepper, and wrote that he had many trees of the white kind in his territory. Then the druggist gave up his argument, so as not to go against the Governor.[1] With this we may end our discourse on pepper, for to say for what it is used is nothing new. The Indians calling it cold is a thing to laugh at rather than to treat seriously. I say, and often repeat, that I know not how to prove that fire is hot. The proof is that it burns.

RUANO

And the physicians of that King who is your friend, what do they say, for they are learned?

ORTA

They, like the Portuguese, say that it is hot in the third degree. As you now know that there are three different trees,

[1] The druggist was right, and Orta was mistaken. The white is simply the black pepper without the external layer of the pericarp.

it is not necessary to place black in place of white pepper, for neither Galen nor Avicenna knew this, nor do you need to know more than that the white pepper burns more and is more aromatic; and when you have to use it, always take the black rather than the long pepper which is a different plant. They do not take the long because it is their custom to use among themselves the black rather than the long. And because you have not any black pepper to see, you can see here these smooth seeds which in this country are called *pimenta canarim*, used for reducing swellings and for toothache. It is a very good medicine and is given to those who have cholera. I do not tell you the appearance of the tree, it not being necessary as you are going to Portugal. We drink it over some conserve, for it will not kill you, and it will be over conserve of peaches. They come, very good, from Ormuz.

Ruano

They are fresh and good, and it is not to be wondered at, for last week they gave them to you, so that they must be of this year. Tell me—they say that they are poison in Persia, and that when transplanted to Egypt they lose their poison.

Orta

Those you are eating are from Persia, for all the fruit of Ormuz comes from there, and as regards these I never heard, nor is it within the memory of man, that they were poisonous. I have talked to Persian physicians about it, and they say it must be understood of the fruit which Dioscorides calls *persea*, of which they could not give me any account. But they consider this[1] to be very good. It is also in Balagate, where I saw it and partook of it without fear.

[1] The *περσέα* of the Greeks was an unidentified Egyptian tree; their *περσικός* or *περσικόν*, the "Malum Persicum" of the Latins, was our Peach; and the *Persea gratissima* of modern botanists is the "Alligator" or "Avocado Pear" of tropical America.

FORTY-SIXTH COLLOQUY

Ruano

I know the conserve very well, though I now know better what you have told me about the pepper; and a modern author also writes what you said, that there are three distinct trees, but I feared he was a person to whom entire faith should not be given.[1]

[1] The species of *Piper* known in India are :—(1) *Piper nigrum*, Linn., of Southern India and the Indian Archipelago, our "Black Pepper," the undecorticated berries of which are the *kala maricha* or *mirich* of the people of India, and the *filfil aswad* of the Arabs—that is, "*black pepper*"; and the decorticated berries, their *safid maricha* or *mirich*, and *filfil abead* respectively, that is, "*white pepper*"; (2) *P. longum*, Linn., of the Circars, our "Long Pepper," and the Sanskrit *pipala*, the Persian *pipal*, and Arabic *dar-filfil*; (3) *P. Betle*, Linn., or "Betel Vine" of Southern India, Ceylon, and Pegu, the *vetila* of the Malays, and *vetili* of Southern India, whence we derive our word "Betel," and Linnaeus "Betle," and the Sanskrit *tambala*, Arabic *tambal*, and Hindustani *pan*, which, with the Malayan *vetila*, means "leaf," that is, *the* Leaf *par excellence* [compare *opium*, *the* Juice, and Chinchona bark, *the* Bark]. Its leaf, with some areca nut and slaked lime folded in it, is used universally in India as a masticatory and digestive. The leaf of *P. Chaba*, Hunter, of the Moluccas, is also used in India along with "Betel Leaf." *P. Cubeba*, Linn., the source of "Cubebs," the *kubaba* of the Arabs, is also a native of the Indian Archipelago. Pimenta or Pimento, "Allspice," is the berry of the West Indian and South American Myrtle-bloom, *Eugenia Pimenta*, DC.

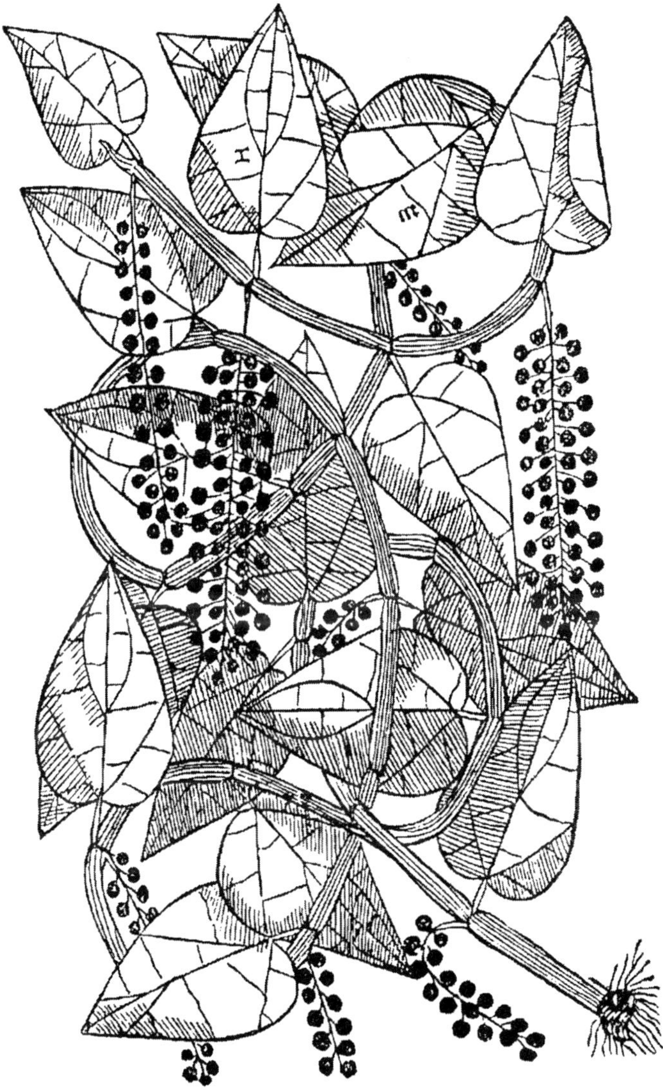

PLATE XX.—PIMIENTA.

FORTY-SEVENTH COLLOQUY

ROOT OF CHINA

RUANO

I WANT to take back with me to Portugal some of the root or stick of China,[1] for it is not a forbidden drug. Will you describe to me its appearance, and for what illnesses it is serviceable in your opinion, and tell me all its signs and the mode of administering? If they use it in Portugal, being a colder climate, the medicine would be weaker. I also want the best way of preserving this root so that it may keep fresh; and which is the best—this or the GUAIACAM of our so-called Indies? Do not be annoyed if the reason that this medicine is more used by you is that it is nearer.

ORTA

This stick or root grows in China, a very extensive country which is supposed to march with Muscovy. Laguna calls it the most eastern India, which may be excused as all unknown lands are called Indias. I will not here give the reasons why

[1] The root of *Smilax China*, Linn., or "Chinese Sarsaparilla," the *kub*-[pronounced *koob*]-*chini* of the Indian bazaars; for which the roots of the Indian *S. lanceaefolia*, Roxb., *S. glabra*, Roxb., and *S. ovalifolia*, Roxb., and also of the Asclepiad *Hemidesmus indicus*, R. Brown, are substituted throughout India.

it is supposed to border on Muscovy, because it is a matter of little profit, and is unconnected with your questions. As all these lands and China and Japan have this *morbo napolitano*, it pleased a merciful God to provide this root as a remedy with which good doctors can cure it, although the majority fall into error. As it is cured with this medicine, the root was traced to the Chinese, when there was a cure with it in the year 1535.

Ruano

How did you know the use of the root, for the ships of China do not come nearer than Malacca, and the Portuguese who go to China do not converse there with the Chinese?

Orta

I came from Portugal in the year before, and brought little property (as happens with many). But amongst what I did bring were five quintals of *guaiacam* which, in the time it needed shelter, was not well lodged, and they would take it from me, or those who wanted it wished to take it. Arriving in this land, I found that many perished of swellings and other sores of the *sarna castellana*, and for many of them they did not apply the remedy of ointments. Arriving in this land, I was much welcomed for having brought this remedy, for some persons had already been cured by it, their treatment having been very successful, so they had looked forward to some arriving from Portugal, and I sold what I had brought for 1000 *crusados*. It pleased God (I having brought little merchandize, and besides this I gave away some, and much was stolen from me in embarking and disembarking) that with all who used it the remedy succeeded very well. When my supply came to an end they bought the root, already boiled, at 5 *crusados* the *arratel*, and because it cost so much, it pleased God that it should be provided. At that time the people lived in the hope of the arrival of the ships from the kingdom, to see if they would bring much of the root, little, or none. At this time a very honourable and rich man was cured who, being in Diu, told my master Martim Affonso de Sousa, who was there taking possession of the fortress which Sultan Bahadur, the King of Cambay, had ceded to him, how he had been cured by the root of China, which restored him to complete health, not requiring any special diet except that he was not to eat beef, pork, fish, or green fruit. In China fish is conceded, because they are great eaters of it. As this became well known, people had a strong desire to have this root. For all men are inclined to eat and drink, and much more in this land owing to their laziness. Then they began to take the GUAIACAM as part of their diet, as they did in Spain, so that when the ships of Malacca came, a small quantity of the root which I saw on board was valued at 10 *crusados* the *ganta* (which is a weight of 24 oz.). In subsequent years it was cheaper, valued at 30 *reis* the *ganta*. From this time onwards the root of the Spanish Indies was contemned, like a Castilian who comes to kill by starvation the people who live here, so that the ships corresponding to that in which I came brought a quantity of the root from Portugal, expecting to make a good profit, but they could get no price for it, and little by little it was destroyed. Now, sir, I have reason to hold that this *guaiacam* is better than the root of China. It is certain that the other is efficacious given with reference first to the quality and character of the patient, then to the nature of the illness, the season, the country, the heat, the cold, the sex and age of the person who takes it. Be not surprised that I praise it, for I have heard no one else praise it, so many writers praising the *guaiacam* every day. Among them there is a German writer who composed a book on his labours in a very copious style and very pure Latin, which might all have been written on one sheet of paper. Of this other root of China, Vesalio and Laguna say many evil things, that it is rotten and without virtue and very dear. I do not care whether it costs much or little, whether it is dear or cheap, but I consider what Mateolo Sinense says to be good, that it is sufficient proof that this medicine is valuable that the Emperor Charles V. took it and benefited by it. It is certain that, given with the condition above mentioned, it is very profitable to all.[1]

RUANO

What quantity of this root do they prepare for one person?

ORTA

If the illness is very bad they cook an ounce of this root in 4 *canadas* of water and use half the water. The rest is kept in glass or in a glazed jar. They take off the froth at the boiling, because it is good to put on any sores; sometimes we put it on swellings when it is being boiled, and it is very good for allaying pain. At other times we make fomentations with this hot water to apply to swellings. At others we put wet cloths on the sore, which is a very good cleanser. The Chinese are accustomed to give larger doses of the root in their country, and some people here wanted to imitate the Chinese, cooking 2 oz. or $1\frac{1}{2}$ oz. of the root, but it was found to do harm, for it is very heating. I took the same root, with sudorifics, for a sciatica I had, without suspicion of *morbo gallico*.[2] But because I took sudorifics and drank hot water, as is usual at the beginning, when this root took effect I was filled with erysipelas and pains, from the great heat engendered in my liver. It became necessary to bleed me, and to give me barley water and sweetened rose

[1] There are two West Indian Guiacs, *G. officinale*, Linn. (the Lignum-vitae of Jamaica), and *G. sanctum*, Linn. (the source of the Lignum-vitae of the Bahamas and Hayti). In medicine it is used as an ingredient of the compound decoction of Sarsaparilla. Guaiacum resin is diaphoretic and alterative, and used in gout and rheumatism. No doubt it found its way to Lisbon, and thence to the East Indies. It is quite different from the root of *Smilax China*.

[2] Syphilis: unknown in Europe before introduced from America by the returned sailors of Columbus, at Cadiz; and unknown in India before introduced by the sailors of Vasco da Gama, at Calicut. Such were the first-fruits of "the augmentation of the Indies."

water, and in that way I was restored to health. Many persons took example from me afterwards, and refrained from hot water and such a quantity of the root as they give in China. For that land is very cold while this is hot. Only in taking it here, when a sudorific is needed, apply heat in the morning, and something to cause perspiration morning and evening. If the weather is very hot, we do not give the root, but more soporifics. This is the largest quantity we are accustomed to give, namely 1 oz. boiled in 4 *canadas* of water, and boiled until half is gone. To others they give less of the root with less boiling.

RUANO

And are there not rectifications with some medicines?

ORTA

Yes, sir. For they order to rectify when the disease is slight or the compress less hot, by giving 1 oz. of root boiled in 4 *canadas* of water or 1½ or sometimes 3, but to that quantity we seldom reach. We also take care that the root shall be good and of proper weight, and free from rottenness. For these conditions it should be white, or better if it is vermilion. With regard to rectification, the Chinese put celery into the boiling water, and I give it, there being reason in the Chinese practice, but I am not accustomed to give the root without rectification. When the patient suffers from headache or neuralgia I give rosemary or roses, or celery if the liver is torpid. At other times for ulcers of the kidneys and bladder I give liquorice. Here is a ptisic[1] which I give, mixed with the same quantity of barley as there is of root, with little boiling.

RUANO

What moved you to give the root in such a heated illness, *em membros esperituaces*?

[1] The Goa edition has *fisico*. Count Ficalho thinks it a misprint for *tisico*.

ORTA

It moved me to see a patient covered with swellings on the head and other parts, discharging matter, and not to provide remedies, but let him go on suffering. So he was treated and it succeeded well, and afterwards many others were treated successfully. But now no one takes the root without being first rectified with some other medicine. Though I have a right to boast that I was the first that used this, others following my example.

RUANO

Tell me whether it is best first to purge the patient who takes this medicine, and whether any accidents have befallen those who take it, and whether the root does most good at the beginning of an illness, or in the course of it, also whether it is most efficacious in great or in slight illnesses.

ORTA

The general rule is to give syrup and purge the patients before they take it, and if the case is very bad we make the syrups soluble. And as in most cases there is inflammation, we add *turbit* or *agarico*, and order the syrups to be watered with water of the root. After purging, the rule is to begin to administer the root, and, after fifteen days, if it is necessary, we give another slight purgative, and sometimes another at the end of thirty days. If in that time it does not cause looseness, we crystallize every day with water of the root, honey in rose water, oil, and *cañafistula*, and this according to the necessity of the case. These slight purgatives, which we give occasionally, are nothing more than manna and *cañafistula*, or rhubarb diluted in water of the root or of endive, or an infusion of prunes or liquorice, or barley water. We then give water of the root in less quantity, or mixed with endive water and fumitory if it is to be had, or bugloss. If the patient is very feverish the root is given up and he is given other treatment more suitable to his case. Sometimes this root does good in twenty days, at others it takes a longer time, but usually the pain increases for fifteen days and from that time it goes down. But I saw the case of a youth whose pains increased very much for twenty-five days, and on the thirtieth he was quite free from pain; I therefore say that we should never despair. I have seen others who took it many times and at last got benefit from it, and others who did not. My advice to your honour is to give the root of China in Portugal, trusting to God for the root, increasing the quantity that we give here because the country is colder, doing as we do here, and when the need is great, administering a chicken cooked in water of the root, or bread mashed in it, according to the nature of the case.

RUANO

Do they drink it hot or cold, and do they eat it with salt or not?

ORTA

It is seldom given hot as it was given at first. In the morning feed with fowl or mutton seasoned with salt, saffron, and dry coriander, or sometimes it is given roasted, according to the need of the patient. Wine is always forbidden, only when the root is given for weakness of the stomach, for in that case the root with wine does much good, for it takes away the loathing and secures good digestion.

RUANO

In using *guaiacum* we always give salt, because it is an enemy to arid humours and inflammations. Many men have told me that they also give it, but I do not know how they would use salt with this root.

ORTA

Salt may be used moderately, for it is not necessary to be very exact in physic, leaving much to the good judgment of the physicians. I think that the use of a little salt cannot do any harm to the arid humours, nor to salt inflammations, for I have always done well in using it, and I trust in God that I may continue to do so. In making this cure the Chinese also use bread with honey.

RUANO

Have you seen any one who has taken it many times and in great quantities?

ORTA

I knew a friend of mine who took ointments and *guaiacam* and this root also, and each time he got worse. He went to Malacca and got worse there, and thence to China, where they gave him root of China in a boiled fowl, which did him so much good that he never was taken ill of it again. For this root is better for chronic diseases than for new ones, and for great swellings and very bad sores. Marvel not that he should have found health at the end, for the tardiest cures are for the oldest griefs. This aphorism applies to the diet and also to the cure. Above all, hear well what is given and done, for there are many who die consumed by high fevers.

RUANO

That is well said, but I should like to know whether there is any other way of taking this root.

ORTA

In Balagate I saw some people who took the root in the way I have described. They mixed it in warm water to take night and morning, each time one *drachm* and a half of root pounded. With this they said that they benefited, and they did it by advice of their physicians. Others take it in the morning with a good slice of conserve made of powdered root with honey (or sugar if the fever is high), and on this they drink water of the stick or root. This conserve is taken with the powdered root, in quantity according to the advice of a good physician. It may be rectified according to the need of the patient, for you know better than I do how the treatment should be regulated. Now it may be seen how much root is required for this conserve. We usually require for a cure of thirty days 30 oz. I cured two persons who had the testicles much swollen for a long time. One was completely cured. The other got well with local remedies only. I would advise you to vary the remedies, and I will tell you more if you are not tired.

RUANO

From this time for a thousand years I should rejoice to listen to you. I have said as much.

ORTA

In China they eat this root cooked with meat, as we eat turnips. For it is very tender when fresh, and it seems to me that it would be a very good thing to take distilled water from it. I do not know whether they will want me to distil and bring it; but now it has been ordered to be brought, and for this I ordered the alembic.

RUANO

This is based on reason, for the distilled water would be a very good medicine. For what illnesses is it found efficacious?

ORTA

For any infirmity connected with *morbo napolitano* or for humours engendered by it. It is also good for paralysis, and for shivering fits, of which I cured Nizamoxa with it in a short time, for arthritis, exema,[1] sciatica, gout, scrofula,

[1] CIRAGRA. Scales like wax formed on the skin.

indigestion, swellings produced by melancholy or by white tumours,[1] old hurts, stone and ulcers of the bladder sometimes, for with this root the stone is got out which never could be removed before. What will surprise you most is that I knew a learned physician, well practised in curing others, who fell ill sixty leagues from this city, where he lived, and where he cured people in the hospital of a city belonging to the King our Lord. He fell ill of a lymphatic disease, and was down with it for four months. Finding that the fever would not leave him, and being more a friend to himself than to any one else, he took to drinking asses' milk. But he got no better; on the contrary he became worse, with inflammation of the stomach. He then came to me to be cured, and I had him in my house. I found some swellings on the liver and excrescences, and I was convinced that it was lymphatic, accompanied by some arid melancholy. He confessed to me that he desired that I would tell him as if the patient was any one else, and not deceive him, for those who care most for the patient always think worst of him. I cured this man in a few days, and he was without fever but with a pain in his stomach. For this I gave him conserve of ginger which made him better. This doctor could never be quite cured without giving him root of China rectified with a little distilled cinnamon water, and so at last he was perfectly well.

Ruano

Certainly you have told me many things relating to the good practice, and I do not wish you to finish so soon. Tell me the name and appearance of the wood or root of China.

Orta

I say that it is a shrub, three to four *palmos* above the ground with a root one *palmo* long, a little more or less. It

[1] "Freima como gesa."

has a thick root and another thin, as here you see these roots which come from there. When the root is taken up it is very tender, and it is eaten in mouthfuls, raw or cooked. When they eat a moisture exudes like sugar cane. From the root there come out some small fibres the size of writing pens, and as the root grows so the fibres shoot out, and from these shoots grow some leaves with the shape of young orange leaves. This shrub is called LAMPATAM by the Chinese. This is what I know about the shrub and root. I saw a small plant of it in Goa, but it withered before it could grow up. Before we finish the history of the root, I will tell you what happened to me in former times. Before the root was ever seen in India, there was a merchant of stones, a lapidary as we call him, or a jeweller to give a more honourable title. This man had universal paralysis in all his body, arms, legs, hands, and feet, so that he could not move a ring to look at it. He had been in this state for six months without getting any better. He asked me to advise him whether it would be a good thing to take *guaiacam* and I told him that at least it would do him no harm. I cured this man, first giving syrup and purging, and also at the end when he was getting better. Having paid me very well, he finally gave me a ring and a diamond for which I got 50 *crusados*, also a watch as a remembrance. He also gave me some advice. He further said that when he felt very sad and thought he was about to die, it was only a feeling of having sinned against God and against his health. The root you take to Portugal should be put in Martaban jars, for they are glazed inside and the root will not be injured.[1]

[1] Made of glazed clay. Duarte Barbosa says they were of porcelain. Linschoten describes them as being of immense size. Pyrard de Laval wrote in praise of them.

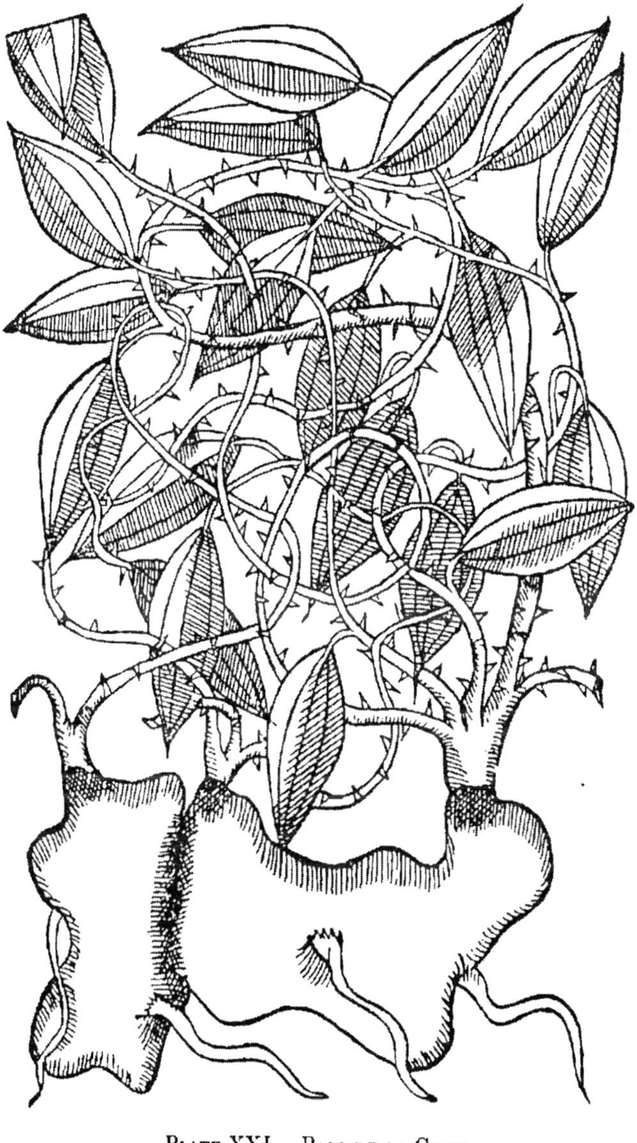

Plate XXI.—Palo de la China.

FORTY-EIGHT COLLOQUY

RHUBARB

RUANO

I WANT to know the appearance of the tree yielding rhubarb, and what the leaves and fruit are like, and whether this root we see is true or false; for certainly I would give much to see the true one.

ORTA

It is many years since I saw, in the treasury of Cochin, a box full of rhubarb which had all turned into powder. They told me in Cochin that the Chinese boiled or distilled the roots, and used the water as a purge. After this I talked to many persons about it, but was never satisfied, because none were eye-witnesses. We hold it for certain that all the rhubarb that comes from Ormuz to India first comes from China to Ormuz by the province of Uzbeg which is part of Tartary. The fame is that it comes from China by land, but some say that it grows in the same province, at a city called Çamarcander.[1] But this is very bad and of little weight. Horses are purged with it in Persia, and I have also seen it so used in Balagate.

[1] Samarcand.

It seems to me that this is the rhubarb which, in Europe, we called *ravam turquino*, not because it is of Turkey but from there.

RUANO

I would ask you whether none who come from China have spoken the truth.

ORTA

At last I came to know this, told by the merchants who come, not from the port of Canton, but from the interior to sell, and thence go to China and some to this India, where their goods arrive so damaged by the sea that we do not care to use them, better arriving from Ormuz.

RUANO

How are you certain that there is no rhubarb except in China?

ORTA

Because those who arrive from Ormuz confess that it comes from China by Tartary or Uzbeg to Ormuz and Persia, and it is therefore called *ravam chini*, and the Moors call it only *ravam*, but all confess that there is no other but that from China. So that the rhubarb or *ravam indico* does not come from Barbary, but that which is brought to India is taken on to Barbary, coming first from China to India.

RUANO

Speaking the truth to you, my belief is that the best rhubarb is sold in Castile, at Medina[1] or Seville. That which is sold in Portugal is much dearer.

ORTA

The rhubarb that comes to Persia or Uzbeg goes thence to Venice and to Spain. Some comes to Venice by way of

[1] Medina del Campo, then a great market for Indian goods.

Alexandria, a good deal by Aleppo and Syrian Tripoli, also to Venice. All these routes are partly by sea, but chiefly by land, and the rhubarb is not so much powdered for it is more rubbed in a month at sea than in a year going by land. The rhubarb that comes to India in May is there until September and is not yet fit for use. Then better and fresher comes from Ormuz, and is bought for India and to send to Portugal. That which remains in India goes to the interior, for India's interior land is not a place where drugs are subject to rot, so those who want to preserve them send them to Bisnaguer or to Balagate. I must beg you to excuse me for not discussing rhubarb, because I do not know much. I trust in God that it will be much better known, as China now has so much intercourse with the Portuguese.[1]

[1] Rhubarb, that is "Rha barbarum" [compare "Rha barbarum monochorum" of the older Pharmacists, *Rumex Patientia*], is the root of various species of *Rheum*—*R. palmatum*, Linn., *R. Rhaponticum*, Linn., and *R. officinale*, Bail., N.O. Polygonaceae, natives of Tartary, Thibet, and Western China, and known in commerce as "Chinese," "East Indian," "Turkey," and "Russian" Rhubarb, according to the route by which the drug is imported into Europe. In India the natives use also the root of *R. Emodi*, Wallich, *R. moorcroftianum*, Royle, and *R. webbianum*, Royle, of the Himalayas. *Ravaud-chini* is the Indian name of the foreign, *i.e.* "Chinese" rhubarb, and *ravaud-chini-hindi* of native Indian rhubarb.

FORTY-NINTH COLLOQUY

SANDAL[1]

Ruano

THE sandal is very necessary as a cordial, and being cold it has a good scent (a thing which happens in few medicines), and for this reason what the Arabians say of the appearance and properties of the sandal appears wrong to Mateolo Senense. They tell me in this country that the red sandal is avid for more cold because it has no scent. In order to understand more about this, I should like to know its names in the countries where it grows and in Arabia, and whether it is used as a medicine by the natives.

Orta

The sandal grows in Timor, where it is in greatest quantity

[1] Sandal-wood, or White Sandal-wood, the Sanscrit *chandana* [*i.e.* "refreshing"], Arabic *sandali-aswad*, Mahratti *gandha* [*i.e.* "fragrant"], and *sri-gandha* ["holy incense"], and Hindustani's *safid-chandan* ["white Sandal-wood"], is the wood of *Santalum album*, Linn. It is also the σάνταλον of the Greeks; but their σαγάλινα, also identified with it, was probably teak-wood, the *sag*, *sugun*, and *saguan* of India. Red Sandal-wood or Sanders is obtained from *Pterocarpus santalinus*, Linn., f. *rakta chandan* and *Caesalpinia Sappan*, Linn., *bukam*, both of the N.O. Leguminosae.

and is called CHANDAM, and it is known by that name in all the lands round Malacca. The Arabs, as persons who have a scent for the commerce of those lands, corrupt the word and call it SANDAL; and all the Moors of whatever nation call it so. The Canarese, Deccanis, and Guzeratis call it CERCANDA. The trees grow in Timor where there is the greatest quantity, and the woods do not become exhausted either on one side of the island or the other.

RUANO

And is all the sandalwood only in these islands?

ORTA

It grows in other parts as I will tell you, and though the red sandal is not in Timor, it is met with in Tenasserim and on the coast of Coromandel. I have not been able to ascertain the appearance of the tree up to the present time, but it is certain that all the red sandal comes from it which is little used here, and only for fevers, and some is sent to Portugal and the west. The red sandal is used in pagodas where there is a demand for very large poles, so that the larger the pole the fewer there are in a *bar* (4 *quintales*) and the higher the price. A very great quantity of the white and yellow sandal is used throughout India. For most of the people, whether Moors or Gentios, anoint themselves with sandal soaked in water and pounded on stones, prepared so for this need. With this they anoint their whole bodies, and let it dry, for when cold it smells well. For this land is very hot, and the natives are very fond of pleasant scents.

RUANO

Mateolo Senense says that it grows in both the Indies, both this side and beyond the Ganges.

SANDAL

Orta

The red sandal does not grow beyond the Ganges,[1] but this side. The natives call this river Gamgua, but the white and yellow sandal grows beyond the Ganges.

Ruano

How do you know that this red kind is sandal and not brazil, for neither of them has scent?

Orta

It is true that neither has a good scent, but the brazil is softer and more dyed. It was in this way that a friend of mine, a merchant, was a loser. For he bought red sandal for brazil, and the dyers found that it gave no dye, and he found no sale for it.

Ruano

Does not red sandal fetch more money than brazil?

Orta

The red sandal is worth more, though little is used, and of brazil much is used, so that when much sandal comes it is worth little. I must resume the account I was about to give of the places where the white and yellow sandal grows. I say that it is in Timor, an island which has many ports on both sides, and that the port of Mena is the best of all. Matomea is another port where there is much yellow sandal. There are many trees, but little pith, and the scent is in the centre. Another port called Camanace has bad sandal, much wood and little or bitter pith. Of this kind also is that of Cerviaguo, another port so called. Merchants trying it, ask whence it comes and take much or little. At Verbali, a port of Java, there is yellow and white sandal with a very strong scent, but

[1] He has just said that it does, in Tenasserim.

this sandal lasts only a short time. If it is a year without being sold, it is necessary to cut the pole and draw out more of the scent. It is also found in Macaça[1] where there is a wood of sandal, but it is so bad that there is no market for it.

RUANO

Are there two kinds of sandal in Timor, or is it all white, and which is most valuable?

ORTA

The most esteemed and best scent is that of the *amarelo* (yellow), but in the place where the sandal is best, that is in Timor, there is little yellow; in fifty sticks there is one. When there are many more the price rises. In the other yellow sandal I mentioned, the scent lasts only a short time, which is not the case with the Timor sandal, so that little comes. Talking, the other day, with a merchant who knows those lands well, he told me that in parts much exposed to the sun there is much yellow sandal, but in both kinds the trees are alike, so that we cannot tell the difference of one from the other. The natives who work on these trees may know it.

RUANO

Tell me the appearance of the tree, whether it has fruit or flowers.

ORTA

The sandal tree is about the size of a walnut tree. The leaf is very green, in shape like that of the mastich, the flower dark blue, the fruit the size of a cherry, and droops. It is first green, then black, with no taste.

RUANO

Now I want to state the doubts caused by what is written

[1] Perhaps Macassar.

in the works of Arabs and Latins. The ancient Greeks did not know it. Rasis does not say what it is, only what it is used for. Serapiam prefers the citron[1] to all, and you also affirm that. He says that the red comes next, and other things which are not doubtful, except what he says about Syria, and a quotation from Galen who did not write about it.

ORTA

Serapiam was wrong in both statements. As there is traffic between India and Syria it is not much to say that it is there, as long as he does not say that it grows there. What he says about Galen is also wrong, though the Arabs have often said the same without ever having seen Galen. They hear some Greek say that Galen knew the medicine and so believe it. Nor does Avicenna say anything about sandal, so that there can be no doubts which have not been well explained—nor does Avenrrois.

RUANO

Antonio Musa says that we owe the sandal to the Portuguese, that it is brought to Calicut where it is collected, being the principal market in India. Yet you say that it is in Timor, and the red in Tenasserim, lands bordering on India.

ORTA

The city of Calicut, where all kinds of merchandize were bought and sold, was celebrated in these parts. Thither goods were brought from elsewhere, and the Chinese came with their trade, including the sandal which they sold there, whence it was taken to the west. I have told you already that the Chinese factory, called CHINACOTA, in which the Chinese reside, is to-day still established there. But as the people of that land committed treason against the Portuguese when

[1] Dark yellow.

they first came to India they went to Cochin, and laid Calicut waste. So little by little its trade fell off, having been the chief city full of rich Moors who brought wealth to it. This is the reason why Antonio Musa said that sandal grew in the plain of Calicut. There is no plain in Calicut, but palm groves along the beach, and hills inland. That which came was in the Portuguese ships from Malacca in great quantity, and was taken to Cochin and Goa, and from those ports it was distributed to Malabar, Canara, Bengal, the Deccan, and Guzerat. A small portion goes to Ormuz, Arabia, and Portugal, as I told you.

RUANO

They usually call the yellow sandal MACHAZARI or MAHAZARI, or some name like that, and for this reason I want to know whence the name was derived. For the Friars say that in some books of synonyms they have MACHAZARI or ODOLIFERI; and that Serapio says that when the name SANDAL is used *par excellence* it means the yellow kind. The same Friars say, in another place, that the yellow is not met with in Europe, only in the pith of many of the sticks. Many others say the same, including Sepulveda. This Sepulveda says further that the best is half the powder of red and half of the white, and he also says, praising it, that he has seen the yellow sandal. Give me your views on all this as one who has been an eye-witness, and you need not give me any further evidence than your own sight.

ORTA

There is no doubt that the yellow sandal has the best scent and the highest price. It is so in many places, and I have seen much here, and many others have seen it. As it is bought here better than in Portugal they do not take it there, owing to the little care taken by the Portuguese druggists who do not ask at the India House that it may be brought there, and also those are in fault who make these drugs for the King and do not send them to Portugal. As to the names MACHAZARI and MAHAZARI, I think (failing better judgment) that it means "brought from Malacca," or it may have been written MAZAFRANI, which means the yellow or saffron. However that may be the *citrino* or yellow is the best of all. The delivery of half red and half white will not be *citrino*. It is better to receive all white, for the white is nearer in its nature to the yellow. Both are found in one and the same land, while the red is a long distance from where the white grows. I would also have you to know that this sandal wood tree is found in other parts. I saw it in Amdanager,[1] where it was brought to be sown. This Amdanager is a city of the Deccan where Nizamoxa resides, where I have been many times. I saw there, at a pleasure house where there were many fruit trees, trees of sandalwood, and many of ours, some yielding fruit. But this sandal stick has no smell on the tree, and many say that the sandal has no smell until the bark is removed and it is well dried.

RUANO

Is there sandal in any other parts?

ORTA

The native negroes say that it is in the island of San Lorenzo[2] and some parts of the coast of Melinda. But I afterwards learnt that there is a scented wood there, as there are many with us, which has not the signs of sandal. The Malayalims also say that they have a scented wood which is like white sandal, and they anoint themselves with it for fevers. They call it SAMBARANE.

[1] Ahmednagar. [2] Madagascar.

FIFTIETH COLLOQUY

SPIKENARD

Ruano

SPIKENARD was of great price and much praised in ancient times. For it is said in the Gospels that that unguent could be sold for more than 300 *dinheiros*[1]; and 300 *dinheiros*, calculated according to the reckoning of Budeu,[2] is 40 of our *crusados*.[3] This was, in those days, a great sum which the unguent cost. Now, owing to the great abundance of perfumes, natural and manufactured, it is not worth so much. Of the scents are the soft Pastilhas and Caçoleas, the delicate Pivetes, the Ambar, and Almisque, Algualia, Linaloe, and many other perfumes. So, for this reason, and for its use in physic, it is well that we should discuss it, the more so because some writers doubt whether we have the true Espique.

Orta

We do not lack the true spikenard, but we have medicines which are not so falsified as they were at first, from the great

[1] John xii. 3, 1 lb. for 300 pence or *denarii*.

[2] Guilherme Budeu, *De asse et partibus ejus, Libri quinque*, 1533.

[3] £4.

quantity that goes from the east to the west. For we have heard the sayings of Pliny[1] on these medicines, and we must not be surprised that they should be falsified, for the high prices conduce to it. Now that the navigation is better known, and that there are more ships, both to Portugal and to other parts of the West, it is not to be wondered that we have so much, and so cheap, without being falsified. And they buy these medicines better and they are grown better now, because more are cultivated.

RUANO

Tell me where they are, and what name they have in their own land among the natives, also what they are called by the Moors, for I know well what the names were among the Greeks and Latins. When this is done we will examine what the writers say who have written on the subject.

ORTA

They call it ESPIQUE in the country where it grows, among the Gentios CAHZCARA, and it is found in Mandou and Chitore and some parts of Bengal on the river Ganges, which the Indians call Ganga, a very famous river, and held to be so holy that the Bengalis, when they want to die, order them-

[1] xii. 12 [26]. The true Spikenard, the "Nardus Assyriae" of our older Pharmacists, the "*Aspic*" or "*Espic* de l'outre mer" of the French, and our "Valerian Spikenard" [all these names being given to it to distinguish it from "Nardi Spica," or *Lavandula Spica*, "Sweet Lavender," yielding "Oil of Spike," the French "Huile de Spic"; and again from "Nardus rustica" obtained from *Valeriana celtica*, *V. Phu*, *V. dioscorides*, etc.] is the fibrous root of *Nardostachys Jatamansi*, DC., N.O. Valerianaceae, of the southern slopes of the Indian Himalayas. Its Sanskrit names are, where fresh and soft, *pisita* [compare νάρδου πιστικῆς πολυτελοῦς of St. Mark xiv. 3 and 5; John xii. 3], and when dry *jatamansi*, meaning "matted-hair plant"; and its Arabic and Persian name *sumbul atibhindi*, meaning "Indian fragrant Arsenic," from, it is said, being poisonous to sparrows! In India its name of *jatamansi*, in its many corrupted forms, is also given to the roots of *Acorus Calamus* and *Cyperus rotundus*, and the combs of *Celosia cristata*; and to the oils of the grasses *Andropogon Nardus*, Linn., "Citronelle," *A. citratum* seu *Schoenanthus*, "Lemon grass," and *A. muricatus*, Retz., "Kuskus."

selves to be taken there and their feet to be put in the water, which water is very good, as I have proved.

RUANO

Do the Gentio people of other nations hold that river in veneration?

ORTA

Yes, in great veneration, for a river which is said to be a branch of it, running north-east to south-east, is also called Ganga.[1] It is in the territory of Nizamoxa, and every year his subjects go to wash themselves in it. As some of them are poor and unable to undertake the journey, the King ordered, tyrannically, that he who wished to go and wash in the river was to go, and that the rest were each to pay him 180 reis, being half a *pardao* of gold. For this he caused the number of persons to be counted, and the rich were to pay for the poor. Thus a sum of money is got together. Of this I can bear witness, for I saw it collected, and they paid me out of it the sum the King owed me. On the banks of the Ganges in Bengal and Orissa there are certain pagodas to which the merchants of Orissa and Guzerat go in pilgrimage and to bathe in the river Ganga. They offer great gifts to the pagodas, and thence come bathed and taken by the devil, which they call being sanctified.

RUANO

This is a new form of tyranny. What is the name in Arabic, and do the Moors and Gentios use it in medicine?

ORTA

Avicenna and all the Arabs call it CEMBUL, which means ESPIGUA in Arabic, and so the ESPIQUENARDO became CEMBUL INDI, which is as much as to say ESPIGUA DA INDIA. What

[1] The Godavery.

we call ESPIGUA CELTICA is in their language CEMBUL RUMIN, as we might say ESPIGUA of the land of the Rumis. Mateus Silvaticus calls it CENUBEL and SO-BEL, as one who does not know Arabic, or it may be that the names were corrupted little by little. As to what you asked if it is used as medicine, I reply in the affirmative as regards the Moors, and much more so as regards the Gentios.

RUANO

It is now necessary that we should examine the writers on this subject that I may get rid of some doubts. Dioscorides, the most ancient writer, says that there are two species, one Syrian and the other Indian, not so called because they are found in those regions, but because they grow on one mountain with two aspects, one facing Syria and the other India. Afterwards it is said that, among the Indian unguents, there is one called GUANJETICO from being found near the river Ganges, and it grows on a mountain. They report that the most luxuriant is on the loftier part, but of less virtue, and that the scent is like that of the *cypress*. Tell me your opinion on these points.

ORTA

I do not know any spikenard in this land except that which I mentioned to you, and that which comes from Chitore or Mandou, countries which border on Delhi, Bengal, and the Deccan. All these lands, and others beyond them, are in India. I do not understand the statement that there are Syrian and Indian ESPIGUA on a mountain with two aspects, one facing east the other west, for the mountains are in India, and those in Syria are far to the west. Further, it does not all grow on the mountains, but in many parts where they sow it. It does not grow unless it is sown, except in very small quantities. A growing root throws a short shoot over the ground, some with a length of 3 palmos, others much shorter. Presently a shoot comes out at the top of the root, and these they take for sale to Cambay, Surat, and Guaga and other seaports where the Arabian and Persian merchants buy them, for we only consume a smaller part. The people of the country also use a great quantity, and I bought some at Diu for the King our lord. There is some in powder made from the hairs of the same ESPIQUE, and the above merchants buy it all. They tell me that they wash their hands with the powder. We do not find in this land that one is better than the other, but those who come to sell say that they grow in the valleys and in the mountains, and that those from the mountains are the best. All the others grow near the ground, which is what generally happens. Of one thing you may be sure, that if Dioscorides saw this ESPIQUE which we see and order, he would say that it was the true ESPIQUENARDO, and certainly he would be astonished at those modern writers who say that they have not cinnamon nor cassia lignea, confessing that, like espique, they come from India. They had much better say that what they talk about is not this India, but another which we do not know. It is certain that many of the medicines are worth little because it is found that on the voyage it is doubtful whether they are themselves.

RUANO

You do not speak unadvisedly, because Mateolo Senense is of your opinion, rebuking Menardo and Fucio for saying that this is not the true ESPIQUE. He says that Pliny describes it as small and black, and that one kind, growing near the river Ganges, is all spoilt by rot. He further says that the price of it is 90 *livras*; if this had been a quintal of the true spikenard it might be true, for it is worth in Diu 25 to 30 *crusados*, which are 90 *livras*. They say that those with large leaves are worth 30. But, in his annotations of Pliny, Hermalao Barbaro says they are not *livras* but *dinheiros*, for they have this note made with x, meaning that the value is x *dinheiros*. This is more reasonable as the true price of this medicine, as of that of pepper and many other drugs.

ORTA

In this land I have not seen any other ESPIQUE but this which is sent to the West, and all comes from near the Ganges. This is what is used by physicians of India, Turkey, Persia, and Arabia who come to live in this land, curing the kings and princes. It is not to be wondered at that the prices are high, for the roads by which it comes are unknown. Pliny may be right in what he says, but not in the assertion that the ESPIQUE of the Ganges is to be condemned altogether, for there is no other, or if there is it is in such small quantities that it has never come to my notice.

RUANO

Laguna says that the ESPIQUE sold in the druggists' shops is not ESPIGUA but the root, and this Dioscorides does not contradict, saying that it is like ESPIGUA. They say further that the ESPIQUE of India is to be suspected, because they make a poison of it called PISSO. They say that this PISSO not only kills internally but also when it is applied externally, and they add that it comes from Syria. What have you to say to that?

ORTA

I say that I have not seen any composition called PISSO, nor have I heard of it. I affirm that, the Nizamoxa wanting to try a little of the liquor called NAPELLO, he gave it to a prisoner condemned to death. It would seem that PISSO is more poisonous when given to drink. But I have never heard of it, and its fatal external application appears to me still more fabulous. Those who say that spikenard comes from Syria mean that it arrives by way of Aleppo, whence a part reaches

Venice and is consumed in Europe. In this way Sepulveda is to be understood, who mentions ESPICA ALIEP as much as to say ESPIGUA of Aleppo; for Aleppo is always the head of Syria and the principal step from India to the West, and now more than ever. Sepulveda mentions a species called SATIECH or SATIACH, and this means SATIGUAM,[1] which is a well-known port in Bengal near the mouth of the Ganges. This medicine, which is highly celebrated and much used, is not found to be false; only when it is old it loses some of its scent. There is no necessity for us to talk more of the spikenard.

RUANO

What city is Aleppo? Is it by any chance Haran?

ORTA

The Bishop Don Ambrosio[2] came to this land by Arabia and Turkey, moved by zeal for our faith. He knew Arabic and read it very well. Conversing with him in the convent of St. Domingo, to which order he belonged, he told me that Abraham, when God called him from Ur, a city of the Chaldees, went to Aleppo, the chief city of Syria, and took great quantities of cattle. He gave milk for all the necessitous and poor to drink, and they came for it every day. These poor people when they came asked for *yalep*, which means a question: "order where shall we eat now." This was the reason why that name was given to the land. The Bishop said that this was the tradition of the old people of Aleppo, who believed that it was inhabited and ruled by Abraham.

[1] Chittagong.

[2] Friar Ambrosio de Routecalli, a native of Malta, was sent to India by Pope Paul IV. with powers of a Legate. He was some time at Goa in the convent of St. Domingo, with the fame of a good theologian, orientalist, and mathematician. He died at Cochin.

RUANO

Could I converse with that Bishop?

ORTA

No; because when he was about to embark for Portugal, he died at Cochin.

RUANO

Certainly I should have rejoiced to have had a conversation with that Bishop.

FIFTY-FIRST COLLOQUY

ESPODIO

Ruano

THE Espodio enters into so many preparations made by the Arabs that it makes us doubt what the composition is, the name of which is taken from the mouths of the ancient Greeks; or whether the Latins use a different Espodio, taking the name from the Arabs. For this reason I want to know from you what the Espodio may be which is used by the physicians.

Orta

There is only one Espodio[1] in the world or Pomfolix or Tutia, and for want of it the Greeks took another medicine

[1] The Spodium and Pompholyx of the Greeks and Romans were substances formed in the furnaces of copper-smelting works, Pompholyx being an imperfectly combusted form of Spodium, both being a combination of copper and cadmium. "Antispodos" was obtained from the ashes of the burned twigs and leaves of the fig, olive, myrtle, quince, and other trees. [Pliny xxxiv. 13 (33 and 34).] Tutia is known in India in three forms, *safid tutia*, the "white" sulphate of zinc; *nila tutiya*, the "blue" sulphate of copper; and *hira tutiya*, the "green" sulphate of iron. Tabashir, the Sanskrit *tabakshira*, meaning "bark-milk," is the siliceous concretion formed within the stems of "bamboos."

and called it ANTISPODIO, which means false ESPODIO. But the Arabs do not mention this ESPODIO except under the name of TUTIA or POMFOLIX.

RUANO

Well, how did this distinction arise of calling one medicine ESPODIO of the Greeks and another ESPODIO of the Arabs?

ORTA

From Davus of Terence,[1] who disturbs everything. On this occasion the Davus was Gerardo of Cremona who translated ESPODIO in place of TABAXIR, the two not having any resemblance whatever to each other, being as different as black from white. And he is not alone in this mistake, for all those who have translated the Arabic books into Latin say TABAXIR instead of ESPODIO, the blunder not being that of the writer but of the translator.

RUANO

Does it not seem wrong, wanting a name, to impose another in its place?

ORTA

If the name means something quite different from what appears in the book, the mistake may lead to many errors, and if physic is given under such a mistake, it is dangerous and becomes a greater error.

RUANO

Tell me, what is this TABAXIR according to the doctors and people of this country?

ORTA

The name TABAXIR used by the Arabs is derived from the Persian, whence the Arabs, Avicenna and others, took it. It

[1] Davus was the slave of Simo, in the *Andria* of Terence.

means milk or juice, or moisture, and it is known by this name all over Arabia, Turkey, and Persia.

RUANO

If the medicine is used in India, what is it called in these parts?

ORTA

The people where it grows call it SACARMAMBUM, which means "sugar of mambu," because the canes are called by the Indians where it grows MAMBU. Now the people of this land call it TABAXIR because the Moors, who come to buy it from Persia, Arabia, and Turkey, call it by that name. It is worth much when there is a dearth of it, and little when there is plenty to sell, as with all other merchandize. The ordinary price in Persia and Arabia is its weight in silver.

RUANO

What are the trees or canes they raise, how are they gathered, and what country is the principal source of these canes?

ORTA

The trees are lofty and large like ash trees,[1] and in Bisnaguer and its territory they are smaller, also in Malabar. For the most part they have their branches straight, but some of them, having a good appearance, are twisted, and these serve for the canes of the palanquins used in this country. From knot to knot these canes have a length of 1 *palmo*. The leaves are little longer than our olives. Between the knots these canes generate great humidity, like starch when it is much coagulated. It is white, sometimes much, at others

[1] Acosta says that some of the canes are so large that they are used on the rivers. The natives sit on the canes naked, with a small paddle in each hand, and so cross the river. He says he saw two men on one cane on the river of Cranganore, in the part where the current was strongest.

little, like that which forms in canes for writing, and which the boys call *ladras* in Portuguese. As to what you said, you will see that the roots of our canes are not destroyed, as some Arabs assert.

RUANO

Have you seen the TABAXIR in the canes, and how is it that some are black or ash-coloured ?

ORTA

I have seen it many times, those of Bisnaguer and Batecala, and some of Malabar. The people of the country, namely, the carpenters, when they work at them to make some woodwork, find thick juice or pith which they put on the lumbar region or reins, and forehead if there is headache, and if the master carpenter does not take it. Some of them are black or ash-colour, and they are not the worse for that, because there is much in the cane and the moisture reaches to the heart of it. It may be taken for certain that the reason they set fire to the canes is to reach the heart. But afterwards the truth comes out, because sometimes they do not set fire to the canes, for many come which have never seen fire, whence it appears to be the truth that there is much moisture in the canes, and so I was told by natives.

RUANO

The Arabians and Latins have much to say of this simple; therefore, will you kindly take a little trouble to explain to me, declaring what they say, and pointing out where they are wrong and where right ?

ORTA

Rasis, in writing of the TABAXIR, does not say of what it is made, but only of what use it is. Serapio says that SATAXIR is more certainly ESPODIO, and states what use is made of it, quoting Rasis, and Rasis quotes Galen. In this he says well, for Galen never wrote any such thing, nor any other Greek. But this could not stop Serapio, who referred to Dioscorides and Galen, who never said anything of the kind. He also says, quoting the same Galen, that the taste is bitter. It is manifest that this is wrong, for it is sweet; it being for this reason, as I have said before, that the Indians call it "sugar of mambu." Calling it SATAXIR instead of TABAXIR is some mistake, for Serapio had written TABAXIR, and time must have corrupted the name. To write more directly, ESPODIO is an error of the translator. Avicenna says that they are roots of burnt canes, which is wrong. The Belunense says that ALCANA is the name of the tree whence comes the ESPODIO; and in his so-called "Vocabulary" he gives a description of the tree, and says that the roots of the cane are the TABAXIR. Thus Avicenna errs in both translations. Avenrrois says that it is the charcoal from the knots of the burnt cane of India, from which it is clear that he had not seen it, or he would not have called anything so white coal.

RUANO

And what is your opinion of the mistakes of these men?

ORTA

It seems to me that the navigation not being so much used was the reason that information was false and meagre. Valerio Codro says much that is bad about the Arabs, because they make ESPODIO from the roots of canes, the ESPODIO being metal or made from metal. In this he is wrong, for the Arabs, as I told you, do not know that name, but call it TUTIA in conformity with the Greeks. Antonio Musa says that Avicenna used the ESPODIO of canes because he could not get that of metal. You know well that TUTIA never fails to be metal, and they do not use it to drink. He further says that we do not use this ESPODIO, for what we get is counterfeit and false. He adds that modern writers are not wanting, such as Menardo and others, who say that nothing is used for making ESPODIO except metals. In this they are quite wrong, for Dioscorides explains how ESPODIO is made in Book 5. Avicenna is free from all these errors, for he only treats of TABAXIR, and is not answerable for having had bad translators who confuse the names, making ESPODIO to mean two things. In fine, it is said that we use ESPODIO made of the canes of Avicenna, of burnt coral, of burnt ivory, or of burnt bones of elephants. You see, sir, how many errors we have to excuse. With the Greeks you would use the true ESPODIO of metal. With the Arabs you would use the ESPODIO brought from India. With the Latins you would take a medicine to be applied internally and externally, in conformity with the intention of the writer who prepares the composition. But it will be seen that to cool the heart, the brain, the liver or the kidneys, or to check diarrhœa, it will be well to use the TABAXIR of India. Many doctors and druggists and composers of medicine will inform you on this subject. Nearly all treat of one material. Those who say that less harm is done by taking ESPODIO made of our canes are in error, for this is not a cordial medicine like the ESPODIO, but has a cooling effect like TABAXIR. If it is said that it is made of coral or burnt ivory, if this was the intention, Avicenna and the others have said well. Those who say that it is made of elephants are, I declare, worth nothing at all. When an elephant dies the Gentios eat the flesh, and the ivory goes to Europe for sale.

RUANO

You have spoken very well. But, to conclude, I want to be informed how the physicians in India use this medicine, as well those in the service of the King, as those of Persia, Arabia, and Turkey. For with that I shall rest satisfied.

ORTA

Those in this country who understand physic use this TABAXIR against over-heating, external or internal, and for fevers and for dysentery. The physicians of Nizamoxa and those of Arabia, Persia, and Turkey use it in the same way, and make lozenges with the seed of sour fruit trees. In this way I cured, by advice of Nizamoxa, a Portuguese, native of Matosinhos, named Sancho Pirez, who was such a favourite that he was visited daily. The Nizamoxa would let no one treat him but me, because he feared that the other doctors would kill him because he was a favourite.

RUANO

He must have been much beloved. Was he a Moor or a Christian, and was he rich?

ORTA

In secret he told me he was a Christian, and he ate with me things forbidden to Moors, and spoke ill of them. He was not circumcised, though all supposed that he was, but I have seen, and he was not. But he pretended to be a Moor and died worth 6000 *crusados* of rent. It is true that, with this rent, the people who served him were paid. Certainly, whether or no, the devil caught him first in the combat of Calabarga,[1] I had a promise from him to come with me; and I was eager to tell the secret to Don Affonso de Noronha. He made many gifts to Portuguese churches and charities in his will.[2]

[1] Kulburga.

[2] He had been an artilleryman under Nuño da Cunha, and passed into the service of Nizam Shah, who raised him to be a general of cavalry. Bahram Nizam Shah died in 1553. His successor, Hosein, employed Sancho Pires in his war with Adil Shah, and he attacked the fortress of Kulburga, being slain in the breach. Orta obtained a pardon for him from Affonso de Noronha, governor from 1550 to 1554.

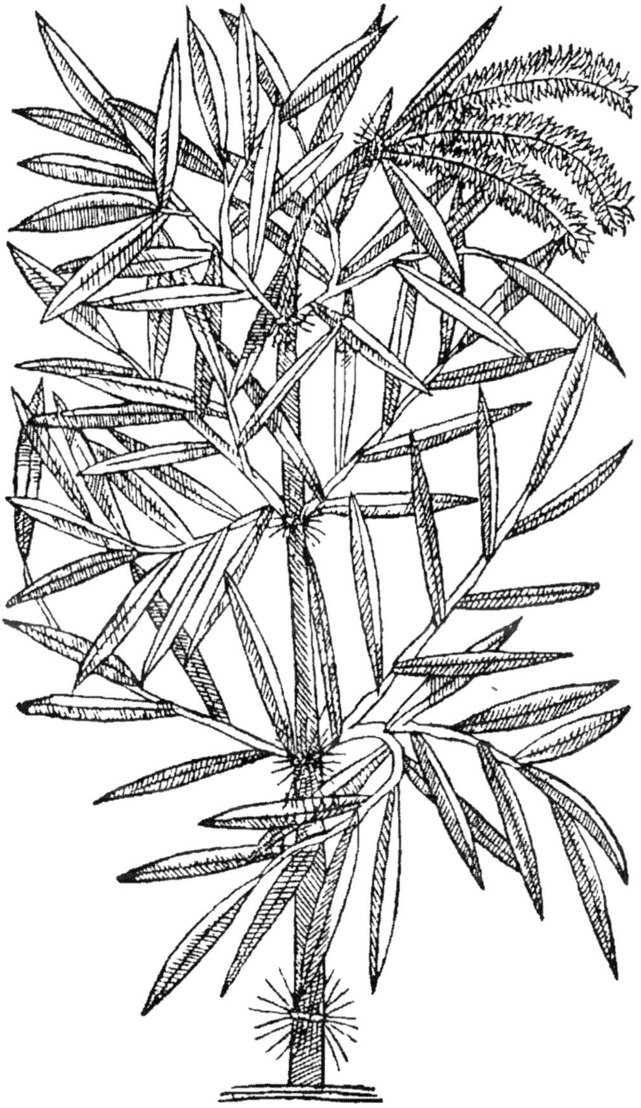

Plate XXII.—Spodio.

FIFTY-SECOND COLLOQUY

ESQUINANTO

RUANO

IN Portugal they say that ESQUINANTO[1] (the medicine used in pharmacy) comes from India, and in Spain also it is said to come from the East. I should like to know its name in the country whence it comes, and in Arabia. For I agree with you that the Greek or Latin need not detain us. I also want you to tell me of the land whence you know it comes, and whether it is much used by Indian physicians.

ORTA

Its native place is Muscat and Calaiate, places in Arabia, where it is as plentiful as the common grass in Spain on which the cattle browse. There they call it CACHABAR, and some call it HAXISCAÇULE, which means "grass for washing." In Persia, which borders on these cities, it is called ALAF, which means grass, and it may be so called *par excellence*. In India it has no other name than the grass of Muscat. In Portuguese,

[1] See footnote at end of this Colloquy, p. 420.

and in Greek and Latin you already know the names. In our land they call it PALHA DE MEQUA, nor are they far wrong, for this land, though a long distance from Mecca by sea, is not so far by land, so that the Arabs of Muscat and Calaiate reach it in a short time. Nor do they err much in calling it pasture for camels, because there are camels in that land. There are so many that feed on the herb and the flower; mules, asses, and the horses called Arabs, of great price, and many cows, goats, and sheep browse on the same grass, which is abundant in that land. It comes to India as a medicine recommended by the faculty, but most of it arrives in the ships bringing horses, for their fodder, and as it gets wet and bruised it is thrown into the sea, and a fresh lot is strewn under the horses. Some sailors also bring it in bundles for sale. I saw many bales in Diu offered at a low price, to be sent to the kingdom with other drugs. I told you that it was called CACHABAR, but I do not deny that it has other names in parts of Arabia. Avicenna calls it ADHAR, and Serapiam ADHER, and this name is used by Arabian and Persian physicians in India, and they call the flower FOCA. Few or none of these flowers come to India. I have not seen any, and they are of no account where they grow, the natives being wild and ignorant. Mateus Silvatico calls them AZQCHIR and ADCARAM, being names corrupted. In this country the natives do not use the medicine, only we use it and the Arabs and Persians; but in its native land it is a common medicine.

RUANO

It remains for us to examine the authors. Beginning with Dioscorides, as the oldest authority, he says that they have it in Africa, in part of Arabia, and in the region of the Nabateos, whence the best comes. He also states that in Arabia it is called by some BABILONICO and by others TEUCHITES. The worst of all comes from Africa, and the flower is used in pharmacy. Do you know whether it is found in those parts?

ORTA

I know that in these parts all point to Arabia. As to the land of the Nabatea I am ignorant whether it is there or not. I say that Nabatea is a province of Arabia, near Judea, so called after Nabatoch, a grandson of Ishmael.[1] Doctors tell me that in Jerusalem and Galilee it comes from Cairo, and the supply in Cairo comes from Muscat, but often the medicinal herbs are not known by the natives, who have little curiosity. This is the reason why I did not ask whether they had it in Babilonia. Dioscorides says that the worst comes from Africa, but he does not say in what part of Africa. What he says about the flower is, I confess, true; for the doctors are not curious enough to send for it. I am also in fault, for the use of it is lost, owing to making no demand for it. I know well that ESQUINANTO is a word corrupted from the Greek, meaning a flower, as you well know.

RUANO

Others call it JUNCO ODORATO, and nearly all the Greeks, with Cornelio Celso, call it JUNCO REDONDO.

ORTA

It is something like a rush, but it does not grow so high. Celso has JUNCO REDONDO, or JUNCO TRIANGULAR, others JUNCO CHEIROSO, to distinguish it from the common rush which we use. Avicenna also says that it has a sweet scent, and that one kind is from the land of Agiami, that is, of Damascus. But I do not know whether it is in those parts, as you say.

[1] Nebajoth, son of Ishmael, Genesis xxv. 13.

ESQUINANTO

RUANO

Avicenna also says that ESQUINANTO has a black fruit, quoting from Dioscorides. Is what Dioscorides says false?

ORTA

It may be that the text was faulty, or that Dioscorides was mistaken.

RUANO

Serapiam says, quoting from Bonifá, that the ESQUINANTO is a herb with roots, having many fine branches under ground, like the root of CHULEM, but much longer, and that it has a fruit like the flower of canes, though softer and smaller. He says that it grows in meadows, and that when it is dried it becomes white.

ORTA

I say that it is not a plant[1] but a herb, as he himself says, being lower. It does not smell like a rose, but it has a good smell when it is fresh.

RUANO

Mateus Silvatico says that it may be kept for ten years.

ORTA

I say that in this land, near the sea, it cannot be kept long, though in other countries it may be preserved for a long time.

RUANO

Antonio Musa says it grows in Apulia.

ORTA

That may be true, if he saw it.

[1] Meaning a bush.

FIFTY-SECOND COLLOQUY

Ruano

The Friars, after describing it, say that it is not the flower but the grass and root, and that the grass they sell in the shops as Esquinanto is not it, and that many believe that the root of Calamo aromatico is the root of Esquinanto, while others think it is the root of Galanga, and that Junco aromatico and Calamo aromatico cannot be very different, owing to the similarity of the names.

Orta

It may well be that all the details given by Dioscorides do not square with Esquinanto, yet Esquinanto is the same as it always was. So the learned physicians of Nizamoxa call it, and the flower Foca, confessing that these names are Greek, the physicians being Arabs. I do not know what further proof you want, for Dioscorides could not know about it so well as those of Muscat which, by land, is not very far from Mecca. As to the writer who says that the *calamo aromatico* resembles the *esquinanto* one with another, the facts are, the former is like a rush, has leaves like lilies, and the *calamo* is much warmer, and has a much longer root. Also the Esquinanto belongs to Muscat, and the *calamo* to India, whence it is taken as merchandize to Arabia. As for the *galanga*, it belongs to China, and is 2000 leagues from Muscat. Likewise the roots and leaves are very different. Here in Goa we have the *galanga* sown. Esquinanto is abundant in its own country without being sown. *Galanga* and *calamo* are cultivated. Those who mistake *galanga* and *calamo* for *esquinanto* are deceived by the prices, for they will cost him ten thousand times more than the *esquinanto*. He would do good by curing us at the will of these men who write, and they would do well to adopt a new plan in trying to cure us, which would be to use no medicine that is doubtful in the opinion of Fuchio. For I observe that the moderns who write now use medicines in their Arab practice, though they are very doubtful about them.[1]

RUANO

Do not be so angry that men should say they are doubtful, it is when they are perverse and obstinate, without giving good reasons, that they are at fault. Let us now pass on, and converse about the tamarinds, for they are so medicinal that a discourse respecting them will be pleasant.

[1] "Esquinanto" here is undoubtedly one, or more, of the scented grasses of India and Southern India, such as the *Andropogon Schoenanthus* of Linnaeus, or *A. citratum* of De Candolle, and *A. Iwarancusa*, Blane, the *alapu* of South-Western India, and *ivarankusha* of India generally. The latter word is Sanskrit, meaning [literally "fever-goad"; compare *ankus*, the elephant-"goad," and our word "anchor"] "febrifuge." It is the *zakhura* and *izkir*, meaning "forage" for camels, of the Arabs, and *gorgiyah*, meaning "camel-hay," of the Persians. All we can say of the σχοῖνος of the ancient Greeks and Romans ["schoenus"] is that it was an aromatic rush; and of the "juncus odoratus," or "squinanthum," "squinanth," "sweete rushe," and "camel's meate" of our older Pharmacists, that it was some fragrant sedge or rush, which they catalogued along with their "juncus holoschoenus," "the matrushe," or "bulrushe," and "juncus mariscus," or "juncus laevis," "whereof watching candles are made." The "Squinancywort" of modern florists is the "Woodruff," *Asperula odorata*.

FIFTY-THIRD COLLOQUY

TAMARINDS

RUANO

THE taste of the tamarind is pleasant, and its medicinal virtues are beyond price, so that it will be well that we should know the names in the Indian languages and in Arabic, and you should tell me what the tree is like and how it is used by the native physicians.

ORTA

There is no deception in the statement that it is a medicinal fruit well known to all, and met with in many places. In the mountains the best are grown, and those that last longest are from Cambaya and Guzerat. In Malabar they are called PALI and in Guzerat AMBILI, as well as in all other parts of India. The Arabic name is TAMARINDI, for TAMAR, as you know very well, is TAMARA, or, as the Castilians call it, DATIL, so that TAMARINDI means the TAMARA of India. This was because the Arabs could find no other name so appropriate, and not because the trees or fruit are alike. The tree is large, like the walnut or ash or chesnut, and the wood is strong, and not porous or soft. It is very well covered with leaves, like fern leaves, on all the branches. The food which covers the fruit is green, and becomes grey when dry, and is easily torn off. Inside there are stones,[1] and being thrown out we use the marrow, which is sticky. These tamarind pods are like a finger curved in the shape of a bow. When they are green they are more acid, but not so much so as not to have a pleasant taste. I use many of them, eating them with sugar, and they agree with me better than acid syrup. Also I may say that I have derived much benefit, in my pains, from an infusion of tamarinds, putting 4 oz. into cold water, or endive water, after standing for three hours, putting in the tamarinds with a little sugar. It has the effect of discharging some of the choleric humour, and throws out and cuts the phlegmatic humours. The people of the country take a purgative of this tamarind with cocoa-nut oil, which is certainly a good medicine, causing neither trouble nor molestation. The Indian doctors used the pounded leaves for application to parts affected by erysipelas. We also use it in our food in place of vinegar, because it is a more agreeable acid when it is ripe. They take them to Portugal with salt, and in the same way to Arabia, Persia, and Turkey, because they say it makes them last longer. I have often had them in the house with their pods on, and they keep very fresh. But the natives preserve them in salt; also making a very delicious conserve of the tamarinds with sugar, and made fresh and without salt. I believe that tamarind is a very good digestive and purgative, very pleasant to the taste. Girl, bring the tamarind in conserve.

RUANO

I shall be much pleased to taste it.

SERVANT GIRL

Here is the tamarind.

[1] *Caroços*, "fruit-stones," which is an incorrect description.

FIFTY-THIRD COLLOQUY

RUANO

It is a very nice conserve, with a pleasant taste. Kindly let me have some to take; I like it best with red sugar of Alexandria, and I should not consider it inconvenient should it be necessary to throw in some rectified scammony.

ORTA

It may be: just at the proper time, and by advice of a good doctor. Further, I caused the tamarinds to be distilled, and used the distilled water as a digestive: but I do not make much now, because I found this sweet water. Pardon me if I wearied you by talking of this more than is necessary.

RUANO

On the contrary, I should like you to spend another hour on this subject, and as nothing can be so well stated but that some doubt will arise, I wish to submit some questions, that the truth may become more manifest. As the ancient Greeks did not know this medicine, we will examine the Arabs and Latins. Mesué, who is so much admired by his Arabian imitators, says that they[1] are from the wild palms of India, and Avicenna does not say what they are, but that the electuary is better when new. Serapio, quoting Bonifá, tells us that in Cesarea, in the lands of Amem,[2] they are found, and that they have a leaf like a willow. He further states, on the authority of Aben Musuai, that the outside of the tamarinds we use comes from India, and that the fruit is red. What do you know of this?

ORTA

I say that there are none in Cesarea, nor in the Amem or Jamen which are the lands of Syria.[2] But the first thing

[1] Tamarinds.

[2] Yemen.

he says is true, that the outside (which is the pulp) comes from India. This is not to say that the stones are not used in medicine. When Mesué says that they are the fruit of wild palms, he does not know what he is talking about, for in all India there is no fruit of palm trees, except the *tamaras* coming from India, which are used dry, in great quantities, as well as when mashed without stones, and some are like those we call datiles.[1] It is true that in Cambaya I saw some fine palm trees, though they were very different from tamarinds, which are taken to Arabia for merchandize.

Ruano

Laguna, the translator of Dioscorides into ordinary Castilian, mentions his author's silence about tamarinds, and accounts for it, if we rely upon the Arabic vocabulary, by its being supposed to be a kind of date coming from the East Indies. For this reason it seems to him that tamarinds do not differ from Thebaic dates, seeing that they come from the East and have the same properties. He further says that, according to some, the tree of the tamarind is a kind of wild palm having its leaves long with sharp points like a willow leaf, also that sometimes white stones are found inside, of various shapes, and that the red ones are looked upon as the most perfect, being thick, fresh, and tender.

Orta

It is not a kind of date, nor has it the form of a date tree except that both have stones, nor are there date palms in this country that yield fruit. Yet there are fine palms in Guzerat, but they give no fruit whatever, neither tamarinds nor dates, but, as I told you, there is a trade with that land. As to his saying that tamarinds are like Thebaic dates, it seems to me that he speaks without sense. He has mistaken the true

[1] Dates.

tamarinds coming from this country for dates. The buyer thinks that what he buys comes from the country where he makes the purchase, as happens with other drugs; for instance we call good cinnamon of Aleppo, though it comes from India. The Arabs who trade in this land call them *tamaras de India* because they have stones, and not because they are like dates. The tree which produces them has not a leaf like the leaves he describes, nor are the stones yellow, nor are they bright and the colour of earth, nor are they of different shapes, but of one size and shape, like a lupin, round and massed together above. They are not thick, fresh, and tender, but as I say. They are not used because the tamarinds come massed together, bring few stones with them, and are not well known.

RUANO

Valerio Codro makes additions to Dioscorides, saying that XIFERHX is tamarind, and that FENICO BALANO is different.

ORTA

He may be right in this, but I do not know them.

RUANO

The Friars say that the true ones seldom come to Europe, and that the good are LEIROM according to Mesué, which they select for the confection of ALIFRACOST.

ORTA

That the tamarinds come adulterated is a foul and base falsehood, for they are so cheap that in Portugal they can yield a very good profit. The tamarinds which Mesué calls *alcairo* are better. The reason is that they come from India to Cairo, and thence, by Alexandria, to Venice, not because they have them in Cairo.

TAMARIND

Ruano

What name had Cairo formerly, and why is it so called now? I ask this, not being connected with medicine, because it is a very famous and ancient city.

Orta

Cairo was called Memphis by the Greeks in ancient times, where there are now the famous pyramids, and where José was a captive. Now the vaults are to be seen where the stores were kept. The Moors call them *Meçera*.[1] And because a Queen named Alcaire increased the size of that city in one direction, the whole city got the name of Cairo.[2] That city, since the Turks have been at Constantinople, continues to fall off in population. To return to the tamarinds I say, what Antonio Musa states, that it will be very unreasonable to dislike the Arabs owing to the tamarinds. Truly they have reason, for I use them, and not *caña fistula* or manna, in very coleric fevers, and this because, being sweet, they increase the heat. This is not wanting in authority, for Avenrrois so orders it. This receipt is much used by the doctors of this country, who do not care to give sugar in burning fevers. So says Antonio Musa, who is clear that the *mirabolano* of Pliny and Dioscorides is not the tamarind, for it has no stones and the tamarind has. He also finds fault with Menardo for reprehending Mesué, and says that the Fenico balano has restringent properties and the tamarind is purgative. He also has not much ground for reprehending Mesué, who says it is cold in the second degree, while Avenrrois places it in the third, for this may be a clerical error, some books of Mesué also saying that it is in the third. With this I say that the account of the tamarinds to your honour is finished.

[1] Misr, from Mizraim, son of Ham.

[2] El Kahirah, the "Victorious."

FIFTY-THIRD COLLOQUY

RUANO

Do not think me tiresome if I ask about a thing that your gardener told me.

ORTA

If he told you that they sleep at night protected from the cold by the leaves, he spoke the truth; for at night I have seen them close together among the leaves, and in the day they open and come outside the leaves.[1]

[1] He alludes to the movements of the sleep and vigils of the leaves of *Tamarindus indica*, Linn.

The names of the *Tamarindus indica* current in India are :—Sanskrit, *tintrina*, meaning literally "to become wet," *i.e.* to "excite salivation," and *amlika*, meaning "sour" [compare *Phyllanthus Emblica*, footnote, page 320], and *chincha* [the name also of the "Baobab-tree" of Africa], meaning "tart," or "biting sour"; and Arabic *tamarihind* [in Persian *tamarihindi*], meaning the "Indian date-tree." With the "Mango-tree" it enters largely into the "folklore" and "proverbial wisdom" of the people of India; one of the "old saws" of them being, "The full flowering of the Tamarind is prophetic of prosperity, of the Mango of drought and adversity."

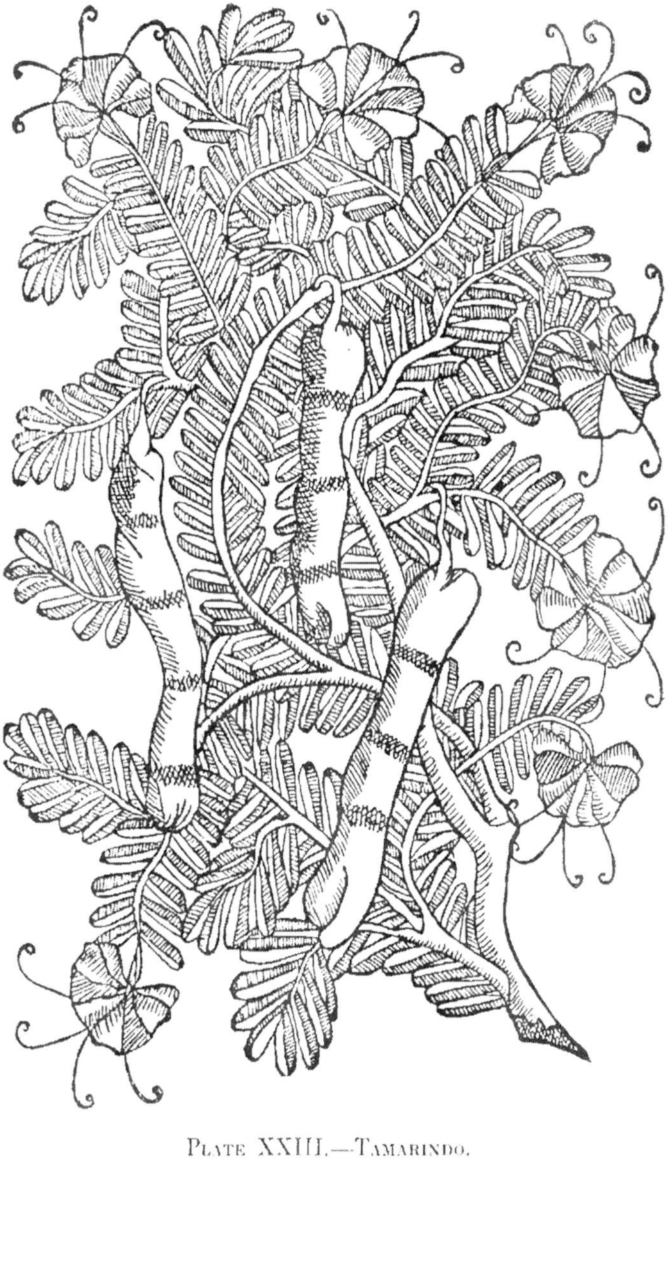

Plate XXIII.—Tamarindo.

FIFTY-FOURTH COLLOQUY

TURBIT [1]

RUANO

I HAVE often heard it said, in very notable cities of Spain, that the manufacture of DIAFINICAM and other confections had been abandoned for want of TURBIT. Others said that it was not genuine, not being black or gummy. Others said that the TURBIT of the Arabs was one thing, and that of the Greeks another, one being called TURBIT and the other TRIPOLIO. These names, they say, are taken from Dioscorides, and to make out their point they amend ancient texts, attack Pliny, and blame Teodoro Guaza.[2] It is certainly a pity to see

[1] The true Turbith of the Arabs, or "Indian Jalap," is obtained from *Convolvulus Turpethum*, Linn. [*Ipomoea Turpethum*, R. Br., *Operculariu Turpethum*, Mauso], a Bindweed closely allied to *C. Scammonia*, Linn., the source of Scammony [the σκαμμώνιον of Hippocrates], and *Ipomoea Purga*, Hayne, *Exogonium Purga*, Bentham, the source of Jalap. It is quite different from the "Turbeth barbarum" of the old Pharmacists, obtained from one of the Umbelliferous *Seseli*; and from the ἄλυπον of Dioscorides and Pliny ["alypon"], the *Turbith blanc* of the French, obtained from *Globularia Alypum*. The word Turbith is the Arabic *turbud*, a corruption of the Sanskrit *tribet*, or "three-virtues," *i.e.*, that it is anti-costive, anti-bilious, and anti-phlegmatic. In Western India it is known also by the name of *chivaka*.

[2] Theodoro Gaza, a translator of Theophrastus.

what trouble was taken by the most learned Lioniceno, Menardo, and others to find this TURBIT in Dioscorides or Pliny, which they could not do, only finding the horn of Amaltea or the city of Platam.[1] Other more modern writers want to reconcile the Greeks and Arabs, leaving themselves at variance. I pray you of your kindness to disentangle these questions, also giving me the names in Arabia and in the countries where the plant grows. If it should be what I have seen, it will be to me a thing of great price. I also wish to hear how the native doctors make use of it, and whether they use much or little.

ORTA

I will tell you, Sir, all I can, for I know this simple very well, and have seen the fruit it yields when it is green, and the flowers. From thence you will yourself see how these modern writers are to be answered, and whether it is the TURBIT of the Greeks or not. I say that what we call TURBIT has the same name in Arabic, Persian, and Turkish, though Andreas Belunense, in his amended text, calls it TERBET. All other educated doctors of those nations call it TURBIT and not TERBET. The Guzeratis, where it is in greatest quantity, call it BARCAMAN; while the name for it among the Canarese of Goa is TIGUAR. It grows on the surface of the earth. I mean that the root does not go down deep and is small. The trunk is the size of a finger, sometimes larger, and it crawls along the ground like ivy. The beginning of the root or branch is the good part, and as it goes tapering away it becomes full of leaves, not having the appearance of TURBIT, nor is it good, nor does it form the gum except at the root which is the proper stick, and this becomes like the real TURBIT. The leaves and flowers are like those of the French mallow, and the flowers do not change three times a day as some say. The taste of the trunk, branch, and leaves is insipid when it

[1] Things that no man hath ever seen.

is gathered. It grows in maritime countries, but not very near the sea. I have seen it two or three leagues from the sea in a place where the sea does not reach, though some say that it does. It mostly grows round Cambaya, Surat, and Diu, and Bazaim. It also grows at Goa, but the doctors do not think it good, preferring to use that from Guzerat. From thence it is taken in great quantities to Persia, Arabia, and Turkey, and a small quantity goes to Portugal. I sent forty quintals when I made the drugs for the King, and heard it was considered a great quantity. The physicians of Nizamoxa in Balagate ask for it, which is a sign that it does not grow in that country or is not good. It may be that it is in other parts of India, but it does not sow itself, and perhaps, if the natives were more curious, they might find it. Some people say that it is in Bisnaguer, which is 150 leagues from Guzerat. Some doctors here in Goa told me that they took it from Guzerat to Bisnaguer, and others that they have it in Bisnaguer, but that it is not so good, and also in Goa, where it is not considered good and is not used, but only that of Guzerat. It is true that what was seen by Mesué, Serapio, and Avicenna was from Guzerat, for the ships that go to the west always bring it as merchandize. I tell you that the plant does not have a branch divided, nor a high part, but all is full of leaves and flowers, in the way I have said.

RUANO

Before you discuss the statements of Grecian and modern Latin writers, I want you to tell me how you know these things that you have explained to me, not because I do not put entire faith in what you say, but that I myself may be able to give reasons to those you do not know.

ORTA

You have reason in what you say. Well, I knew because

when that invincible Captain Martim Affonso de Sousa came to Diu with forty men, by order of Sultan Bahadur, who was the most powerful king of Mourama, and, with so much force and risk, occupied that his city of Diu, so renowned throughout the world, I was with him. After we received the *prazme* of the King to build a fortress, I was idle, looking upon the opulence and traffic of that city. One day being in the bazar (as we call the market or fair) in the afternoon, sitting at the door of one of the merchants they call Banians, a woman came past with a sack of dried TURBIT for sale. As I was an expert in medicines, and had heard that they were brought there for our ships, I asked the Banian what it was. He replied that it was TERUMBU, and that we and the Moors gave it that name, but that the Maratas (who are Gentios) call it BARCAMAN. I then asked for what it was bought and its use. He said it was of use to purge the stomach, that it was considered a good medicine, and was taken by the merchants in their ships to Arabia and Ormuz. He praised it much and asked me if I wanted to buy it, and he showed me its gummosity and whiteness. As I knew that our people valued it, I bought it, each maund for a *tangua*, which is 60 reis, one maund being 27 *arrateis*. But he gave the woman very little. From what I afterwards heard from some other Banians he doubled the price to me, twice over.

RUANO

I am satisfied that this is the TURBIT we use, and that they so call it; but how could you know that the signs of its good quality were that it was white and gummy unless you knew it from our books?

ORTA

I must tell you that I knew that from our books, but not from what the Banian said, for, to tell you the truth, those are not the signs which identify the TURBIT only. For it is possible that it may be TURBIT without gum so good as to be gummy. For the gum collects, because those who gather it twist and prick it when it is green, causing it to press out and exude gum. It is a sign by which we distinguish the good from the bad. I knew this afterwards, for I have a relation who is a doctor in Baçaim, one of our cities, 50 leagues from Diu by sea, and he told me that he had been gathering it with the Indians many times. At first they twist, cut, and prick it, and then, after some days, they gather it, finding it full of gum. He arranged that they should not twist nor cut one, and afterwards he found that there was no gum, and that in some others there was very little.

RUANO

Is the gummy kind as good as the other, and are they from the same plant?

ORTA

You are quite right to ask that. The gum forms within, but some TURBIT is gummy without the twists and blows upon it, as it gummifies more easily. It is our arrangement that we make, according to which the Indians twist the plant, and that no doubt is the truth.

RUANO

How is it that the black is worthless and the white good? What are the conditions of its goodness?

ORTA

The custom of the native druggists, who are called GUANDIS, is to dry it in the sun. They say that it turns black if it is dried in the shade. Then they bring it to our druggists, and they have learnt this way of drying this medicine from experience. It may be that that which becomes black from being dried in the shade is the best, but up to the present time I have not tried the experiment.

RUANO

Do the Indian doctors use this TURBIT to reduce inflammation?

ORTA

Yes, sir! and on this point I wish to call in a native doctor who seems to me to be the best, that I may question him in presence of your worship. Girl! call Malupa.

SERVANT GIRL

He comes here every morning to cure those negresses, and he is coming up. [*Enter* MALUPA.

ORTA

Malupa! tell this gentleman, who is a doctor, how the TURBIT is used in this country, if it is mixed with ginger, and from what part the best comes.

MALUPA

Yes, we use it to reduce inflammation, sometimes mixing ginger with it when there is no fever. We also mix ginger with other purgative medicines, but we also give them without ginger. The best TURBIT is that of Cambaia, whence it is brought to other parts of India. I have shown the TURBIT to this gentleman here present. I may tell you that we sometimes cure with that grown at Goa, but not unless it is gathered near the sea. They tell me that it is at Bisnaguer, but not good.

ORTA

What you say is very true, for the Nizamoxa asked me for this TURBIT of Cambaia and I sent for it. But after all it may be in the interior, and is not found owing to the want of curiosity among the natives. I myself saw in Balagate the bugloss which is wanting here, and the fumitory. You can go with God, Malupa, and I have told this gentleman already what you know of these medicines.

MALUPA

Dr. Orta knows better than all of us, for we only know the Gentios, but he knows Christians, Moors, and Gentios better than us all. I kiss the hand of your honour.

[*Exit* MALUPA.

ORTA

This Indian has told you the true rectification. For Rasis does not rectify with ginger, but with oil of sweet almonds, for fear of the excoriation it may cause.

RUANO

Now let us examine the writers, commencing with the Arabs, for with them we are on more certain ground as you have said, the Greeks having learning and the invention of good letters.

ORTA

Do not go on too fast. I do not say anything against the Greeks as the inventors of good letters, but they were also the inventors of many lies, had many bad habits, and were effeminate in their customs. The Romans, besides what they received from them, received many more things. But after all I do not speak evil of them as regards what they wrote of things in their own land, but only what they wrote about things of which they were ignorant. They filled their books at their own sweet wills, as is shown by what they wrote of the things of India, which are so fabulous. I affirm that as regards India the Arabs are better authorities and err less than the Greeks. Now let us have your contradictions that we may the better reach the truth.

TURBIT

Ruano

Mesué says that it has leaves like the *ferula*[1] except that they are smaller, and that those of the plants give milk; and that there are wild and cultivated, small and large, black and white and citron colour; also that it grows in the driest places for the production of the milk; and that it has seven properties, white, hollow, reedy like a cane, gummy, a bark the colour of ashes, smooth, easily broken, and that the thick kind is not good.

Orta

Mesué told the best he knew and had heard, but he is not altogether right. For the leaves are not like the *ferula*, but like the mallow called by the Portuguese *Malva francesa*, and they have no milk. They are not cultivated but are all wild. There are large and small, as he says, white, yellow, and black, but they are not so from their birth. Those which are badly kept are not white. It thrives in damp and dry places, but more damp than dry. They are not dry, as he says, because of their milk. They are not better for being white and gummy. They are not like canes. The bark is not ash coloured nor smooth, but curled and wrinkled and grey. When new it is good, and not fragile except when dry. He also says that the thick kind is not good, in which he is wrong, for it appears to have more virtue.

Ruano

What think you of Avicenna, who says that in preparing the bark it is scrubbed that it may not be ash coloured but white?

Orta

I say that this only serves for selling it, and for nothing else.

[1] Giant fennel.

FIFTY-FOURTH COLLOQUY

Ruano

Serapio says, on the authority of Dioscorides and many others, that it is born on the sea-shore in places where the sea covers it at high tide, and that when the tide goes down it is not touched; also that it has a leaf like that of a plant called *arasidis* only the leaves are thicker, and that the trunk is two *palmos* long, and that the upper part is divided, and the flower changes its colour three times a day, being white in the morning, red at noon, vermilion at night, that the root is odoriferous and that when chewed it heats the tongue, that it is an antidote to poison, like another medicine called Bezedarica. These things and others have the authority of Galen, translated by Albataric and many other Arabians.

Orta

I have already told you that I have seen the Turbit growing near the sea but not so near as that it would be touched by the tidal waves, because its place is two leagues from the sea, which the spray does not reach. The leaf is not like that of the plant called Arasentis or Ahisatis, nor like the mallow, nor the *murtinhos* as Lioniceno would have it. Indeed they are very different from the mallow. The trunk is sometimes two *palmos*, at others eight and ten *palmos*. The flower does not change colour three times a day, but is always a mixed red and white, sometimes white. The root is not odoriferous, nor biting to the tongue. We do not use the root, only the branch with leaves extended on the ground. It is never used as an antidote for poison. What I say to you I have seen with my eyes.

Ruano

Dioscorides says, speaking of Pitiusa, that it is a species of Laticinios, or of the herbs which yield milk which seems to be the Turbit, and this is the opinion of some modern writers. They also say that the TRIPOLIO of which Dioscorides speaks is translated literally in Serapio. Autuario, a Greek doctor of authority, also says that PITIUSA is TURBIT and that it is white and black, and he says that some falsely use TURBIT PRETO for HESULA, and Mateolo Sinense also thinks so. He also says that ALIPIUM is TURBIT, and ALIPIA the seed of it. But this Mateolo does not believe, because the TURBIT has not seed, and while ALIPIUM cures melancholy, TURBIT reduces inflammation. The Friars say the same as the moderns and as Antonio Musa, and also hold that what Dioscorides, Galen, and Pliny say of TRIPOLIO is true. They say that they have the TURBIT of Serapiam and that it appears to be all one. These reverend Fathers affirm further that the TURBIT sold in most druggists' shops is not the TURBIT of Mesué, and that those who have gathered it with their own proper hands have said so because it has not got the leaves of purslane. It is also conceded that the TURBIT of Mesué is not TAPSIA, because in washing and scrubbing it causes swelling on the hands and face. They say further that the TURBIT which they bring from Apulha is the true TAPSIA and has great roots, but that it must not be administered until six months after it is gathered, nor when it is eaten by an animal.[1] These things and others, which I do not repeat because they are not relevant, are said by many modern writers of repute, who, as you have heard their views, can be answered by you as you see fit, for you must receive belief as you speak, from the testimony of your own eyes.

ORTA

I have myself read all the things you have repeated and more. What I answer to you is that the milk-yielding plants are numerous and that most of them are poisonous. The lands of India and of Europe are full of our kind and of many others.

[1] Bicho.

It pleases God, for the sins of our first parents, that some should be poisonous and, by the divine mercy, some of these are of use in other ways, and for those which are pure poisons we know that there are antidotes. I will give you examples in this country and in Portugal. The ESULA, called by the Arabs ALFEBRA, is poisonous, and its juice or milk causes a swelling where it falls, as I have often seen in Portugal. In these parts there are plants which do the same. One is a species of wild *manga*. For this reason the ancients wrote of seven species of milk-yielding plants, besides which there are many unknown kinds. Here in India there are many other plants with which the doctors purge and cure various infirmities. One of these is TURBIT. It has no milk, or if it contains any it is very little; it is not a poisonous plant, but purges without annoyance or trouble. Here the natives take it in chicken broth, or in water in greater quantity than is the practice with us, either in Portugal or here. It does not cause swelling in the face or hands as you say on the authority of the Friars. It is not this species of milk plant but ESULA, and ESULA is not this TURBIT, nor is TRIPOLIO, nor PITIUSA, nor HISIATIS, nor ALIPIUM, nor is ALIPIA its seed, for it has no seed. Its leaves are not like either FERULA or BELDROEGA or MURTINHO. It does not grow so near the sea that the waves cover it, nor do its flowers change three times a day as these Greeks affirm. So that for these and many other reasons it is clear that TURBIT was unknown to the Greeks and, properly speaking, to the Arabs, although they use it for their people when it is brought from India. This led them to search among the Greek writers for some plant that was like it, for they look upon the Greeks as very high authorities. The reason is that the Greeks were the first writers on human things, though the first who wrote on divine things were the Hebrews. This is why Serapiam translated literally the chapter on TRIPOLIO from Dioscorides, thinking that there could not be a medicine on which the Greeks had not written. Certainly he would have done better to have written a chapter from his own knowledge, and time is better spent in research, as is now done, and proving it to be a medicine of this land. He says that the TRIPOLIO and other medicines are species of milk plants, but he did not know, nor did Dioscorides know everything, for he often says "as is the common fame."

RUANO

Laguna holds that PITIUSA is black TURBIT and that ALIPIUM is white TURBIT and good.

ORTA

I have already told you that none of those medicines are either black or white TURBIT. The ESULA is a very strong laxative and TURBIT is not. ALIPIUM purges melancholy, and TURBIT only reduces inflammation. The TURBIT has not a fragrant root nor does it bite the tongue, nor is it like FERULA nor BELDROEGA nor MURTINHO, nor does it rise from the ground but extends along it. For these and many other reasons it is clear that it is not identical with any of these medicines mentioned by the modern writers.

RUANO

And the TURBIT that is brought from Apulhas. Is not that TURBIT?

ORTA

No! only some milk plant. Some say it is the true TAPSIA because it has great roots; and the TURBIT we use in this country has very small roots, so that we only use the stick.

RUANO

These reverend Fathers say that it must not be used until six months after it is gathered, and that it must not be used if it has been eaten by any animal.

FIFTY-FOURTH COLLOQUY

ORTA

The last statement is clearly true, for in this country things are subject to putrefaction to such an extent that rhubarb and other medicines can only be kept during the four winter months, June, July, August, and September. But the other statement that it must be kept six months before it can be used is wrong, for it is gathered in November, December, and January, and if it was kept for six months it would be rotten. It is true that in the interior the medicines are not subject to corruption, as they are near the sea-shore. You may be assured that the Greeks were totally ignorant of this drug, but I would not say worse of them, nor that they are more wrong than the moderns, who say that they do not know the root which, in our time, is sold for TURBIT. Yet this is true in a way, for it is the branch not the root, and we do not call it TURPETUM but TURBIT. Now let us get on, for to-day is Saturday and we have to go to *Madre de Dios*.

RUANO

I have many other things to say, that I may have much noted down to relate in my country, where to hear them will be much appreciated. For this you should tell me of the cities and countries where the TURBIT grows, namely Baçaim and Diu, which are territories of the King of Portugal.

ORTA

Diu is an island which includes a city and a good port, with a very considerable trade, and concourse of many merchants, Venetians, Greeks, Rumes, Turks, Persians, and Arabs. It was given by the great Sultan Bahadur to Martim Affonso de Sousa, being Chief Captain of the Indian Sea. The grant to him was that he might build the fortress in any part of Diu that he chose, and this he did, so that it could be defended by land and by sea. Afterwards, owing to many treasons that they practised upon us, they lost the city and the whole island of which we have been possessed of for many years. It is a very great and strong city, which, with a small besieged force, we defended against the Grand Turk, with a great power, in 1539. Afterwards, in 1546, it was besieged for seven or eight months, the walls being beaten down, the few Portuguese in the garrison defending it gallantly. At last the Governor, Dom Joam de Crasto, entered the island and city and drove out the Moors, killing a great number. He then resolved to build another and a larger fortress. As these events have already been well described in Latin and in Portuguese I will not write more about them for, as I said, they have already been described in a better style. One thing I will add, that Dom Joam Mazcarenhas, who was captain of that fortress, did deeds worthy of a very valiant captain during the siege, showing much industry, ability, and fortitude, as well as patience when it was necessary. His merits deserve the highest praise.

Ruano

Speak now of Baçaim, for it is a greater thing, and not so well known.

Orta

Baçaim is a very great city, and under its jurisdiction there are many lands and cities. It gives a rent to the king of more than 160,000 *cruzados* with its land and fortresses, afterwards granted to Francisco Barreto. The said lands are called Manora. They include, in one part, an island called Salsette where there are two pagodas or houses of idolatry under ground. One is under a very lofty hill built of stones in greater quantity than in the fortress of Diu, and which may be compared, in Portugal, with a town of four hundred houses. This hill has a grand ascent, and on arriving at the hill it is found to be a great pagoda worked and cut within the rock, where the Friars of San Francisco afterwards built a church called San Miguel. There are many pagodas of stone on the ascent, and near the summit there are other stone houses with their chambers, and still higher are houses cut in the rock, and in them there is a tank or cistern of water, with pipes to lead down the rain water. Altogether there must be three hundred houses, and all contain idols sculptured in stone. But they are very heavy and dark, as things made for worshipping the devil.

They have another pagoda in a part of the island called Maljaz, which is a very grand thing, also cut out of the rock. Within there are many other pagodas very dark and dismal. All who enter these houses say that it makes their flesh creep, it is so dreadful. Another pagoda, the best of all, is on an island called Pori, which we call the Isle of the Elephant. On it there is a hill and in the upper part of it is a subterranean house worked out of the living rock, and the house is as large as a monastery. Within there are courts and cisterns of good water. On the walls, all round, there are sculptured images of elephants, lions, tigers, and many human images, some like Amazons, and in many other shapes well sculptured.[1] Certainly it is a sight well worth seeing, and it would appear that the devil had used all his powers and knowledge to deceive the gentiles into his worship. Some say that it is the work of the Chinese when they navigated to this land. It might well be true seeing that it is so well worked and that the Chinese are sutis. It is true that, at the present day,

[1] This is one of the earliest descriptions of the Cave of Elephanta. Linschoten was there forty years afterwards. The first description of the cave, fuller than the one in the text, was by the Viceroy Dom Joao de Castro, 1545–1548, in his work *Roteiro de Goa a Diu*. It is reprinted in the recent Life of Dom Joao de Castro by Manuel de Sousa Pinto (Lisbon, 1912), pp. 29-31. But Orta probably saw the cave before Joao de Castro; as the first visit of the latter to India was in 1538, while Orta was there in the time of Nuño da Cunha, 1529–1538.

this pagoda is much defiled by cattle getting inside, but in the year 1534, when I came from Portugal, it was a very fine sight. I saw it at the time when Baçaim was at war with us. Soon afterwards the King of Cambaya ceded it to Nuño da Cunha.

RUANO

What sort of people inhabit this land of Baçaim?

ORTA

The Moors were originally lords over it but now there are few there, only those who trade by sea and are called NAITIAS, what we call mestizos, descended from Moors who intermarried with the people of the land. The GENTIOS are of many castes. Those who till and sow the land for rice and other crops are called CURUMBIS,[1] and by us labourers. Those whom we call gardeners, who cultivate flower and fruit gardens, are called MALIS.[2] Clerks and accountants are known as PARUS.[3] They collect the rents of the king and of others, and of estates, and are great negotiators. Others bear arms. Others are called BANEANES, being those who observe the precepts of Pythagoras very exactly. In each inhabited locality there is a caste despised and abhorred by all, who have no intercourse with the others. These eat anything, even dead animals. They are given food to eat in common, without touching the givers. Their work is to clean up filth in houses and streets. They are called DERES or FARAZES.[4] They are also employed as executioners. There are also merchants called COARIS, and in the kingdom of Cambaia they are known as ESPARCIS.[5] We Portuguese call them Jews, but they are not so. They are Gentios who came

[1] *Kumari* cultivators who clear patches in the forests, cultivate them for a year, and then clear another patch.

[2] Gardener caste.

[3] Parvu, clerk caste.

[4] Pariahs.

[5] Parsees.

from Persia, and have special letters of their own and many peculiar superstitions. When they die the bodies are taken out of another door, not the one generally in use; and they have sepulchres where their bodies are put and left until they corrupt and dissolve. They look to the east. They do not circumcise nor are they forbidden pork. Beef is prohibited. By these things we see that they are not Jews. Nor are the Jews who are in the territory of Nizamaluco, and who remain there, held to be Jews. They take strange oaths which, as they do not refer to the subject we are considering, I will not relate to you.

RUANO

Do not tantalise me, but tell me briefly.

ORTA

A cow is used for taking the oath, and they put water on one side of the cow and fire on the other. Then they take a knife in their hands, and say certain words which mean that as this cow is killed with iron, and is encircled by water and fire, so may it be to him and so may he suffer if he swears falsely. It is a thing worth noting as with these men so with others, that no one changes the business of his father, and all those of the caste of shoemakers are so.

FIFTY-FIFTH COLLOQUY

THURE AND MYRRH[1]

RUANO

WHY do they write of two kinds of incense, one of Arabia, and the other of India? It is necessary to know the tree of each, what it is called in the land where it is found, and to know whether it is a medicine used in that land.

ORTA

In this land there is no incense, but it is sent to the King of Portugal from here, for giving alms to many religious houses. But in India there is none. It is brought from Arabia where it is called LOVAM, corrupted from the Greek word OLIBANO, which is much used by them. They also call it CONDER, namely Avicenna, for CONDER or CONDROS means resin, and ÇAMAC means gum in Arabic. For this reason they call the gum of Arabic ÇAMAC ARABI. It is true that Serapio calls it RONDER, but that name is corrupt. For I have spoken with many Arabs about it and they all call it LOVAM, except a few who call it CONDER, but none

[1] See footnote at the end of this Colloquy.

call it RONDER here nor in its own land where it grows. I have asked the Portuguese, who have lived for a long time in this land, and all say that it has no other names, and that the tree is also called LOVAM. These men say that the best is from the hills, while that from the plains is bad, as it gets mixed with the resins of other trees and is used as pitch for the ships. The trees are the property of the king and no one can gather from them without his license. The merchants come from Adem, Xael, and other parts of Arabia and contract with the king for the quantity they are to be allowed to gather, and the price they were to pay for the incense, being good and what we call *incenso macho*. The price is not high. The best that is brought from Arabia to India is worth 2 *cruzados* the quintal.

RUANO

Do they call it *macho*?

ORTA

No. The good incense is called MELETO in India. The bad kind has another name and grows on the plains and not on the hills. Sometimes the two kinds are mixed to send to India, and fetch a much lower price. It comes mixed with the bark of the tree. The tree which produces this gum is not very large, and the leaves are like those of mastich. In India they do not adulterate or falsify it, as they would gain so little by doing so. The Indian doctors use a great deal of the incense for unguents and perfumes, and it is eaten for many diseases of the head, and for dysentery. But the greatest quantity is taken to China for sale, where it is largely used. Those who trade with it there make large profit, and it is also used in the parts bordering on Malacca, where the myrrh is also in demand, called BOLLA in India.

RUANO

Dioscorides, Avicenna, and others say that it is in India, and that it is black; but I know, as you have said, that there is no incense in India.

ORTA

This word INDO is often taken for black. For instance Mesué speaks of MIRABOLANOS INDOS, which are the black kind, as I told you in discussing them.

RUANO

Now tell me about the myrrh. You have said such a few new things about incense, that it is fair that you should now say where the myrrh comes from and what it is like.

ORTA

Much comes to India from Arabia and the land of Abexim [1] which is in Ethiopia, but the truth about this gum has never been ascertained, nor what the tree is like; only a merchant who traded between Melinde and Mozambique told me that the Bedouins brought it by land from Brava and Magadaxo, and that they came, according to what they said, from Caldea, so called by these Bedouins. These men were mountaineers, and spoke pure Arabic, which is said to be nearer the Caldean language than the ancient Syriac. This was told me by an Abexim priest and an Armenian bishop. I asked him why Pico Mirandolano, in his *Apologia*, said that MAGO in the Caldean language signified one who knows. He said that the Holy Scripture was written in the Caldean language, and that he would tell me what MAGO means. He said that MAGOXI, in that Caldean language, means a man of letters or learning, and that these were the MAGOS that came to worship our Lord. He said that these men were not kings but men of

[1] Abyssinia.

great learning, as well in the stars as in other things of nature. This bishop further told me that the star which guided these MAGOS was not a natural celestial one, but an elemental one, such as we call a comet. Tell me what appears to you as regards these stories. None of them seem to me to be good, or what is said by those who rule the Holy Mother Church of Rome.

RUANO

What you say appears good to me, but I should rejoice if you would tell me something more about the incense, for our Castilians say that it is in the Western Indies of our King.[1]

ORTA

I only tell you what I know to be well authenticated, or told me by persons worthy of credit. As to that which you say of writers on the Western Indies I have read it, but as I have not seen, I cannot say whether it is true or not. You can find out in Castile and write there, if it pleases God that you should return, for here you cannot find out much.

[1] Tus, or Turis, meaning incense in general [from *θυσία*, an "offering," a victim "offered" in sacrifice], here refers to Frankincense [*i.e.* "pure" "Incense"], the Hebrew *lebonah*, "milky," "white" [cf. Lebanon, "the White (*i.e.* 'snowy') Mountain"], the Greek *λίβανος*, and Latin *olibanum* [as if "oil of Lebanon"!], is the product of the Burceraceous trees, *Boswellia carterii*, Birdw., the *mohr-modow* of the Sumalis; *B. bhaudagiana*, Birdw., the *mohr add* of the Sumalis; and *B. freriana*, Birdw., the lemon-scented *yegaar* of the Sumalis. The two first make up the bulk of the Frankincense of commerce, while the last is reserved, chiefly in South-Western Arabia and North-Eastern Africa, for special use. See Sir George Birdwood's paper, "The Genus Boswellia," *Transactions of the Linnean Society*, 1871, xxvii. 111-148, and plates 29-32; and his articles "Incense" and "Boswellia," *Encyclopædia Britannica*, 12th Ed. *B. serrata*, Roxb., *B. thurifera*, Colebrooke, supplies a semi-fluid oleo-resinous exudation, the *salika*, contracted to *salia* [meaning "running," "drooling"], of the people of the country, and "Indian Olibanum" of Anglo-Indians. The *thus terrae* of Pliny and our older Pharmacists is the "Ground Pine."

Myrrh is obtained from the Burceraceous *Balsamodendron Myrrha*, Nees, of North-East Africa and South-East Arabia; and in the bazaars of India is always mixed up with "Indian Bdellium" produced by the *Balsamodendron roxburghii*, Arnot, of Assam.

FIFTY-SIXTH COLLOQUY

TUTIA[1]

RUANO

THEY take TUTIA from here to Portugal, according to what they told me there. Avicenna also says that there is TUTIA in India, and Serapio tells us that a species of TUTIA grows there. Will you tell me what this TUTIA is, and in what part of India it grows?

ORTA

The parts of India we know do not yield TUTIA nor ESPODIO, as the Greeks call it, nor the copper or other metals of which this TUTIA is composed. But if you want me to believe, you will tell me what is the TUTIA they use in India, Portugal, and Spain, and what is the mineral, whether it is only the ANTISPODIO mentioned by Dioscorides, or another like it.

RUANO

Where does this TUTIA come from, how is it made, and whither does it go?

[1] See Colloquy LI. *supra*. Marco Polo found an impure "Tutia"—sulphate of Zinc—in Kerman, obtained from a zinc mine there. Teixeira gives a similar account.

FIFTY-SIXTH COLLOQUY

ORTA

A rich merchant of these lands, who is very curious in such matters, told me that he learnt from native merchants of Persia that it was certainly made in Guirmon,[1] a land of Persia not far from Ormuz, and that it consists of the cinders of a plant called GOAN, yielding a fruit also called Goan, and that they take the crust or bark, eating the pith and outside, but not the inner bark. From this tree, yielding this fruit, they make this TUTIA out of the cinders. The city of Guirmon is very celebrated for the best cummin seed in Persia. It is brought to Ormuz and other parts of Arabia, whence it finds its way by Alexandria to Portugal. In many ships this TUTIA is found, brought as merchandize, and I saw it sent to Portugal for the King. A Portuguese druggist told me that the TUTIA was what was used in Spain and France under the name of ALEXANDRINA, not made there but brought from Persia, and this is one of the ANTISPODIOS of the Greeks.

RUANO

I am not surprised at these counterfeit things, for I saw that a doctor brought you some counterfeit copperas, and told you that the Indian surgeons used it with good effect, because it was a good caustic.

ORTA

The Indian doctors know the use of mineral substances in medicine. I have seen them melting and pulverizing metals such as steel, iron, and mercury. The King of Cranganor in Malabar was given pulverized mercury to drink for a long time, as he had a previous disposition to leprosy of which I cured him, and he is now much better. He was cured by the Portuguese method.

[1] Kerman.

FIFTY-SEVENTH COLLOQUY

ZEDOARY[1]

Ruano

YOU know well what doubt there is about the drug called Zedoaria and the other called Zerumbet. Avicenna devotes two chapters and Serapio one to Zerumbet, and Rasis gives a chapter to each. Tell me about them, giving the names and the uses in their native countries.

Orta

I had the same doubts as you profess for a long time. I found that Zedoaria, being the most famous, was what we call Zerumba, a drug used at Ormuz and taken thence to Turkey and Venice. The Zerumbet is what we call Açafram da Terra, which on the face of it seems like our Ruiva seca of which I have spoken to you as the Croco Indiano. After much enquiry and research I concluded that this was a mistake owing to the different effects produced by the Açafram da Terra and those by Zedoaria and Zerumbet, so called by us. For Avicenna makes a chapter for Zedoaria and for

[1] *Curcuma Zedoaria*, Roxb., the *Curcuma aromatica* of Salisb. The *C. Zedoaria* of Roscoe is the *C. Zerumbet* of Roxb. See also Colloquies XVIII. and XXIV. *supra.*

ZERUMBET, but that which we call ZEDOARIA Avicenna calls GEIDUAR, and I know no other name, for it is not known elsewhere, except on the confines of China. This GEIDUAR is a medicine of great price and only known in the hands of the Gentios called Joques,[1] or others called by the Moors CALANDARES. All these are mendicants who pass lives of begging and travelling and visiting pagodas. These and kings or great persons have this GEIDUAR, which we corruptly call ZEDOARIA.

RUANO

And how do you support this which you so boldly assert?

ORTA

The physicians of Nizamoxa told me. Wanting to give something, for the bite of a rodent, to a man in the retinue of Nizamoxa, they ordered me to petition the king. I said to them that the druggists must have a remedy, and I showed it. They answered that what I showed them was ZERUMBA and not GEIDUAR. Giving it against the bite it did the workman good, restored the circulation, and proved its efficacy.

RUANO

And what was this ZEDOARIA like?

ORTA

It was the size of an acorn and nearly the same form, and the colour was light. I asked the King for an *arratel* of this medicine and he replied that he could not give me so much, giving a bit weighing half an ounce. I showed it to the druggists of Chaul and Goa, and they all told me that they did not know that medicine, and did not use it. I sent it to Portugal with an Armenia stone, and all were lost with the ship on board which they were. God be praised. Afterwards

[1] Yogis.

I found some in the hands of a Joque, but I did not buy it because I did not know it well. If I find any doctor here with some, I will buy it, and show it to you.

RUANO

Is this GEIDUAR useful in other ways?

ORTA

Mula Ucem, a learned physician, with whom I conversed when I was at Juner curing the sons of Nizamoxa, told me that it was efficacious for thirty-six things, telling me of many of them. I saw him apply it to a knee from which a merchant was suffering pain. At least you may believe that it is much esteemed as a medicine, principally as an antidote for poison.

RUANO

Well, Avicenna does not make so much of it.

ORTA

Avicenna did not know, and was very doubtful about this medicine. He devoted two chapters to what was dubious, and in chapter 752 he has "GEIDUAR what is it," and adds that it may be ALGEIDUAR. Dioscorides never alluded to it. So that you will see that Avicenna was doubtful about this medicine. The Belunensis, in his exposition of Arabic names, seems to have scented this, for he makes mention of ZEDUAR and of ZEDOARIA and ZERUMBET. From this you will know that ZEDOARIA is a corrupted word and that GEIDUAR is the true word. Now I will tell you what ZERUMBET is, and you will see the end of the contradictions it is your habit to make. For I am on foot, it being true that I have feet and can walk and never die. I say that the Arabs, Persians, and Turks call the ZERUMBET ZERUMBA, the Guzeratis and Deccanis CACHORÁ, the people of Malabar ÇUA. Most of it is in

Malabar, namely Calicut and Cananore. It grows in the woods, but is planted or sown and thrives in many parts. Some people call it *wild ginger* and with some reason, for the leaf is like that of ginger, except that the ginger leaf is longer than the ZERUMBA and more open, and the ZERUMBA leaf is larger. What is collected is dried in slices and sent to Ormuz as merchandize for Arabia and Persia, then to Jidda and Alexandria, to Venice and other places. The trade is profitable. They also make a conserve of it with sugar, which is better than ginger. This is an account of it, by which you will see that it is not a tree as some have falsely stated.

RUANO

Now it is necessary to face the doubts which have arisen from this account. I say that the statement of Avicenna is that the ZEDOARIA has certain slices like those of ARISTOLOGIA, and that that plant is better, growing near the NAPELO or water cress of Pero Jogral, and the NAPELO is believed to have virtue against poisons. In chapter 745 he says of the ZERUMBET that it is a herb like the CIPERO or JUNCA AVELANADA except that it is more fragrant, and in another sentence he says that it is a tree. But he does not say all that Serapio affirms about ZEDOARIA. Serapio in chapter 172 says that ZERUMBET is ZEDOARIA on the authority of Isaac Aben Amori, that ZERUMBET has round roots like ARISTOLOGIA, and that it is like ginger in colour and to the taste, and that it is brought from SENI. Now consider this, and tell me how it appears to you.

ORTA

Avicenna only saw ZERUMBA or ZERUMBET as I told you, and as these are sent by the strait of Mecca,[1] they are made into round slices or compressed, which may have given rise

[1] Red Sea.

to the idea that they were two species, namely ZEDOARIA and ZERUMBA. He never saw the leaves, nor a picture from India, but only the form in which they were brought from India, as roots like those of ginger. Now the round and pressed ZERUMBAS have different prices, and the small ginger roots are worth less than the larger ones. He who says that it is better when it grows near the NAPELO repeats a very fabulous tale, for the NAPELO is scarce and ZERUMBA grows wherever it is sown, but the greatest quantity is in Malabar, in the woods. What they sow in those parts is very little, and the woods are not suited for the growth of NAPELO. If the name of NAPELO is in the language of this land, I never heard it used by the people of Malabar who live where it grows. Avicenna says of the ZERUMBET that it grows like it, and another amended version says that it is *lignum* or tree, by which you will see that Avicenna did not know, for it is not a tree but a legume. Serapio has not written positively of the ZEDOARIA, for this is a mistake of the translator, who did not know the difference between ZEDOARIA and ZERUMBET. At the end he says that these roots come from Seni, and that in India these roots do not grow, but in Seni, and that the few in India are brought from China, as I have already told you. So that there is ZERUMBA in India and ZEDOARIA in China.

RUANO

How do you know that China is the same as Seni?

ORTA

For many reasons that you will know. At present it suffices to know that RABAN SENI means "root of China," and so it is, for the good RUIBARBO is not in China. On this point you need have no doubt.

FIFTY-SEVENTH COLLOQUY

Ruano

Antonio Musa put great dishonour on Zedoaria by calling it barbarous. He could not use greater abuse. Serapio, talking of Zerumba, must have meant Zedoaria, for what Simao Genovense says is a proof that they are different medicines, for Mesué, describing the electuary of *gemis*, makes mention of Zedoaria in one place, and further on refers to Zerumbet in another place. He further says that some others said that Zedoaria was Arnabo or Zarnabo, which with Paulo and Aecio is Arnabo. Others allege that it is Bem Album and Rubeum, others Carpesio, so that I do not know what I can say about it.

Orta

The Carpesio is not any of these medicines, and there is no such drug in this land except what comes from the Strait, which fetches a good price. Of the others there are plenty here and they are very different. The clear Carpesio is not the same, for one is a root and the other grains. Zarnabo it cannot be, for it is a large tree, as Avicenna says, and further it has little scent. Zarnabo or Arnabo is a very large tree, while the Zedoaria or Zerumba is a legume. With this you can answer Fuchsio, Mateolo, Ruelio, and the Friars, who all say almost the same thing.

FIFTY-EIGHTH COLLOQUY

Treating of some things which came to the notice of the author and of medicines already discussed; increasing them by some other medicines and fruits; also speaks of a kind of wild rice, of betel, of the city of Badajos, of the caña fistula, and of cirifoles, a medicine praised for dysentery; also of the city of Chitore, of ivory, of the mangosteens, patecas, root of China, and of a stone much praised as an antidote which is found in the skin of the porcupine.

Licentiate Dimas Bosque [1]

AMONG friends all things are common, so that I have the privilege of amending the things which have been discussed by them, much more so when I am asked to do so and to tell you about the things with which we have both practised, to see if we can discover truths not known to all. You thus give me leave to make such corrections as seem to me desirable, and so I come now, but it is necessary that you should give permission anew.

[1] The Licentiate Dimas Bosque was a Spaniard, a native of Valencia, who graduated at Salamanca and also studied at Coimbra. In 1558 he went to India with the Viceroy Dom Constantino de Braganza as his medical adviser. He acquired the small island of Santa Cruz, near old Goa, in 1561. He accompanied the Viceroy in his expedition to Jafnapatam at the north end of Ceylon. He wrote a detailed description of the Dugong (*Halicore Indicus*), published in a volume of the *Historia de Sociedade de Jesus*. The date and place of his death is unknown.

FIFTY-EIGHTH COLLOQUY

ORTA

You are excused from asking it anew, because you before had done me a great favour in this.

DIMAS BOSQUE

I want to say that there is a kind of rice brought from Java to Malacca called PULOT which, when boiled only in the steam, sticks to the hands and is so moist that it is as if it had been boiled in butter.

ORTA

I am not surprised at the first effect, for it must have been boiled with the steam like *cuzcuz*;[1] for the same thing happens with the other rice, and to that which goes to Portugal when boiled in the same way, with salt water in the absence of fresh. The other kind, which becomes greasy and damp, I have never tried, for I am not very fond of rice.

DIMAS BOSQUE

Well, I asked every one in your house and they will tell you; further, I have tried it, and you may take my word for it.

ORTA

That I will do willingly. But tell me what the gardeners of your island say about the betel. Do they tell you anything new?

DIMAS BOSQUE

I never get more from them than that it is very much liked, and that when it is collected it is not well to touch it much with the hand, and that it does not like either much heat or much cold.

[1] A kind of *pâté d'Italie*.

Orta

It seems to me that you have reason for this *betre.* It does not thrive in the interior, and is brought from the skirts of the sea to Balaguate. I know that it will not grow in China because the climate is too cold, nor in Mozambique nor Sofala owing to the heat, and in all these lands they require much of it.

Dimas Bosque

I also find in one of your colloquies, written by the way, that the city of Badajos, so written by the Spaniards, ought to be written *Guadajoz*, city of nuts. But I find it stated by a learned modern author, named Gaspar Barreiros, that the Moors corrupted the name, for it was originally called Pax Augusta, and as the Moors have no P they put B in its place and called it *Bagus.*

Orta

I have seen that writing, and the author seems to be a man of learning and judgment, yet I think his derivation strained, and mine seems to me to be best for it is adopted by the Moors and it is common fame. I may be wrong, for all men think their own ideas better than those of others, though it is true that the Moors have no P, but B serves them as P when it has two points and is then pronounced P.[1]

Dimas Bosque

With reference to what you say about *canafistola*, they now have it in Malacca, and there is much in Malacca and Siam and in all those parts. These things do not fetch much

[1] ب is B, but the additional two dots پ belong to the Persian alphabet, not the Arabic. Orta had spoken of the etymology of Badajos in the Thirty-second Colloquy.

Gaspar Barreiros' book was published in 1561. It gives several derivations for the name Badajos—from the river or country of nuts (Lebrija), from the land of supplies (Fray Joas de Sousa y Marmol), and from a corrupt form of the Latin *Pax Augusta* turned into Bagus or Badaxus.

there, for they are nothing more than curiosities, and are often found in the city of Chitore. I do not know whether you know what *sombreiro* means, for so a chronicler of India writes it, and people rejoice in any new thing.

ORTA

Derivations of names are much better ascertained in the countries where they originate, and where we know the languages well, than in strange countries where we scarcely know a word much less know the derivations. I must tell you that CETRI means *sombreiro*, called by some CHATRI. Talking to some Guzeratis on this subject they told me that CHITOR signified a bird so called, and, more properly speaking, a drawing or picture. It seems to me that this derivation fits better, but as in this I know little, it must be as your worship pleases, for truly the city is a picture, according to what people say who have been there. I have not seen it.

DIMAS BOSQUE

These things that I have said to you up to this time are but the flourishes of fencers. What I now speak of is a medicine very good for dysentery. You know that one of the principal cures we have to effect in this land is that of dysentery. For though there are many medicines which cure it, yet we sometimes meet with it so old and chronic that none of them will take effect. I have since seen a simple medicine which even then effects a cure, and I have brought it here for you to see.

RUANO

What you say of it we also say, and we have already tried it.

DIMAS BOSQUE

Have you ever heard of *Marmelos de Bengala*?[1]

[1] The Orange-bloom, *Aegle Marmelos*, Correa; the Sanskrit *bilva*, *i.e.* "destroyer" of disease, and *shriphala*, "holy-fruit"; Arabic *safarjalihindi*, *i.e.* "Indian Quince." Cultivated throughout India, along with *Feronia elephantum*, Roxb., the "Wood-Apple Tree."

ORTA

Yes, I have heard, and sometimes I have seen it in conserve. It seems to me to be very styptic, and the physicians of Guzerat use this fruit when young and tender as a conserve in vinegar, which they call ACHAR, and also in conserve of sugar as we use it. It always preserves that styptic taste however ripe it may be.

RUANO

Now that you both agree that it is a styptic and good for the dysentery, it will be well that the doctor should state the names and appearance of this fruit or tree.

ORTA

This fruit had the name of *Marmelo de Benguala* because the conserve was brought in a ship of mine, and I saw it mixed with others when they came. Now I see, with that name, that it is good for dysentery. I having praised it to my very good friend, a man of rare intelligence, who often went into the woods to hunt for game, he told me that this fruit need not be called *Marmelo de Benguala* for that there are many such trees on the mainland, and some in this island. If you want to know the name of this fruit I tell you that in Bengal and in all that part the name is CIRIFOLES and BELAS. As I knew that it was called BELI in Baçaim I asked the native doctors what was its proper name, CIRIFOLES or BELAS. They replied that CIRIFOLES was the vulgar name and BELI the name in pharmacy and in their writing. The tree is the size of an olive or larger; the leaves are like those of a peach tree and with the same scent, of little use, and last but a short time. At first they are tender, the colour dark green; the bark thin at first, and afterwards it becomes thicker. It is dried when the fruit is ripe, because then it becomes as hard as a cocoanut. At first the fruit is the size of a small orange, but it increases till it reaches to the size of a large quince. They draw out the inside, which is very tough when ripe, and cutting it into large slices make into a conserve with sugar, as I have said. When it is younger and more tender they eat it in ACHAR or salted. This is all I know about this fruit or medicine. The Senhor Licentiate may now say what he knows of this medicine, and what has happened to him with respect to it, for he also is among the number of physicians who are lovers of the truth.

DIMAS BOSQUE

When the Viceroy Dom Constantino was in Jafanapatam, owing to the continual labour of fighting and the heavy rains to which the soldiers were always exposed, a great many sickened with dysentery; and their cure always fell into my hands, as there was no other doctor in the fleet. All the medicine we brought with us had been expended in the island of Manar, on the illnesses of two royal ships, where in forty days I cured over 300 men. Having nothing left with which to treat the sufferers from dysentery, whose sickness gave so much work to the army, I was forced to try experiments with these *Marmelos*, of which I heard from the natives. I cured many cases with them, ordering plasters to be made for the belly and stomach. I also ordered a marmalade to be made which did not taste bad, and had a pleasant acid flavour. I ordered the sick to eat it roasted, with sugar. During the time that the dysentery lasted I caused clysters of the boiled bark to be applied, which had an effect not very different from the *balaustias* which we use. Thus with this drug called MARMELOS cures were effected in the absence of other medicines. I cannot refrain from relating an incident which happened to me in connection with these *marmelos*. Augustinho Nunez, son of Lionardo Nunez, Chief Physician of those kingdoms, had many of his soldiers in the sick list. I ordered two *marmelos* to be roasted by a negro, and taken to a sick soldier. The kernels of these *marmelos*, bursting in the fire, the negro that was roasting them was so burnt that it looked as if it had been from a pan of gunpowder, on the breast, arms, and face, and it did not leave a place that it had not reached. It appears to me that this fire did more, for the material was more styptic and compact, the fire burning more in iron and stone than in the tow. This is what I saw of the medicine in question, respecting which I can bear testimony.

ORTA

There is much reason in what your worship has said, for that inside part, when the fruit is not very dry, is so glutinous and sticky that those who eat it cannot get it clear of their hands.

RUANO

I am taking some jars of the conserve of these *marmelos*, if I can find any.

ORTA

I will do you the service of seeking for one. I said that I brought you some letters from Malacca in that *catur* which arrived from Cochin yesterday, and the news is that the ships of Malacca are coming.

DIMAS BOSQUE

They bring me letters and news from my estate. I rejoice that Dr. Ruano is here, because he can see the *doriões* and *mangosteens* before he goes. For they will bring me here each sort of fruit made in wax.

RUANO

This year I have not yet seen a pear in the kingdom, and when the ship arrives I shall rejoice much to see these fruits.

ORTA

It is a very beautiful fruit, the size of a very large pine-apple, and with the same appearance, though its nibs are finer, like those of a hedgehog.

DIMAS BOSQUE

In my letter it says that there are others much larger than these, and they are called *elephant's head.* They have four divisions on the top, which they call *peitacas,* and between them a leaf like a small lance, divided down the middle into two threads, and others which twist round the inner sides. The tree is very green, very large and well tufted. They say it does not yield fruit for forty years. The fruit when ripe has the green more clear.

ORTA

A man who is married at Malacca told me that it gives fruit every four years, and that he has seen it.

DIMAS BOSQUE

It may be so, and we cannot get at the truth too clearly. He writes to me, with reference to the *doriam,*[1] that the pulp inside is like cream. Here you see the *mangosteen.*[2] It is also dark green and the size of a small orange.

ORTA

Well, they are planted here, so they may yield fruit, and we shall see by experience if God spares us long enough.

DIMAS BOSQUE

I am also reminded that, in your chapter on ivory, I saw that you speak of mineral ivory mentioned by Andreas de Laguna. One of two things occurs to me in this: either you have not seen this author, or some friend of yours has not seen

[1] See Colloquy XX. *supra.*

[2] See Colloquy X. *supra.*

the whole chapter, and you do not wish to find fault with him. This may be and you do not know the title, for you are wrong as regards the name. You call the author *Tordelaguna* when his name is Andreas de Laguna.[1]

ORTA

Laguna makes a statement so contrary to reason that I was ashamed to denounce it, it being false on the face of it, nor does he refer to any author as his authority, so that the error is self-evident. What I have to say about the mistake in the name is that I cannot blame myself because I did not read the title well, and also because I knew a pharmacist in Alcala who called it *Tordelaguna*, a man who knew a little Arabic, and was a great herbalist, and I therefore thought that it was right. I am glad to be corrected. For he was my friend, and I would not rejoice over such a mistake as he made.

RUANO

If we begin to expose errors, there was Leonardo Fuchsio, a learned man, who said that there was no true ivory in the world.

ORTA

There are some lies so gross that it is not well to expose them, nor do they merit exposure; better to let them pass on, giving them a dozen strokes like the clock at noon. This man wrote many years ago, and I do not mention him by name, for he knew little of physic, and still less of things to save his soul, being a heretic condemned for Lutheranism. His books were put in the condemned catalogue. A monk of the order of preachers told me that he knew him in Germany, and often conversed with him, but could never convince him, for which

[1] Orta did this throughout, and Count Ficalho has substituted Laguna in consequence of this correction.

reason I came to abhor his works. Though medicine is not the science of the Christian religion, still I abhor the author, and his statement that there is no true ivory was very disgraceful, when there are so many elephants in several parts of India and in Ethiopia, and they have been taken to Portugal. I suppose the Lutherans have some ivory in hell which they are taking care of.

RUANO

It seems to me that you may excuse Andreas de Laguna, for you showed me here, a few days ago, horns which made roots, and I saw them with very large roots.

ORTA

It is true that I showed them to you, and there are many in this land, as it is moist. But the ivory is not buried, at least not in that way.

DIMAS BOSQUE

Have you written about this fruit called *ananaz*, for it is certainly the king of fruit as regards taste, and more so as regards scent?

ORTA

Oviedo has written on this fruit, he who wrote of the Western Indies, as one proper to that land, so that it was not necessary for me to treat of it. In the province of Santa Cruz, called by us Brazil, they know better how to describe it.

DIMAS BOSQUE

About the *patecas*, I find it written that they are not the *balancias* of Africa. In this I think you are mistaken. For here men born and bred in Azamor, Tangiers, and Arzilla have told me that the *balancias* of India are the same as the *datecas* of Africa.

ORTA

I did not say that the one fruit was different from the other, for, to judge of that, it was necessary to know both the fruits, and I never saw them in Portugal. But I say that I may be mistaken in this, for the plantation of these *patecas* is very different from the ground where the melons of Portugal are grown, and the *albudiecas* and *sandias*[1] of Castile are different from the *patecas* of India. I must refer it to persons who have seen both.

DIMAS BOSQUE

You also have to further increase what you said of the *Pao de China*. Where they sow it, the plant binds itself to the trees like ivy.

ORTA

I believe this, and I wish that it could be written from testimony of eye-witnesses.

DIMAS BOSQUE

This medicine that I want you to speak about now is very necessary, because it is an antidote against poison, and is brought from Malacca by an educated man, a friend of yours who you know very well.

ORTA

If it is the man you met the other day when we went to visit that gentleman, I know well what medicine it is. I had rather not write about it until you have first told me that you have seen it, and what you have to say about it. Then we shall have two witnesses touching the public voice and fame this medicine has in the parts of Malacca, which will give us authority.

[1] Water melons.

FIFTY-EIGHTH COLLOQUY

DIMAS BOSQUE

I know that you saw it, from the signs that you gave.

ORTA

I have not seen it, but your man told me that it was a stone against poison, that it belonged to you, and that if I came to your place he would show it to me. He further showed me the appearance of the stone, and said that it was held in great esteem in Malacca. This stone is found in Pam,[1] a country bordering on Malacca. It is found in the skin of a porcupine, and the natives put a high value upon it.

DIMAS BOSQUE

Do you know in what esteem? Do you know that another was found, sister to this, and sent to the Count of Redondo, Viceroy of India? In the land of Pam, where they find the bezoar stone in great abundance, or at least in greater quantity than here, it is more esteemed than the bezoar concerning which you have already written.

ORTA

I do not remember having read of this stone of the porcupine's skin, and therefore I desire to gain some knowledge of it.

DIMAS BOSQUE

Well, I will give you reason and experience.

ORTA

You promise me much.

DIMAS BOSQUE

You have told me, treating of the bezoar stone, that the Moors of Persia say that it is found in three places, namely

[1] Pahang.

Coraçone,[1] the Isle of Cows near Cape Comorin, and Pam near Malacca. You add that the pasture on which the cattle and sheep feed is all of the same kind in those parts, and that for that reason the sheep and goats secrete in their stomachs a stone which is useful against poison. Now the stone of the porcupine's skin is found in the same land, and the natives know its value. It is conformable to reason that they are not deceived. As for experience, I gave it to two persons who had been given poison. Being very ill, I gave them the water of this stone and they were soon much better. Now you see how I have kept my promise. They gave you an account of the stone and its use against poison, and I give you my experience as the result of an experiment.

ORTA

I have nothing to reply to all this except that it is well said. God prolonging my life, I will try the same experiment many times, for poisoning is very common in this land.

DIMAS BOSQUE

Now I wish to show it to Dr. Ruano. See, it is here.

RUANO

The colour of the stone is a clear vermilion. It is bitter to the taste, and to the touch it is like French soap. It is necessary that you should tell us how you made experiments with it.

DIMAS BOSQUE

I put it in water and gave it to drink. The patient said that the taste was bitter, and that it left the stomach strong and comforted.

[1] Khorasan.

ORTA

All this is true, for the man who was treated told me that the taste was bitter, and gave his stomach much content. I am well satisfied with the stone, and should like to know more about it.

RUANO

And I want one to take back to Portugal.

ORTA

If I had one I should wish to give it, but this does not seem likely, for there are not many. When the time comes to find out all, it will be discovered. Certainly the physicians of this land are much indebted to you for making the experiment, for the more medicines we have against poisons the more they are wanted. It also seems to me that if this stone was taken to Rome it would be highly valued.

THE LAST COLLOQUY

On betel and other things, in which some mistakes throughout the work are amended, which have been left through forgetfulness.

Ruano

IT appears to me, sir, that we forgot to speak of betel.[1] The people throughout this country are so accustomed to eat it. But I have never seen your worship eat or taste it, and the people in this house tell me that they never like to be seen to eat it. It seems that either you are very pertinacious, or you only remain in the faith of a Portuguese.

Orta

Well! my memory is so weak that you must pardon this instance of forgetfulness, as well as many others that I may have passed over. As for my not eating it, this is no proof that it is not very good, but it is a proof of my obstinacy, as you said. I tried the betel when I came from Portugal, in Pangim, which is a small fort at the mouth of the river, and it embittered me as it embitters all who eat it. It is mixed with *areca* and a little lime, and with this mixture it is said to

[1] See Colloquies XXII. and XXIII. *supra.*

make a very pleasant juice. But my first trial left me with such an abhorrence of it that Nizamoxa, who eats it, never could induce me to do so. Many women have it in their mouths (though they may be Portuguese), for no woman converses with a man who has not chewed it in her mouth.

RUANO

Do they mix nothing with it besides what you have said ?

ORTA

They mix CATE with it and important persons add camphor of Borneo, some LINALOES, and ALMISQUERE, or AMBRE.

RUANO

I should not think that they throw camphor in, for it makes men impotent.

ORTA

Yes, they do; and the great Sultan Bahadur, King of Cambay, laughed, saying, "You may tell the Portuguese that this camphor makes men impotent." I replied that camphor in small quantities, mixed with other medicines, does not have that effect. In our colloquies that treat of camphor, *areca*[1] and *cate*, you have discussed these other medicines, so I will not speak of them to you now, but only of the betel. When it is made with this mixture, it is so agreeable to the taste, and makes such a pleasant scent, that every one chews it continually. Very little time passes when those that can afford it are not chewing it. I say this because in the countries far from the sea it is very dear. It costs the Nizamoxa 30,000 *cruzados* a year, for all the fruit they give you is this, and when you want to take your leave, with this you have to depart. Each person spends what he can afford on his betel. Sometimes the King presents it with his

[1] See Colloquy XXII. *supra*.

own hands, or else by others called XARABDAR or TAMBULDAR. I know two persons who hate this betel: one is myself, the other is an Arabian physician of Nizamoxa named Mula Ucem. We cannot chew it.

RUANO

I have seen many people who never use it.

ORTA

That is true. But if those people could eat it, they would like it. I cannot eat it, because I have no appetite for it. The Indians prize it so much because the betel has veins and nerves along the length of the leaf. They take the leaf in one hand and tear it with the thumb nail, which is not round like ours but with a sharp point in the middle, which they grow for this purpose. In that way they double the leaf and mix the lime with it in small quantity, and the *areca* in pieces ground, and the leaf folded three or four times for chewing. They spit out the first juice, which is the colour of blood. Some persons do not do this, but chew it all, and afterwards take other leaves made in the same way. It is usual when some person is taking leave to give him leaves in a purse of taffeta, with some grains of *areca* and *cate*, and a little mashed-up lime. This lime does not do harm because it is in small quantity, and also because it is made from burnt shells for the most part. I have said that according to the person who gives or receives, so is the number of leaves—the most when a prince takes leave of any person, or when he himself departs. A guest does not go until he has been given his betel, and with it he goes, as it is a hint to depart.

RUANO

This thing is a great deal used and chiefly, it seems, in this land. Is it used in all parts, and what is the time most taken for chewing it?

LAST COLLOQUY

Orta

Chiefly when men go to have an interview with some person of quality, they approach chewing it in their mouths, so as to give out a pleasant scent. Among these people it is so detested to smell bad or musty that common people put their hands before their mouths so as not to give out an unpleasant smell when in presence of a person in authority. Also a woman having to treat of love never seeks her man without first chewing a little. Every one chews it after meals, for they hold that the two things should be done alternately. Many Portuguese say that when they eat fish they presently alternate it with betel. Many say that habitual chewers smell musty or bad if they do not chew, owing to indigestion of food in the stomach. When they chew this goes off. Those who have lost father or mother do not chew for several days, nor when there are fasts. The Moors, and those called Moalis, who are the followers of Ali, fast for ten days. For they say that the sons of Ali, besieged in a fortress, died of thirst. About this they tell a thousand interesting stories, some of them laughable, and they at that time do not chew betel. You ask where it is used. I reply in all parts of India known to the Portuguese, and this applies to all parts near the sea, for in the interior it has to be brought from the sea. It is true that in Dultabado, the chief city of the Deccan, and in Bisnaguer they have it, but in small quantity. For the parts of Persia and Arabia it only goes to Calaiate, eighty leagues from Ormuz, and is imported thence, but it is very dear for those who are able to buy it. Others chew areca with cardamom or cloves.

Ruano

I should like to know the appearance of the tree. I have seen the leaf. I would also wish to be told which is the best, and whether it is of any use in physic.

BETEL

ORTA

The name in Malabar is BETRE, and in the Deccan, Guzerat, and Canara, PAM. The Malays call it CIRI.

RUANO

Why is the Malabar name adopted rather than the others? It would be more reasonable to call it *Folium Indum*, or we might call it PAM, as it is called in Goa.

ORTA

We call it betel because Malabar was the first part of India known to the Portuguese, and I remember in Portugal that they did not say they came to India, but to Calicut. This was because Calicut was the place whence all the drugs and spices were taken up the Strait of Mecca.[1] It was a very rich place, and now, in revenge for what we did in Calicut, all that business is lost. Although the King of Calicut is emperor, he has less power than he of Cochin, because we helped him at first. This is why all the names you see that are not Portuguese are Malayalim. For instance BETRE and CHUNA, which is lime; MAYNATO, washerman; PATAMAR, a runner; and many others. As for calling it "Folium Indum," as you suggest, it is not so called in any language; besides, the "Folium Indum" is quite different. Avicenna gives chapters for one and the other separately.

RUANO

I am much surprised at this, for I always held that "Folium Indum" was the most appropriate name for *betre*.

ORTA

I had your mistake when I first came to India, and after some days I went to see the Nizamoxa, who is vulgarly called

[1] Red Sea.

the Nizamaluquo. Wanting to have a prescription for the stomach, I gave a receipt. He laughed at me, saying that "Folium Indum" was what he chewed every day, for he understood that word "Folium Indum" in Portuguese. Then he showed me Avicenna in Arabic, in which there are two chapters different from each other, the 259th "Folium Indum" and the 77th on *betre*, in which he also showed me "Folium Indum." As in the chapter of "Folium Indum" there is mention of it, we do not put it here. I only know that Avicenna calls the *betre* TEMBUL, which appears to be rather a corrupt form, because every one calls it TAMBUL and not TEMBUL.

RUANO

You refer to a king, but you have no other proof. Yet it is usually said that the word of a king is proverbial; but I do not care to say that a king never lies, but that they never have to lie, being kings.

ORTA

I have the two different chapters of Avicenna, and I can ask any Arab or Ethiopian what they call *betre*, and they will reply *tambul*. Avicenna says the same, that it comforts the gum between the teeth, and the Indians always chew it for that; further, it is comforting for the stomach, and for this also they chew it.

RUANO

I do not know what to say to such strong proofs as regards Avicenna, and I, therefore, want to see the book, for, as they say, seeing is believing.

ORTA

Here is the book with the commentaries by the Belunensis.

RUANO

So it says; yet I feel a doubt in saying that it is cold in the first and dry in the second degree.

ORTA

The text is corrupt. All learned Moors say that Avicenna was mistaken, and that in this he had false information. For people often err as regards these degrees, holding pepper, cardamom, and *cebolla* as cold. As for the *betre* being cold and dry in the second, I have held it to be so myself, judging from the taste and scent, as I have found it useful for several things in physic.

RUANO

I said that I wished to hear about the shape of the leaf and the seed, and about the plant, and which is the best.

ORTA

The shape of the leaf, as you see, is more compressed and narrow towards the point than the orange leaf, and when it is ripe it is nearly yellow. Some women like it best when it is not so ripe, because it excites and then settles well in the mouth. In Maluco this *betre* has seeds like the tail of a newt, and they eat them, finding them good to the taste. This seed was brought to Malacca, where they eat it and find it very good. They plant it and have a place for it to climb over. Some people, to secure more profit, do the same with pepper and with *areca*, making very graceful arbours of the climbing plants. It should be well cared for, kept very clean, and well irrigated.

RUANO

You have explained very well. I wanted to know if you held it for certain.

LAST COLLOQUY

Orta

I say that all who have written to the contrary, ancient and modern, are wrong. Musa and the Pandecta,[1] who have *malabatrum*, are far from the truth. In the colloquy on "Folio Indo" you will see that what all these authors say is false. Let us go and I will show you the *betre* in the garden.

Ruano

You might also tell me some things that have been forgotten, or that need to be declared.

Orta

In the chapter on aloes, I say that the aloes and many other medicines go to Ormuz, and thence by Aden to Cairo. This needs correction. It should be that what goes to Ormuz is taken by Bussora to Cairo, and that which goes by Aden is taken to Cairo and Alexandria, and not by Ormuz, for that would be to go twice over the same road. I said this without considering well. I also remember that the *arbore triste*, when water is distilled from it, wetting a cloth in the water, is good for the eyes.

Ruano

Tell me whether there is much good cinnamon in the island of Mindanao.

Orta

In truth there is, and also in the island of Aynãs, which borders on China, and which is where China obtains its *betre* and *areca*. You may add this to the chapter on cinnamon.

[1] Mathaeus Sylvatico who wrote *Liber pandectarum medicinae.*

RUANO

Tell me also whether you have not something further to say about *Jaca*.[1]

ORTA

I wish to say that the *castanhas* of *Jaca* are useful to stop diarrhœa, as has been experienced by myself and other persons. It is not much that you can add to that chapter. You may do so where I speak of the tower of Babylonia, and say that it is neither Bagada nor Baçora. I have very sure information that the tower of Babylonia, according to the people of the country, was on an adjacent hill, but on this hill there neither appear stones nor bricks nor anything else; yet the traditions are that it was there, though these things do not reveal much nor can they enlighten. Where I speak of *morbo galico*, which the Persians call *bade frangi*, in our language it is *mal franecezi*.

RUANO

Do these men of Persia say nothing more of the bezoar stone?

ORTA

They say that now it is carefully guarded in the land where it is found, and much care is taken that all should reach the hands of the King, and they make parks for it as they do in Spain and throughout Christendom. You have not asked me about the stone of Malacca. Every day I receive more and more praise of it, and I hope to write on the subject if God spares my life.

RUANO

Also with reference to the *olho de gato*,[2] can you tell me of any virtues it possesses?

[1] See Colloquies, XX. and XXVIII. *supra*.

[2] Cat's eye.

Orta

Placed on the top of an open cloth, no fire will make it burn. I have tried it with a candle and found it was quite true. You can try it yourself.

Ruano

I will try it. Further, I beg that we may eat that peacock which you have just had brought in, for they say the flesh does not get bad. This is not a fable, for it is stated by Pliny and other writers, including St. Augustine. Some doctors, in their rules about the plague, praise the flesh of the peacock because it is free from putrefaction.

Orta

That may be true elsewhere. But in this country, as I have told you before, things are so subject to putrefaction that a peacock will not last longer than a partridge, and I have found this by experiment many times.

Ruano

It may be so on the sea-coast, but inland, where it is not so damp, and is colder in the cold season, as every one says, it may be different.

Orta

They eat more peacock in Balagate than in any other part, especially in the city of Juner, which is near the mountains and the cold country; and those who have experimented find that it putrefies more than in Goa. So you may believe that these properties that are attributed to it do not belong to it here. Those who so write about it in Europe may tell the truth, while we speak the truth with reference to this country which we know.

Ruano

Reading yesterday in a chronicle of the King of Portugal which that boy showed me, I found a place where it treated of many mixtures which the writer had seen in his time. I there found that in the kingdom of Delhi there was a deadly poisonous root that bears a fruit which cures every man that is poisoned, and was very wholesome. The root was called Baçaraga, and the fruit Mirabixi. I am very much surprised that you have never written about it.

Orta

We are very little conversant with things in the kingdom of Delhi. We gather what we know about it from people called *Jogues*,[1] and what they tell us on one day they deny on the next. They are a people who live by begging for alms, as I have told you. But I never heard this, though I have conversed with many of them, and none have ever spoken of it. It seems to me contrary to all good philosophy, for the root maintains the trunk, the trunk the branches, the branches the fruit, so that the fruit which is an antidote to the poison is maintained by the root which is poisonous. This is contrary to reason. I say so because the principal base of a serpent's poison is rectified by sixty-three different medicines.

Ruano

If you go by philosophy, we find plants and seeds every day which, in their different parts, have different effects. For instance the Zargatoa.[2] Some trees in India have roots, the water distilled from which is very cold, and the bark and seed very hot. Men from Malacca also tell me that the herb which kills, when arrows are anointed with it, has on one side of the tree that looks to the west an antidote, as they told me, while the poison is on the side looking to the east.

[1] Yogis. [2] *Cineraria campestris*, Flea-wort; *Senecio campestris* of De Candolle.

LAST COLLOQUY

ORTA

These materials for simples need not be treated with such subtlety, nor is so much philosophy needed, for all can be answered. It is not remarkable that any plant should have the root cold and the leaves and fruit hot; but it is another thing for the root to be poisonous, and the fruit fit to eat, and included in the class that may be called nutritive. For poison is totally opposed to this. Poison in itself has not the nourishing but the destructive property. Those who say that the antidote and the poison used for arrows are both from the same tree say what is false. However, this requires investigation, and I promise you that it shall be made when I light upon those *Jogues* of Delhi. Believe me, that if God prolongs my days, I will find out the truth for you, or at least I will make a good attempt to do so.

RUANO

Referring to the way of Ormuz, conversing with the Moors of Persia, they tell me about the *Persian roses*,[1] so called by Avicenna, which we call *açuquare rosado de Alexandria*. Are these roses of Persia held to be soluble? for we find those that are brought and planted to be so.

ORTA

That medicine is much used among the inhabitants of Persia and Ormuz. For a man it is a light purgative in a good quantity, and boiled it is stronger. They give it boiled in 10-ounce doses with a little sugar, which causes five or six motions. A very honourable gentleman told me that it caused twelve, and he had gone for a month without any motion. But in truth the men to whom I have given this medicine never have passed more than six.

[1] *Rosa damascena*, Linn.

PERSIAN ROSES

RUANO

I rejoice much to know this that you have told me. I have a doubt about one thing. The roses have a very superficial value, like all other flowers.

ORTA

Here we have less experience in the roses. In so far as they are styptics and purge fully, it should not cause surprise that there is virtue in the roses, as with all other things that are styptics.

FOOTNOTE ON THE WORD "AMEOS"

(*continued from page* 230).

Without the seeds before you, it is most difficult to identify the various umbelliferous plants producing them, from their current names in the bazaars of India. The principal of these seeds are :—(1) *Pastinaca grandis*, Dalzell, *buphali*, "Hog's Fennel" of Anglo-Indians. (2) *Cuminum Cyminum*, Linn., Sanskrit *jiraka*, "digestive"; Persian, from Sanskrit, *zirah*; Arabic *kamun*, the Hebrew *kamon*, and our Cymen, Cumin, and Cummin. (3) *Carum Carui*, Sanskrit *sushavi*, "curative"; Arabic *carviya*, *i.e.* "of Caria," and *kamunirumi*, *i.e.* "Roman (*i.e.* Turkish) Cummin"; Persian *ziraharmini*, "Armenian Cummin"; Hindustani *methazira*, "Sweet Cummin," and *vilaitizira*, "Foreign Cummin," our Carway or Caraway. (4) *Carum Bulbocastanum*, Koch, *Bunium Bulbocastanum*, Linn., *Carum nigrum*, Royle, the *gunyon* of Kashmir, *umbu* of Ladak, and *shiazira* and *kalijiri* of all Indian bazaars, both these names meaning "black Cummin," and it is the "Black Cummin" of Anglo-Indians; these same Indian names being given in the bazaars to the seeds of *Bupleurum falcatum*, another Umbellifer, of *Nigella sativa* ("Fennel-flower," a Frog-wort), the "Niger seed" of our older Pharmacists, and of *Vernonia anthelmintica*—all of which seeds are used in different parts of India for adulterating the seeds of *Carum Bulbocastanum*. (5) *Carum Ajowan* and *Ptychotis coptica*, DC., *Carum copticum*, Benth., Sanskrit *yamani*, *i.e.* "of the Greeks [cf. Javan]; Persian *nankhah*, "appetising"; Arabic *kamanimaliki*, "Royal Cummin"; Hindustani, from Sanskrit, *ajawan*; Mahratti *omra* and Tamil *omau*—both corruptions of the Sanskrit *yamani*, the "Ajwan" and "Omum" of Anglo-Indians, and ἄμμι [compare *oman*] of Dioscorides, iii. 70. (6) *Carum roxburghianum*,[1] Benth., *Pimpinella involucrata*, DC., and *Apium involucratum*, Roxb., Sanskrit *ajamoda*, "goat-odour"; Persian and Arabic *karafs*; Hindustani *ajmud*; Telegu *ajamodaomamu*, "goat-smelling *ajwan*." The *kharasaniajwan* of the Indian bazaars is "Henbane" seed [*Hyoscyamus niger*], the "Jovis faba," "Faba suilla," and

"Sissili Indicum" of the older Pharmacists. (7) *Pimpinella Anisum*, Linn., Sanskrit *shataphaspha*, "hundred-flowered"; Hindustani and Mahratti *sont*; Persian *badiyan*; Arabic *raziyanaj* and *anisun*; Cyngalese *devadaru*, "divine Cummin," and *mahadaru*, "great Cummin," the ἄννηθον or ἄνηθον, and ἄννισον or ἄνισον of the Greeks, until they came to discriminate between Common Dill and Anis or Anise. (8) *Foeniculum Panmori* or *Anethum Panmorium* of DC. [not to be distinguished from *Foeniculum vulgare*, Gaert., our Common Fennel, Fenel, or Fenoil], the *panmuhori*, that is "bette-leaf-fennel" of the Indian bazaars. And (9) *Anethum Sowa*, Roxb. [not to be distinguished from *Anethum graveolens*, Linn.; *Peucedaneum graveolens*, Benth., our Common Dill]; Sanskrit *shataphuspha* (v. 7), and *madhura*, "sweet Anise," and *mishriya* "of Egypt," the μάραθρον of the Greeks, and the "Sowa Dill" of Anglo-Indians.

INDEX OF PERSONS

CHIEFLY AUTHORITIES QUOTED, WITH TITLES OF THEIR WORKS

ERRATUM

Page 301, line 12. For "*if* she had bracelets of the CHANK shell on her arms" read "*unless* she had bracelets of the CHANK shell on her arms."

INDEX OF PERSONS

INDEX OF PERSONS

INDEX OF PERSONS

INDEX OF PERSONS

INDEX OF NAMES OF DRUGS AND SIMPLES

INDEX OF NAMES OF DRUGS AND SIMPLES

INDEX OF NAMES OF DRUGS AND SIMPLES

INDEX OF NAMES OF DRUGS AND SIMPLES

INDEX OF NAMES OF DRUGS AND SIMPLES

INDEX OF NAMES OF DRUGS AND SIMPLES

SCIENTIFIC NAMES OF NATIVE DRUGS AND SIMPLES

INDEX OF NAMES OF DRUGS AND SIMPLES

INDEX OF NAMES OF PLACES

INDEX OF NAMES OF PLACES

INDEX OF NAMES OF PLACES

INDEX OF NAMES OF PLACES

INDEX OF NAMES OF PLACES

THE END